ANIMAL BREEDING PLANS

ANIMAL BREEDING PLANS

By JAY L. LUSH

C. F. Curtiss Distinguished
Professor of Agriculture

and

Professor in Animal Breeding,
Iowa State University

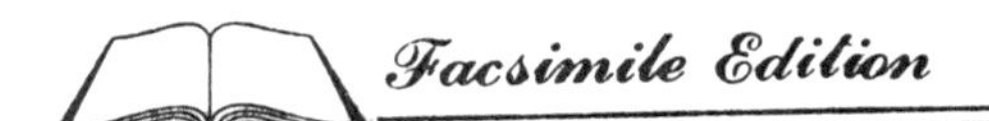

Iowa State University Press, Ames, Iowa

 Composed and printed by The Iowa State University Press, Ames, Iowa, U.S.A.

FIRST EDITION, 1937
Second Printing, 1938
SECOND EDITION, 1943
THIRD EDITION, 1945
Second Printing, 1947
Third Printing, 1949
Fourth Printing, 1956
Fifth Printing, 1957
Sixth Printing, 1958
Seventh Printing, 1960
Eighth Printing, 1962
Ninth Printing, 1963
Tenth Printing, 1965
Eleventh Printing, 1970

Facsimile Edition International Standard Book Number: 0-8138-2345-5

Fifth Facsimile Printing, 1984

PREFACE

This book has grown out of sixteen years of teaching animal breeding to college students who already have had courses in genetics, embryology, anatomy, and physiology of farm animals, herdbook study, history of breeds, and stock judging. The object of this course is to give the students a clear understanding of the means available for improving the heredity of farm animals, more especially of what each possible method will or will not do well.

No effort of mine could keep the book entirely free from statistical terms. After all, a breed is a population, and any attempt at precision in discussing methods of changing its characteristics must necessarily be phrased in terms of the measurements of populations; that is, in terms of averages and of variability. Complete proofs have been presented only where those were simple and brief. In other cases I have sought to present only enough to outline the argument and to show why it is reasonable. That, of course, involves more facts and formulas than most students will use in actual practice but not, I think, more than are necessary to help the student understand why each breeding method he might use is effective in doing certain things and practically powerless to do other things.

Animal breeding is a business; and, therefore, economic considerations of the value and availability of time and materials loom larger in it than they do in the investigation of a purely scientific problem. The work must go on, and decisions must be made in many cases where there is not yet enough evidence to show with certainty what the result will be. The scientist, faced with the problem of deciding what is the truth in such a case, might retire to his laboratory and design an experiment which in due time would reveal that truth. But the man engaged in the business of animal breeding cannot wait for that. Without being entirely certain of what would result from each of the alternatives open to him, he must decide whether to cull or keep each animal and whether to mate it in this way or that. Knowing that the odds are two to one in favor of one procedure as against another may be highly useful to him as a business man, although the scientist may well demand that the odds be higher than the conventional 19 to 1 before he places much faith in a principle deduced from his experimental data. With these needs of the practical breeder in mind, I have sought to state the most probable truth concerning questions which may guide his actual decisions, even in cases where genetic knowledge has not yet established the limits of that truth as closely as is desirable. Such statements have been labeled with qualifying phrases so that the students will be prepared to encounter occasional exceptions to them.

The ideas in this book have been drawn freely from the published works of many persons, I wish to acknowledge especially my indebtedness to Sewall Wright for many published and unpublished ideas upon which I have drawn, and for his friendly counsel.

Ames, Iowa
June, 1945

JAY L. LUSH

TABLE OF CONTENTS

CHAPTER 1

The Origin and Domestication of Farm Animals

The story of the origin and domestication of farm animals, although interesting, has little practical usefulness to the animal breeder who is seeking to better his flocks and herds today. Only living animals can be used for breeding; and if those have the inheritance the breeder wants and can produce fertile offspring, it makes little difference how they came by that inheritance or what their ancestors were. Knowledge of how closely two races of animals are related may be of some help in forecasting the outcome of crosses between them; but such predictions, based on degree of relationship, will have many exceptions. Knowledge of the origin of farm animals, therefore, is useful to the practical breeder only in the same way that ancient history may be useful in training modern people for citizenship. The details matter not at all, but here and there the history may show with dispassionate clearness some general principle of human conduct which will repeat itself in present situations. Also, it may give the student a perspective which will be useful in making decisions concerning contemporary affairs. The present chapter, then, is a compilation of facts which may be helpful in forming a historical perspective from which to view the present general problems of animal breeding. The reader seeking only immediately useful information is advised to omit it or merely glance at it.

Domestication of the important farm animals was accomplished long before the beginning of written history, but long after man had become a toolmaker and tool-user of considerable skill. In terms of human culture it seems to have happened mostly very late in the Paleolithic (Old Stone Age) or early in the Neolithic (New Stone Age), although this varied with different peoples in different parts of the world.

The *origin of the species* of animals which were domesticated extends back into vastly greater reaches of time and is only a special aspect of the story of evolution. Figure 1 is intended to show graphically the contrasts in the enormously long time involved in evolution, the comparatively short time which has elapsed since domestication first took place, and the tiny fraction of time since definite and continuous written history began.

The following comments concerning geologic and cultural time are centered around man and other mammals, since those are the central figures in the story of domestication. There are hundreds of thousands of species in the animal kingdom; but with a few exceptions, such as honeybees and silkworms, all the domesticated animals are included in a few species of mammals and birds. It is an interesting but perhaps an idle speculation to wonder why so few species were domesticated. Did the mental characteristics of the others make domestication impossible? Or did those which were domesticated have among them nearly all the characteristics which man found useful to him? Why did not man domesticate any of the many species of animals which hibernate through

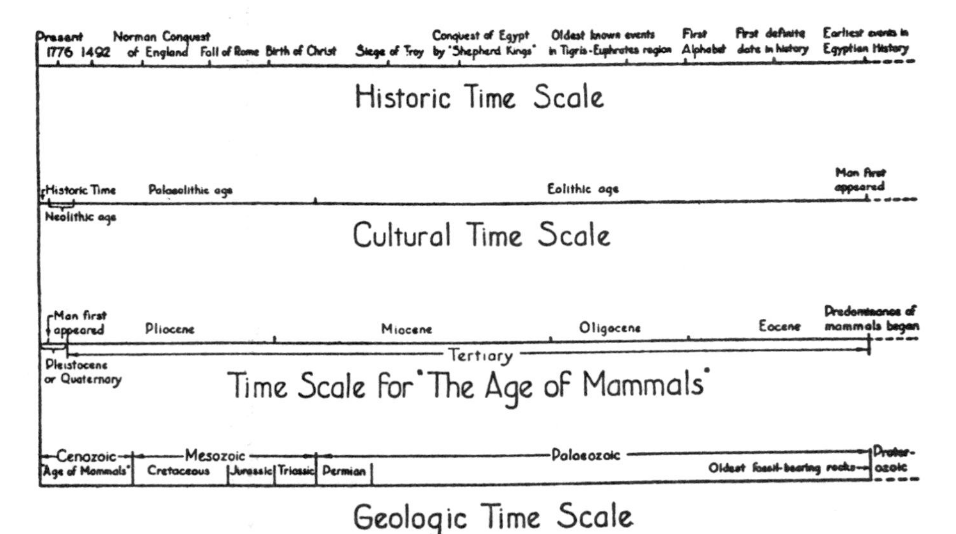

FIG. 1. The comparative lengths of historic, cultural, and geologic time. Each line represents on an expanded scale the small segment at the extreme left end of the line just below it. If the geologic scale were drawn with the same number of years per inch as the historic one, it would need to be more than 20 miles long!

the winter? Those would have had some practical advantages in regions where feed is abundant from spring to fall, but scarce or buried during the winter. Are there perhaps still unrealized possibilities which can be had by domesticating species which are still wild or semi-wild?

Domestication implies several things, no one of which alone is sufficient to define it completely. It usually means tameness, but individual wild animals may be tamed (as trained seals or performing bears are) without our being willing to call them domesticated, and ranch-raised but nevertheless domesticated cattle or horses may be very wild individuals. Domestication implies bringing the animal's growth and reproduction at least partly under man's control, but we call pigeons and cats domesticated, even though their breeding habits and mating choices usually are not controlled by man. Domestication implies that man converts the animal's products or services to his own advantage or purposes, but he does this also with many wild animals, such as the fur-bearing ones. Some of the domesticated ones, such as canaries, many breeds of dogs, and most "pet and fancy stock," serve him only in an esthetic way. As the word implies, domestic animals usually are kept in or near man's own dwelling places, but usually we do not consider as domesticated the mice and rats which live in man's barns or even in his own houses, while we do consider range-raised cattle and sheep as domestic animals although they may never have seen a human habitation. Some of the domestic animals are dependent on man's care for their very existence, at least in many regions. But horses, dogs, and even cattle, have at times run wild and reproduced for several generations without any control or care by man. Strictly speaking such animals are called "feral" rather than truly wild. Domestication has in most cases produced rather large changes in behavior. Some of these are conditioned by the environmental circumstances under which the domestic animal is reared but many of them are hereditary and presumably have been caused during the process of domestication by selection for those individuals or families which were the gentlest, the most cooperative with man, the best trail-runners (in the case of some breeds of dogs), etc.

In the laboratory animals, or the animals used in fish farming and in fur farming, we may perhaps be witnessing the slow process of domestication. Some of them, like the guinea pig or the white rat, have almost as good a claim to be called domesticated as do swine or reindeer or ducks. Others, such as mink and silver foxes, have advanced little beyond the stage of wild animals being kept in cages as in a menagerie.

GEOLOGIC TIME

The Cenozoic Era (the Age of Mammals) began some 50 to 75 million years ago, although the expression of geologic time in years, particularly in the more remote periods, is very uncertain. The first mammals had come from a reptile group called Cotylosaurs through a premammal group called Cynodonts some time in the enormous interval between the beginning of the Permian and the end of the Triassic, but they did not become the dominant form of animal life until the Cenozoic, being overshadowed earlier by the reptile forms. The Cenozoic is divided into five periods, which, as far as the mammals are concerned, are chiefly noted as follows:

EOCENE. The archaic or generalized mammals were replaced by modern types.

OLIGOCENE. The mammals differentiated into many of the orders and families known today. An anthropoid (Propliopithecus) is known from early in this period.

MIOCENE. This period saw the greatest variety and abundance of mammalian forms. It was a period of extensive grasslands and restricted forest areas over much of the earth. Corresponding to that, there was a widespread expansion and development of the grazing forms of mammals at the expense of the browsing forms. It is probable that man's line of descent had already diverged from that of the other anthropoids by the middle of the Miocene.

PLIOCENE. Most modern genera and even some modern species of mammals already were present at the beginning of the Pliocene. Man's definitely human ancestors appeared during this period at a time date of something like 600,000 to 1,000,000 years ago.

PLEISTOCENE. There was periodic glaciation and with it the extinction of many of the great mammals, such as the mammoth, the mastodon, the woolly rhinoceros, the saber-toothed tiger, and many others. Man learned the use of tools and fire and began to domesticate animals for his own use. The period ends with the retreat of the last great glaciation about 30,000 years ago.

RECENT (The Age of Man). Civilization began. The historical aspects of what is known about man and his surroundings constitute the subject matter of archaeology, and archaeology in turn gives way to history when the written records become adequate enough to give a connected account of man's activities.

CULTURAL OR ARCHAEOLOGICAL TIME

Archaeological time is measured in stages of human culture. It does not correspond perfectly to chronological time, since human culture did not advance contemporaneously in all parts of the world. The major subdivisions of cultural time are as follows:

PRE-HUMAN PERIOD (the Eolithic or dawn period). This period begins with the time when man's ancestors can first be called definitely human, something like a million years ago, and extends roughly to the coming of pre-Neanderthal man in Europe about 200,000 years ago. The use of tools was advanced but little beyond picking up and using such stones or clubs as happened to be handy. Probably fire was not used.

Paleolithic Period (the Old Stone Age). This period was marked by the use of stone and bone implements which slowly increased in complexity and usefulness. The use of fire was learned at least by the time of Neanderthal man. No agriculture was practiced; and there were no domesticated animals, except perhaps the dog. The Paleolithic culture in Europe slowly developed to Neolithic culture around 25,000 years ago. What was practically Paleolithic culture still prevailed among the aborigines of Australia and Tasmania and the Bushmen of South Africa when the first white explorers came in contact with them.

Neolithic Period (the New Stone Age). This period was marked by the use of ground and polished bone and stone weapons and tools. Neolithic man made pottery and crude textiles and basketwork. He practiced agriculture in a crude way. He had domesticated animals of nearly all the species we have today, whereas Paleolithic man had few or none. Neolithic man lived in huts or even in wooden houses as, for example, the Swiss Lake Dwellings. The art of metal-working was still unknown or was practiced only on soft metals used for ornaments. The American Indians were in a rather advanced stage of Neolithic culture when the first white explorers found them. The cultures of the Aztecs, Mayas, and Peruvian Indians already were more advanced in many ways than the late Neolithic cultures of Europe. They were working copper, silver, and gold, but those metals were too soft to make useful tools. The Neolithic culture of the American Indian was behind that of Europe in the use of domestic animals.

As long ago as 4500–4000 B.C., the city of Tepe Gawra in the Tigris valley included in its culture such things as gold and lapis lazuli beads, temples, landscape painting, and the firing of painted pottery, although the inhabitants had not learned to smelt copper. The earliest bit of copper known is an ornamental pin made in Egypt perhaps as long ago as 5000 B.C. The Egyptians were working the copper mines in the Sinai peninsula regularly for ornaments by 3500 B.C. and were using copper tools by 3000 B.C.

Bronze Age. The use of bronze began in Assyria by 3000 B.C. and spread west and northwest like a slow wave. It had certainly reached the Danube basin by 2000 B.C. and perhaps Britain almost that early. Apparently there was no bronze age in Africa, except in Egypt and on the northern coast. The other African races went directly from the stone age into the iron age. In the bronze age in Europe, village life and complex social customs had already developed far.

Iron Age. Bits of iron have been found in the Great Pyramid at Gizeh (about 2900 B.C.), but apparently these were only rare curiosities. General use of iron for tools or weapons was begun by the Hittites in

Asia Minor about 1300 B.C. and iron supplanted bronze as the commonest metal for weapons among the Assyrians by 1000 B.C. Then its use spread rapidly.

History and archaeology are intertwined in the long period which we describe loosely as the "dawn of history." The earliest definite date in history is 4236 B.C., when the Egyptian calendar began; but fragments of Egyptian history are known from nearly 5000 B.C. Some time between 5000 and 4000 B.C. the Egyptians began to use ox-drawn plows, and by 3500 B.C. they had an alphabet. The beginnings of history were not contemporary in different parts of the world. For example, the definite history of Britain begins about the time of the Roman conquest; and that of the Scandinavian countries begins about five or six hundred years later. Little is known of South Africa or Australia before 1500 A.D. On the other hand, the history of Greece goes back some 2000 years farther than that of Britain; Cretan history is known from nearly 3000 B.C., and a few events among the Sumerians near the Persian Gulf can be dated at about 3500 B.C.

DATE OF DOMESTICATION OF FARM ANIMALS

All the farm animals were domesticated long before historic times. The evidence about when and how that happened is incomplete and consists of such things as the bones and tools found buried in the trash heaps around ancient camp sites or caves, the drawings or carvings on the walls of caves or on ornaments. Rich sources of evidence are the tools, weapons, utensils, and images which so many peoples placed in the graves with their dead.

Often such evidence is fragmentary and may be interpreted with equal plausibility in several different ways. Even where evidence admits only one interpretation, the dates derived from it are necessarily minimum dates. For example, the evidence leaves no doubt that domesticated horses were widely used in the region of the Tigris and Euphrates rivers as long ago as 2000 B.C. and probably the first of them were brought into that region as long ago as 3000 B.C. Even if the earliest date of domesticated horses in this "two-river land" were established with absolute certainty, it still remains possible that horses may have been domesticated and used a thousand years earlier at some other place as yet unexcavated. Doubtless the thoroughly excavated sites of ancient human camps and cities are only a small fraction of the total number which exist and may be discovered in the future.

One who reads the technical evidence and discussions can scarcely avoid the feeling that they give an unduly large emphasis to details of the shape of horns and skull. This is natural since these parts of the animal are best preserved and best known. It is partly justified because these parts are little affected by ordinary variations in nutrition or other environmental influences. Yet one who surveys the considerable variability in skull and horn shape within comparatively pure breeds today, and who considers the known cases where a single gene substitution can cause large differences in these characteristics, must feel uneasy about placing much faith in genealogies which rest largely on similarities or differences in the size and shape of horns or skulls. Such genealogies are especially questionable when they are based on only a few specimens, perhaps widely separated in time.

It is uncertain that Paleolithic man in Europe actually domesticated any animal, although he may have had the dog. The Paleolithic aborigines who settled in Australia took the dog with them, but no other modern animal. Paleolithic man in Europe used the horse extensively for food, but it is probable that he hunted it as game and had not really domesticated it.

Neolithic man in Europe appears to have had nearly all of our modern domestic animals except the cat and poultry and those animals which were found only in America or in the tropics. Some students of the evidence claim that even in early Neolithic times the bones of the domesticated ox, swine, and sheep were already distinctly different from their wild contemporaries. There is even considerable speculation that the domesticated races of the early Neolithic in Europe were brought from the Caspian region or from Asia Minor in an already domesticated condition by peoples who migrated into Europe then. A considerable amount of care was certainly given to farm animals by the Lake Dwellers and by the men of the bronze age. It is not certain that the horse was really a domesticated animal in Europe before the end of the Neolithic, although the men of the bronze age certainly were riding horses.

The rock-carvings from ancient Egypt and from the pre-Babylonian peoples of Sumer and Akkad, which are among the oldest of what may be called written records, show the goat, sheep, ox, ass, pig, dog, and cat. Caring for these animals was already a well-established part of agricultural practice, even at that remote day.

PLACE OF DOMESTICATION

Domestication took place in the Old World except in the case of the llama, alpaca, guinea pig, and turkey, which are native to the Americas. In the Old World, domestication seems to have taken place largely in central or western Asia, although the evidence points to some domestication in Egypt and in Europe itself. Chickens and elephants were domesticated in India, and at least one center of domestication for swine was in China. Domestication may have taken place independently in several regions. This seems certainly to have happened in the case of swine and sheep and may have happened with other animals. Much of the world, especially central Asia, is still incompletely explored from this point of view. There were no modern mammals (except the dog and man) in Australia. Africa south of the Sahara desert had a fauna rich in mammalian species, but none of the domesticated animals, except the South African ostrich and the African elephant, came from there. Upper Egypt or Ethiopia was one of the centers of domestication for the ass.

SPECIES DOMESTICATED

For many animals it is still disputed whether they descended from a single wild species (*monophyletic* origin) or from two or more wild species perhaps domesticated in different regions or at different times and later interbred (*polyphyletic* origin). The main reasons for this dispute are, of course, the scantiness of the evidence and the different biological views which the writers hold. In many cases domestication was completed so long ago that the original wild ancestor has become extinct, or it may still be living but the domesticated form has been changed so much that we are not now certain which contemporary wild species was the ancestor.

Polyphyletic theories are not as widely held now as formerly. The idea of organic evolution was not generally accepted even by naturalists until well into the last half of the nineteenth century. Many of the naturalists who wrote on the origin of domesticated animals still had in their minds traces of the old Linnean idea of the fixity of the species. Often they had an exaggerated idea of the supposed uniformity of wild species. With this mental background a polyphyletic origin seemed to them the only possible way to explain the tremendous diversity of domesticated forms—for example, the tremendous contrasts between breeds of dogs or of sheep. Modern studies of large samples from wild populations have shown that those populations are not as uniform as many of the older naturalists believed. Some of these modern studies have shown that enough of this variability in wild populations is hereditary that selection, directed toward diverse goals, and other breeding practices, such as inbreeding, could in a few generations produce distinctly contrasting races from a single wild population if man were to control its breeding as he does that of his domesticated animals. Hence it no longer seems necessary to invoke a polyphyletic origin as an explanation for observed diversity. The possibility remains, however, that some domesticated animals may really have had in their ancestry crosses of distinct races or even species which were still genetically similar enough for their crosses to be fertile. This seems to have happened in the case of swine and sheep, although there is plenty of room for difference in opinion as to whether the races domesticated were ever different enough and discontinuous enough to justify calling them different species. The whole question of proper taxonomic terms for domesticated animals is in a chaotic condition. Many taxonomists hold (with Linnaeus) that variations among races of domesticated animals are largely man-made and therefore outside the scheme of nature which is the concern of taxonomy.

METHOD OF DOMESTICATION

Literally nothing is known about how domestication was first accomplished. It is only speculation to guess that hunters first brought home a few young as pets or captured cripples from time to time and thus learned how to care for animals. At least one Egyptian rock-picture shows hunters building a fence across the mouth of a little steep-walled valley into which they have driven some wild animals. Wild elephants are captured in India today by carefully planned drives. Tame elephants are then used to help chain the wild ones and to teach them to work. Hunger is the most generally effective method of taming the most unruly among the wild ones. But it is more difficult to tame the African elephant, even by these same methods. This suggests that temperamental aptitude was an important element in the success or failure of early attempts at domestication.

DETAILS ABOUT DIFFERENT KINDS OF DOMESTIC ANIMALS

SWINE. The European wild boar (*Sus scrofa*) still lives in some of the forests of Europe. It crosses freely with domestic swine, and the offspring are fertile. Doubtless it was domesticated somewhere around the Baltic sea in Neolithic times. A swine race or species (generally known as *S. vittatus,* although some divide it into two groups and give the name *S. cristatus* to one) native to the middle and eastern Asian mainland from western India around to central China, and found also in nearby island lands like Japan, Formosa, Sumatra, Java and Borneo, was separately domesticated in China, perhaps as long ago as 3000 B.C. At least one more center of domestication in Neolithic times was south or southeast of the Alps, where some of the Mediterranean local wild races were domesticated. *S. scrofa* grades into *S. vittatus* by a gradual but continuous series of local races to such an extent that modern writers (such as Kelm, 1938) are inclined to consider them all as one highly variable species—a *"Formenkreis"*; i.e., a chain or circle of local groups or races, each differing only a little from the ones next to it but considerably from those farther away.

Nomadic peoples could not move swine with them easily. They generally regarded with contempt the farmers and settled valley-dwellers who did keep swine. This may have been the origin of the Hebrew and Moslem dislike of swine, which was later fortified by religious precept. Because nomads could do so little spreading of swine and because wild swine by themselves do not usually migrate far, there has been in swine more than in most species a differentiation into local races which vary from one place to another. It may be more accurate to speak of the practice of domestication spreading from tribe to neighboring tribe rather than to speak of an actual spread of one or of a few domesticated forms of swine.

Some of the early navigators report finding native swine on some of the islands in the southern Pacific; but the Polynesians who settled those islands were expert navigators who had come from the general Malaysian region only a few centuries earlier, and probably they brought swine with them. The peccaries, which belong to a different genus, are the nearest American relatives of swine. They had not been tamed by the Indians nor, from their behavior in captivity, does it seem likely that they can be domesticated. The breeds of swine common in the United States probably get most of their ancestry from the European wild swine, but there may have been a considerable amount from the Mediterranean races, and there is clear historical evidence of the introduction of at least a little blood from Chinese swine.

CATTLE. The family Bovidae are the most specialized of the hollow-horned ruminants. They are connected with the other ruminants by way of antelope-like ancestors from which they diverged in the Pliocene or Miocene. Living forms of the Bovidae include the true buffalo, the bison, musk-ox, banteng, gaur, gayal, yak, and zebu, besides what we commonly call cattle. The musk-ox is intermediate in some respects between oxen and sheep or goats. The musk-ox has not been domesticated, although Stefansson reports that it is well suited for domestication. The Asiatic buffalo was in Syria in Neolithic times, but may not have been domesticated until near Christian times. It is an important dairy and work animal of India and lands farther east and is used a little as far west as Bulgaria. It existed in the Atlas region of northwestern Africa even after Neolithic times. The African buffalo has never been domesticated. The banteng, gaur, and gayal are all restricted to southeastern Asia and the nearby islands. The banteng is the common work ox of Java, Bali, and Borneo. It is often crossed with common cattle, but the crosses thus produced are fertile. The gayal may be only the domesticated form of the gaur. The European bison, or wisent, and the American bison have never been really domesticated, despite a few sporadic attempts to do so. The American bison can be crossed with common cattle, but there is much mortality and sterility among the crossbreds. The yak is the bovine species best adapted to cold mountain lands. It is native to the highlands of Asia north of the Himalayas and, although an important domestic animal there, has not found practical use outside that region. Nothing very certain is known about its date of domestication. The yak can be crossed with common cattle and with zebus and with bison, but there seems to be some sterility among the males from such crosses.[1]

There are in the whole world between six and seven hundred million cattle which are commonly grouped under the one species name of *Bos taurus,* although some prefer to give a separate species name, *B. indicus,* to the zebu group. In the Balkans, Asia Minor, central Asia, Korea, Formosa, and in eastern and southern Africa, there is a wide variety of forms intermediate in many respects to the extreme zebu types and the cattle of western Europe. The more extreme types have been separate in their ancestry for thousands of years. Carvings from the Indus valley region show bulls with extreme zebu characteristics from as long ago as the third millenium B.C. The cattle of the United States are of purely European origin, except some in the region bordering the Gulf of Mexico, which have considerable zebu ancestry. Concerning the ancestry of European cattle, the most commonly mentioned species or subspecies are: (1) *B. taurus brachyceros* (or *longifrons*), which was in Europe as a domesticated animal early in the Neolithic and presumably was domesticated somewhere north of the Alps or in northwestern Asia; (2) *B. taurus primigenius,* the urus or aurochs, known in Caesar's time as the wild ox of Europe[2] but domesticated long before (perhaps early in the Neolithic), probably south of the Alps or in the Balkans or in Asia Minor; and (3) *B. namadicus,* which was contemporary with man in India in the early Pleistocene. Other names, common in the early writing but not seen so often now, include *B. taurus frontosus,* the Swiss spotted cattle; *B. taurus brachycephalus,* the short-headed cattle such as the Dexter, Eringer and Zillertaler breeds; and *B. taurus akeratos,* the hornless cattle of northern Europe.

HORSES. Horses were plentiful in Europe in Paleolithic times. That

---

[1] Deakin, Alan, Muir, G. W., and Smith, A. G. 1935. *Hybridization of Domestic Cattle, Bison, and Yak.* Publication 479, Department of Agriculture, Dominion of Canada.

[2] Keller says the last wild aurochs cow in Poland was killed in 1627.

they were used for food is attested by the cracked and dismembered bones, mostly of young horses, around old camp sites like that of Solutre near Lyons. They were probably not domesticated at that time but were hunted for food. The horse is primarily adapted to open grassland country and apparently became rare in Europe during Neolithic times with the increasing forest growth. Formerly it was thought that a distinct type of forest horse remained in western Europe through the Neolithic and was the principal ancestor of the heavy or "cold-blooded" draft breeds. More recent studies (Antonius, 1936) indicate that the heavy horse was developed between 1000 and 1200 A.D. in or near Friesland out of the existing domesticated horses of Tarpan origin. Prawochenski, however, disagrees with this interpretation. Horse bones are rare in Neolithic deposits, but bronze age deposits include bridle bits and other accoutrements thus proving its domestication by that time.

Probably all domesticated horses of Europe and western Asia descend from the tarpan which still existed wild in eastern Europe in the region from East Prussia southward as recently as 1700 A.D. and perhaps in the 1800's. Some writers distinguish between a "forest tarpan" in the more westerly region and a "steppe tarpan" farther south and east, but others think (Antonius, 1936) there was no real distinction. In eastern Asia the wild horse of Przewalskii was reported about 50 years ago and perhaps still exists in the Mongolian desert region. It may have been domesticated separately there. The use of the horse had reached China before historic times, but that is of uncertain date.

The horse first appears definitely in western history before 2500 B.C. when the Neolithic Indo-Europeans from the Caspian basin brought it into Anatolia and later to Babylonia. Presumably the Indo-Europeans from west of the Caspian took domesticated horses westward with them north of the Black Sea at more or less the same time. Certainly horses were in Spain and northwest Africa before any could have reached there from Egypt. The horse first reached Egypt about 1800 B.C. when the Shepherd Kings conquered Egypt from the northeast.

The *evolution* of the horse is especially well known compared with that of other mammals and is used as a classic illustration in many books on evolution and zoology. Most of this evolution took place in the Americas, but horses later became extinct there after some of them had migrated to Asia. There were no horses in the Americas when the white men came. Why they died out is one of the unexplained mysteries of evolution. Conditions were favorable for them at the time of the discovery of America by white men, as is shown by the way the horses which escaped from the early settlers or explorers multiplied. The wild horses of the present western ranges are descendants of these escaped horses and are often used to illustrate the difference between "feral" animals and truly wild ones whose ancestors never have been domesticated.

The early use of horses was for human transport and for pulling chariots in time of war. Their use for pulling loads and for tilling the soil is a comparatively modern development.

ASSES. The ass was in common use in Egypt and Babylonia many centuries before the horse was introduced into those lands. Probably it was originally domesticated in Egypt or on the east coast of Africa or in southern Arabia or around the Persian gulf. Its nearest wild relatives are found in Africa and in Asia Minor. The main (perhaps the only) ancestor is thought to have been the Nubian wild ass, although another variety of wild ass in Somaliland may have contributed something.

It is generally stated that the other Equidae, such as the zebra, the kiang, and the onager, have not been domesticated. However, Antonius concludes that the ancient Sumerians had domesticated the onager and used it extensively, even for crossing with horses to produce mules. Also it is stated[3] that Burchell's zebras were formerly used on stage lines in the Transvaal.

SHEEP AND GOATS. The subfamily Ovinae are all highland or mountain dwellers. Perhaps because of this and the resulting isolation they are mightily given to breaking up into many species, subspecies, varieties, and local races. There is no single trait which distinguishes all goats from all sheep, although there are some things which are characteristic of most sheep and few goats or vice versa. Besides the true sheep and the true goats, there are, according to Antonius, the following groups of Ovinae: (1) The Hemitragus group of primitive or short-horned goats with four teats. Three living species are found in the mountains of India and southern Arabia. (2) The ibexes or steinbocks, which resemble the true goats rather closely and are fertile with them but have contributed nothing to their ancestry. They are all mountain dwellers, and Antonius lists seven species in the mountains from the Himalayas to the Pyrenees and another from the mountains of southern Abyssinia. (3) The burrhel or blue sheep of Tibet. (4) The maned sheep or aoudad, now restricted to the Atlas mountains but extending as far east as Egypt even in historic times. (5) The argali group, which are rather sheep-like and include an enormous number of local races besides the argali itself and Marco Polo's sheep. The argali group probably did not contribute anything to the ancestry of domesticated sheep, unless perhaps something to the fat-rumped races. This is disputed by some investigators (Gromova, 1936) who believe the argali group was

[3] *The Horse Lover.* 7:9. 1943.

important in the ancestry of domesticated sheep. (6) The thick-horned sheep, of which there are an enormous number of local races extending from the northern Himalayas northeastward over Kamchatka and Alaska to California. The Big Horn Sheep of the Rocky Mountain region is an example. Probably this group played no part in the ancestry of domestic sheep. There have been a few accounts of crosses between domestic sheep and Big Horn Sheep in western United States.[4]

Domesticated goats are believed to descend mostly from *Capra prisca,* which is known from Pleistocene fossils from Greece northward, or from *C. aegagrus,* the Bezoar goat or Psaang, which still exists through the mountainous regions of Asia Minor and Persia and in the past has extended from Sind to as far west as Crete. Also mentioned among the true goats are *C. dorcas* (which some think only a form of *C. prisca*) from the Jura mountains, and *C. falconeri,* which is extraordinarily given to the development of distinct local races and lives in the region around Afghanistan. Some of the tame Egyptian goats resemble *C. falconeri* rather closely, but it is not certain that they descend from it in any part. Nothing is known about the place of domestication of the goat, nor about the date of domestication, except that it must have been very remote.

Domestic sheep are thought to descend mainly from the mouflon, *Ovis musimon,* which is still found wild in Sardinia and Corsica and interbreeds freely with domesticated sheep, and from *O. vignei,* the urial, which is found from Turkestan to Asia Minor. Some writers think that a part of the ancestry of domesticated sheep came from *O. orientalis,* the mouflon of Asia Minor and Armenia, or from *O. arkal,* which is found east of the Caspian Sea. There is more confusion and disagreement about the ancestry and classification of sheep than of any other farm animal.

Domestication of the sheep took place so long ago that taxonomic names have even been given to the forms found in deposits from various stages of human culture. Thus, there is talk of a "turbary sheep, *Ovis aries palustris*" and of a "copper-age sheep, *Ovis aries studeri.*" Early Neolithic man in Europe certainly had domesticated sheep with him, but where or when he got them can only be conjectured from what is known about their geographical distribution.

There is an enormous variety of breeds and races of domesticated sheep. The classifications proposed for convenience in referring to these are generally based on the nature of their wool and the length and fatness of their tails. Antonius classifies them for that purpose as follows:

1. Long-tailed sheep

---

[4] *Jour. Wildlife Management* 9:82–83. 1945.

A. The wool sheep of Europe. These include all the breeds which are prominent in the United States.
B. Hairy sheep originally prevalent all over Africa. The black-headed Persian sheep of South Africa is an example.
2. Fat-tailed sheep, usually with a long tail, fat at the upper end but slender at the tip. Karakuls are an example.
3. Short-tailed sheep, such as the Heidschnucke of Germany and other marsh or moorland sheep of northern Europe.
4. Fat-rumped sheep, which were originally native to high Asia east of where the fat-tailed sheep developed.

DOG. Zoologically the dog is essentially the same as the common wolf of the northern hemisphere.[5] The dog was the only domesticated animal found in both the New and the Old Worlds. The polyphyletic origin of the dog, involving the jackal as well as the wolf, was once widely believed, but that belief has almost disappeared now. The dog may very well have been the first animal domesticated. Chaldean and Egyptian monuments show several distinct breeds in existence four or five thousand years ago. The Incas in Peru had bulldogs as well as their ordinary breed (Hilzheimer, 1936).

CAT. The Egyptians domesticated the African wild cat. Thence it was probably introduced to Italy by Phoenician traders some centuries before the Christian era. A European wild cat and perhaps also a Chinese wild cat may have contributed to the ancestry of modern domestic cats.

CAMEL. The single-humped camel, *Camelus dromedarius,* was domesticated probably in Arabia or northeastern Africa, perhaps near the very beginning of Egyptian civilization or perhaps not until near 1000 B.C. The two-humped camel, *C. bactrianus,* was domesticated in Asia somewhere between Iran and the Gobi desert. It had reached Syria at least by 1000 B.C. and perhaps a thousand years earlier. It is now the common camel in China and the regions just northwest, although the other occurs also.

The llama and alpaca are the only living American representatives of the camel family, although that family went through most of its evolution in North America. They were domesticated by the ancient Peruvians from the wild guanaco, *Lama huanachus,* long before the Spanish conquest. The wild vicuna, *L. vicugna,* is a close relative. In the Incan agriculture the alpaca occupied somewhat the place of the sheep in ours, but the llama was used extensively as a beast of burden, as well as for its meat and wool. Llamas sometimes are used for dairy purposes, too.

---

[5] Crosses between dogs and wolves are frequent *(Jour. of Genetics* 43:359–414. 1941).

REINDEER. *Rangifer tarandus* is the Scandinavian wild form which was domesticated by the Lapps and later introduced to Alaska and arctic Canada. *R. caribou,* the caribou of Canada, has not been domesticated. One report has it that reindeer were domesticated in the Yenisei River basin about the time of Chi.st.

ELEPHANTS. Elephants were domesticated at least as long ago as in ancient Carthage. Presumably they were used in India much earlier. Alexander used them in his battle with Darius in 331 B.C.[6] in Mesopotamia.

CHICKENS. Chickens came from southeastern Asia and are generally believed to have descended from *Gallus bankiva,* the jungle fowl of India. It is not certain that more than one species was involved. Domesticated chickens were kept in China at least as early as 1400 B.C., and from India had arrived in Babylon by 600 B.C., in Greece by 500 B.C., and in Rome well before the Christian era. The actual introductions may have been centuries earlier.

GEESE. These came from the grey laggose,*Anser anser,* and perhaps also the Chinese goose, *Cygnopsis cygnoides.* The date is uncertain, but geese were kept by the ancient Romans several centuries before the Christian era, as witness the legend about the cackling of the geese which on one occasion saved Rome from invaders. There may have been several independent domestications.

DUCKS. Probably these all come from the mallard duck, *Anas boschas,* although there are many other species of wild ducks. Ducks probably were not domesticated before Roman times.

GUINEA. The guinea comes from the Guinea coast of western Africa. The wild species is *Numida meleagris.* The guinea was known to the Romans as a domestic fowl but later ceased to be kept in Europe and was reintroduced by the Portugese in the sixteenth century.

PEACOCK. *Pavo cristatus* is found in India and Ceylon. *P. muticus* is found in Burma and Java. Either or both may have contributed to the origin of domestic races. The peacock was a relatively late arrival in European agriculture. It was first domesticated in Persia, or at least first knowledge of it came to Europe from Persia.

TURKEY. The peoples of ancient Mexico and Peru had domesticated the turkey long before the discovery of America by white men. The wild Mexican turkey, *Meleagris mexicana,* is thought to have furnished most of the ancestry, but *M. gallopavo* from the Atlantic coast of the United States and *M. ocellata* from Central America are sometimes mentioned in that connection.

FUR-BEARING ANIMALS. Such animals as the *fox, mink, marten, ferret, skunk, muskrat,* etc., are extensively reared in captivity but prob-

[6] See Armandi's *"Histoire militaire des elephants."* Paris. 1843.

ably should not yet be called truly domesticated. They may represent stages in domestication through which our farm animals passed far back in Neolithic times.

LABORATORY ANIMALS. Among the laboratory animals, the *guinea pig* probably should be called domesticated. South American Indians were breeding it in captivity for food before the white man came. The *white rat* is an albinotic strain of the common Norway rat, *Mus norvegicus*. Probably the albino rat was domesticated before the beginning of the nineteenth century. An albino rat colony in England in 1822 is reported. Fancy races of *mice* were known in ancient Troy perhaps as long ago as 1200 B.C. and probably in China as far back as 1100 B.C., since a Chinese dictionary of that date had a special word for the spotted mouse. Domestic *rabbits* are descended from the European rabbit, *Oryctolagus cuniculus*. It was probably domesticated in the Spanish peninsula or southern France, perhaps as early as 1000 A.D.

BEES. Honeybees are mentioned frequently in the Old Testament and in early Roman writings on agriculture. They were brought to America from Europe. Bumblebees were already in America when the first white explorers came.

SILKWORM CULTURE developed in China, possibly longer ago than 2000 B.C. It did not spread to other countries until the Christian era.

REFERENCES

Encyclopedias, especially the *New Brittanica* and the *Americana* and Volume 3 of Bailey's *Cyclopedia of American Agriculture,* furnish a good starting point for general information. Consult several in order to note the disagreement as to fact and the confusion about specific classification.

The *National Geographic Magazine* occasionally has an article on some domestic animal with pictures and descriptions of many of the living breeds or races. Some of the recent ones are as follows:

November, 1923. The story of the horse. W. H. Carter. 44:455–66.
December, 1925. The Taurine World. A. H. Sanders. 48:591–710.
April, 1927. The races of domestic fowl. M. A. Jull. 51:379–452.
March, 1930. Fowls of forest and stream tamed by man. M. A. Jull. 57:327–71.
April, 1935. Man's winged ally, the busy honeybee. James I. Hambleton. 67:401–28.

Many of the original articles which treat of domestication in detail are in the German language, being written by Germans, Austrians, or Swiss. Among the more recent of these, should be mentioned:

(Various authors). 1936. Neue Forschungen in Tierzucht und Abstammungslehre. Bern: Verbandsdruckerei A.G. (contains the articles by Gromova and by Hilzheimer).

Antonius, Otto. 1922. Grundzüge einer Stammesgeschichte der Haustiere. 336 pp. Jena: G. Fisher.

———. 1936. Zur Abstammung des Hauspferdes. Zeit f. Zücht. B. 34:359–98.

Kelm, Hans. 1938. Das postembryonale Schadelentwicklung des Wild- und Berkshire-Schweins. Zeit. f. Anat. u. Entwicklungsgeschichte 108:499–559.
Klatt, B. 1928. Entstehung des Haustiere. Berlin: Gebrüder Borntraeger.

In the English language the following original articles or books treat of domestication in some detail:

Allen, R. L. 1847. Domestic animals. 227 pp. New York: Orange Judd & Company.
Amschler, Wolfgang. 1935. The oldest pedigree chart. The Jour. of Heredity, 26:233–38.
Brown, Edward. 1906. Races of domestic poultry. 234 pp. London: Edward Arnold.
Ewart, J. Cossar. 1926. The origin of cattle. Proc. of the Scottish Cattle Breeding Conference for 1925, pp. 1–16.
Holmes-Pegler, H. S. 1929. The book of the goat. 255 pp. London: The Bazaar, Exchange and Mart.
Jennings, Robert. 1864. Sheep, swine, and poultry. Philadelphia: John E. Potter & Company.
Lydekker, Richard
1912. The horse and its relatives. 286 pp. London: G. Allen & Company.
1912. The ox and its kindred. 271 pp. London: Methuen & Company.
1913. The sheep and its cousins. 315 pp. New York E. P. Dutton & Company.
Morse, E. W. 1910. The ancestry of domesticated cattle. pp. 187–239 in the Twenty-seventh Annual Report of the Bureau of Animal Industry, United States Department of Agriculture.
Ridgeway, William. 1905. The origin and influence of the Thoroughbred horse. 538 pp. Cambridge University Press.
Shaler, N. S. 1895. Domesticated animals. 267 pp. New York Charles Scribner's Sons.

Among French books, the following is unique enough to warrant special mention:

Chollet, *et al.* (since 1903) Les Animaux. 550 pp. Paris: Bong et Cie.

Among the general references, dealing briefly with man's cultural stages while he was domesticating the animals or with evolutionary aspects of the subject, may be mentioned:

Breasted, J. H. 1938. The conquest of civilization. 669 pp. New York. Harper Brothers.
Matthews, W. D., and Chubb, S. H. 1921. Evolution of the horse. No. 36 of the Guide Leaflet Series, American Museum of Natural History, New York City.
Newman, H. H., *et al.* 1927. The nature of the world and of man. (pp. 332–80 for the story of man and civilization.) University of Chicago Press.
Sumner, F. B. 1932. Genetic, distributional and evolutionary studies of the subspecies of deer mice (Peromyscus). Bibliographia Genetica, 9:1–106.

CHAPTER 2

Consequences of Domestication

The fundamental laws of heredity and the mechanics and physiology of reproduction are the same among domesticated animals as among their wild relatives. The change from the wild to the domesticated condition did not alter these laws nor create any new inheritance. The changes which domestication did bring about were an increased amount of inbreeding and outbreeding and assortive mating and the addition of artificial selection to the forces of natural selection. The changes in environment which accompanied domestication doubtless permitted many differences in heredity to show themselves more clearly than they could in the environment of wild animals, and thus to be more readily and accurately selected. For example, when feed is so scarce that no animal gets all it wants, hereditary differences in ability to fatten could hardly show themselves as clearly as they can in a well-managed feedlot. But there is no direct evidence to indicate that the changed environment directly *created* any new genes or hereditary differences.

Increased inbreeding happened because domesticated animals were more narrowly restricted (by tethering, herding, or fencing) to growing up, remaining, and reproducing in the same region in which they were born. Soon the animals around one community or village would all become related to each other. Most breeders, even among primitive peoples, intentionally avoided the very closest inbreeding; but often the pedigrees were known only for a generation or two, or only in the female line. Under such circumstances, attempts to avoid inbreeding merely made the inbreeding less intense so that more time was required to produce the same amount of fixation and uniformity in the stock. That gave more opportunity for the accompanying selection to discard undesired results of the inbreeding than if the inbreeding had been extremely close. Even when obstacles to exchange between tribes were greatest, some introduction of outside blood went on either by trade or by war. Doubtless these exchanges usually involved stock from only the closest neighboring tribes. Such animals, as a result of previous exchanges, would already be more closely related to those among which they were introduced than animals which might have been chosen at random from the whole species. There would have had to be a large amount of such interchange to prevent this inbreeding from bringing about a situation where almost every tribe or community would have its own type of each kind of animal. Doubtless the intensity of this inbreeding varied greatly from region to region according to the nomadic habits of the people, geographic barriers, or social customs which may have prevented extensive trade with neighboring tribes, and the extent to which their beliefs about breeding led them to try deliberately to prevent this inbreeding by taking special pains to get sires from unrelated or remotely related stocks. This extra inbreeding resulting from restricting the movements, and also from limiting the number of breeding males, may well have been one of the most potent forces leading to the production of diverse races among domestic animals. It does not seem possible now to measure its past importance accurately, since so little is known about actual breeding customs until very modern times. The geographic or physiologic isolation, which in nature divides so many species into small sub-groups between which crosses occur only at rare intervals, is the very same kind of a process in principle, but probably is rarely as extreme in nature as it is under domestication where man has added so many artificial barriers to the natural ones.

Increased outbreeding has certainly resulted now and then from domestication. By the agency of man, breeding animals could be transported far beyond the area in which they were born or over which they could have wandered before they were domesticated. Thus they could be crossed on races more diverse than they would have encountered if they had remained in the wild state. Knights returning from the Crusades brought with them stallions from Arabia. Cattle were brought from Holland to England across water which would have been impassable to wild cattle. In a later day Merino sheep were taken from Spain to many lands, Angora goats came from Turkey to the United States and to South Africa, zebu cattle came from India to Brazil and to the Gulf Coast of the United States, and Shorthorn cattle went from England to the Argentine and Australia. Many other examples could be cited to show how this process went on in historic times. The diverse races of swine from which the Poland-China breed was formed could hardly by any conceivable circumstances have come together anywhere on earth without the intervention of man to transport them.

Presumably this process has gone on more rapidly in the last three or four centuries of exploration and widespread trade than it did formerly, but Phoenician traders were traveling the length of the Mediterranean and skirting the western shores of Europe as far as Britain nearly three thousand years ago and doubtless helped exchange some of the smaller animals at least. Invading armies or migrating peoples usually carried with them much livestock from their native lands; some of these were mingled with the breeding stock of the countries through which they passed. Two thousand years ago Hannibal and Hasdrubal took their armies and livestock, including such large and unwieldy animals as elephants, from Carthage along the northern coast of Africa to Spain and over such mountains as the Pyrenees and Alps almost to Rome itself. Also, Alexander the Great took large numbers of livestock with his armies on the road from Greece to India and had with him men who made careful notes about the strange livestock they saw in the new lands. In the early thirteenth century the armies of Genghis Khan and his sons, with their enormous reserves of cavalry horses, ranged all the way from eastern China to central Europe. The passing of that band of horses (with the inevitably large amount of straying and robbing) must have changed greatly the genetic composition of the local races in the regions through which they passed. With such cases known from definite history, it is only reasonable to suppose that similar exchanges by migrations or wars had been taking place almost since the beginnings of domestication. Here again, as in the case of geographic isolation, there must have been much more of such exchanges in some parts of the world than in others.

A combination of moderate inbreeding alternating with occasional wide outbreeding is an effective plan for producing many distinct families which are moderately uniform within themselves. A population being thus bred is in a more favorable condition for selection to be effective than if matings within the group selected to be parents were entirely at random. In this way domestication made conditions more favorable for the formation of distinct races than exist among wild animals.

Selection means differences in reproductive rates within a population, whereby animals with some characteristics tend to have more offspring than animals without those characteristics. Thereby the genes of the favored animals tend to become more abundant in the population and those of the less favored animals less abundant. Artificial selection differs from natural selection only in the kind or degree of the characteristics which are thus favored. Also, in many cases artificial selection may be more intense, less of the decision being left to chance or to accidental circumstances than in the case of natural selection.

Natural selection did not wholly cease with domestication. More of the weak and sickly than of the strong and vigorous are still doomed to die before they reach breeding age. This will happen whether the breeder consciously aids in this selection or not. Indeed, some of it will happen in spite of the breeder's efforts if he tries to breed a type which is rather frail or susceptible to disease. Among domesticated animals natural selection is merely supplemented by man's selections. In making his decisions as to which animals should leave few and which should leave many offspring, man often strongly emphasizes characteristics which were of little worth in a state of nature. Other qualities valuable in the wild state became useless or nearly so when man began to protect his animals against their enemies, against cold and against starvation. Thus, man's selection may differ from natural selection both in *intensity* and in *direction*.

The practice of favoring for breeding purposes those animals which in their owner's opinion were the most desirable ones must have begun with domestication itself. The recognition that offspring tend to resemble their parents and other near relatives occurred at so early a cultural stage that proverbs embodying this idea are found in practically all languages, even those of extremely primitive people. Primitive man was doubtless quite shrewd enough to put this knowledge into practice on his domesticated animals. Castration is one of the most ancient of surgical practices. Medical literature traces it back at least as far as 700 B.C., and in the Bible there is frequent reference to eunuchs in the time of Solomon or earlier. Eunuchs are mentioned in the Code of Hamurabi; ca. 2000 B.C. Castrated bullocks (balivarda), as distinct from bulls, are mentioned in the Puranas of the Hindus, which deal with the events of the Aryan migrations into northwestern India probably as long ago as 2400–1500 B.C. (Since the Puranas were not put into writing until much later, the words may have been changed during the interval). The general practice of castration must have intensified the selection which was practiced among males.

The Roman agricultural literature of about two thousands years ago[1] contains many bits of advice about the kinds of animals to select for different purposes. Much of this they had copied from still older writings, such as those of Mago the Carthaginian. It is certain that artificial selection has been practiced by man for thousands of years, although there seems to be no way of measuring the intensity with which it was practiced.

Another aspect of selection which was intensified by domestication was that the characteristics favored for one set of conditions might not

---

[1] Harrison, Fairfax. 1917. *Roman Farm Management.* 365 pp. New York: The Macmillan Company.

be the same as those favored under other conditions. This, of course, was true in nature also, contrasting characteristics sometimes being favored for life on the open grasslands or for life in the forest, for mountain or lowland, for tropics or temperate climates, etc. In nature these would often—perhaps usually—be characteristic of wide areas with broad transitional zones between them. Under domestication one man might prefer a certain type of horse or cattle, and his next-door neighbor or the people of the very next village might prefer and select for a distinctly different type. Indeed, as agriculture became more complex the same man might keep two or more types of the same species, each being preferred for some special purpose. For instance, the ancient Egyptians had several breeds of dogs, and modern farmers keep different breeds of cattle for beef and for dairy purposes. So far as one breed or type is concerned, this is nothing but selection directed toward that particular ideal; but, from the standpoint of the whole species, this is assortive mating, which is a powerful tool for producing diversity within a species. This seems certain to have been more intensified under domestication than it was in nature.

SUMMARY

Domestication merely intensified forces or processes which already existed in nature. Increased inbreeding alternating with wider outcrossing, more intense selection devoted toward a wider variety of goals, and mating like to like wherever one man or tribe was breeding the same species for two or more different goals, all had the net effect of tremendously speeding up the slow process of evolution as it occurs in nature, until remarkably large changes were made in animals under domestication during what was a very short period in terms of geologic time. That the changes thus brought about were at the maximum rate possible seems highly unlikely. If further changes are desired, it is probable that the possibilities in most directions are by no means exhausted and that intelligent use of these same processes can result in much faster progress than has been averaged during the long (in terms of human lifetimes) history of domestication.

CHAPTER 3

The Beginnings of Pedigree Breeding and the Formation of Breed Registry Societies

"The virtues of their fathers live on in bulls and in stallions." Horace
"Who would grow spirited stallions for the Olympic prizes or strong bulls for the plow, let him choose carefully the females who will be their dams." Virgil

Emphasis on ancestry in human genealogies is older than history, although human pedigrees may have been used more for social purposes, such as to determine the inheritance of property or of rank in a caste system of society, than because of definite beliefs about the inheritance of physical and mental qualities. Often these pedigrees recorded only the male or only the female line of descent. The genealogies in the early chapters of Genesis are examples of this.

Pure breeding is also an ancient idea as applied to man, not only in those peoples which had a pronounced caste system of society but also in many others where a tribe was warring on neighboring tribes, or a conquering race was trying to live with but keep itself "pure" from a conquered or slave race. Also, many tribes or races with a simpler social structure cherished myths about their racial origins which implied that they alone were the chosen people descended from the sun or the moon or some other deity, while other peoples were of inferior or mixed descent. In most (but not all) human societies there has been a heavy social prejudice against the "half-breed," which in general has meant that the half-breed and its descendants must come up to higher standards of individual excellence than the average "purebred" before they would have an equal chance to contribute to the inheritance of the future race.

The Arabs in their horse breeding more than a thousand years ago were memorizing the genealogies of their horses, but we have no detailed knowledge of how these genealogies were used—if at all—to guide them in making the matings. Probably, like the modern Arabs, they traced the pedigrees only in the female line and used the family name only as an aid to selection,[1] also taking some care to avoid close

[1] Nurettin, Aral, and Selahattin, E. 1935. *Der heutige Stand der Pferdezucht in Arabien.* Zeit. f. Züchtung. (Reihe B), 33:13–38.

inbreeding. The Romans of the time of Varro and Cato made many comments about the kinds and types of animals which should be selected for breeding purposes but apparently made no attempt to memorize or record long pedigrees for their livestock. Varro's comments on the importance of judging the breeding worth of a sire by the quality of his get show that in a general way they were aware of the importance of the progeny test and the use of pedigrees, traced at least to the parents and grandparents, to help them to a more correct estimate of an animal's breeding worth.

"Throughout the Middle Ages the authority of the written word almost completely displaced firsthand observation and experiment in the search for truth."[2] Largely because of this, knowledge of the mechanics and laws of heredity advanced but little. Most of the learning was preserved in the monasteries and what little is known about agriculture in the middle ages comes mostly from the account books, inventories, and fragmentary notes kept in connection with the farming operations of the monasteries. It was not until 1700 that enough was written and preserved to give us a connected account of agricultural practices.

The use of pedigrees in the modern manner began in rural England late in the eighteenth century, and the general formation of breed registry societies began around the middle of the nineteenth century. Robert Bakewell is generally given credit for setting the pattern of modern animal breeding and is sometimes called the founder of animal breeding. Perhaps this is giving too much credit to one man, but at any rate pedigree breeding was established in his time, and his own outstanding success had more to do with making it popular than the efforts of any other one man did.

Robert Bakewell was an English farmer or country gentleman who lived from 1725 to 1795.[3] We first hear of his agricultural efforts when he began to manage the estate at Dishley in 1760. He wrote little or nothing about himself. He was a good farmer in other things besides his stock breeding, having taken a prominent part in the introduction of turnips and other root crops into English agriculture. He was a good observer, a keen student of anatomy and probably a good judge of livestock. According to some accounts he even kept for future reference specimens of the bones or pickled joints of animals which he had bred and which he regarded as nearly ideal. He told so little about his opera-

---

[2] Mees, C. E. Kenneth. 1934. Scientific Thought and Social Reconstruction. *Sigma Xi Quarterly*, 22 (No. 1) :17.

[3] A more complete account of Bakewell's work and of the conditions of animal breeding then is given on pages 176–189 of Lord Ernle's *English Farming, Past and Present*. 1936.

tions that many of his contemporaries thought there was something mysterious about them. Some writers hint that this was done deliberately to avoid competition or censure. This latter point is made because an important element in his procedure was the deliberate and intense use of inbreeding. At that time there was even more prejudice against inbreeding than there is today, and many people thought it almost sacrilegious. Perhaps Bakewell thought it no use to invite criticism by proclaiming openly what he was doing. There is also more than a hint that he kept his operations secret because of certain extreme outbreeding he was practicing which, if known, might have injured the commercial reputation of his stock. Thus, there were rumors of a mysterious black ram used in his sheep breeding which visitors were never permitted to see and whose existence he would never admit.

Bakewell's own breeding work was with the old Longhorn cattle, Leicester sheep and Shire horses. He was so successful with these that his animals came into great demand as breeding stock. He inaugurated the practice of ram-letting. That is, he did not sell his best males outright, but rented the use of them a year at a time. His annual auctions, or ram-lettings, attracted great attention and were a distinct financial success. He is said to have received as much as 1,200 guineas for one year's use of a ram. By this practice of ram-letting, the best sires came back to him each year and any whose progeny had proved them much better than the others could be kept for use in his own flocks or herds. There seems to be no record of how many times he took back for his own use a sire which originally he had thought not quite good enough for that, but no doubt such instances occurred.

Bakewell's success attracted many imitators. From many parts of England ambitious stockmen went to Dishley to work with Bakewell and study his methods. Some of them stayed for as much as six months. Returning home they applied his methods to stock secured from him or to what they thought were the best of their own local animals. The details about these students and what they did are poorly known, but it is certain that the Collings who laid the foundations of the Shorthorn breed were in close touch with Bakewell and that men from Herefordshire were students with him. Enough of Bakewell's followers won distinct success that here and there all over England there soon began to be groups of animals closely related to each other and similar in type. These were the groups from which came the modern breeds, most of which were not formally organized as such until later.

The *principles* which Bakewell used included such things as: "Like produces like or the likeness of some ancestor; inbreeding produces prepotency and refinement; breed the best to the best." His greatest contribution to breeding methods lay in his appreciation of the fact that inbreeding was the most effective tool for producing refinement and fixing type. He was reluctant to make any outcrosses at all when his own stock seemed to him better than that of his neighbors. With his willingness to inbreed was coupled a good knowledge of anatomy and a keen interest in the subject of what types of animals were best suited to his agriculture and should be set up as goals.

The economic setting of the times, of course, had much to do with the increased interest in breeding improved animals. The enclosures of the "common lands" had given the individual farmer opportunity to breed his own stock as he pleased and to reap the rewards of anything he might do to improve them or to build up the fertility of his own lands. The introduction of clovers and root crops to English agriculture had made more intensive animal husbandry possible and had supplied a store of roughages suitable for winter feeding. The times were ripe for commercial appreciation of animals which could utilize the crops of the new agriculture better than their contemporaries and which would produce a quality of product well suited to contemporary market demands. The warfare through the latter half of the eighteenth century, finally coming to a climax in the Napoleonic wars, often made prices high for farm products. Afterward, the industrial revolution and the steadily expanding urban population of Britain made a rising market for agricultural products, more particularly for those like meat which, in the days before refrigeration, could not be imported in the fresh state from the New World. When the improvement which Bakewell and his followers had made in their breeding stock began to be known in other lands, the export of breeding stock to those lands became a considerable source of income to British stockmen. Appreciation of the importance of this was a spur to further improvement in order to keep the foreign customers coming back for fresh breeding stock and had much to do with guiding the policies of breed registry societies.

BREED REGISTRY SOCIETIES

As long as each breed was local the private records of each breeder were adequate for his own purposes. He usually knew at least the sires used by his fellow breeders and knew the integrity of those breeders well enough to have some idea of how much he could depend on their statements or records when purchasing breeding animals from them. But in time the number of breeders increased until many of them were utter strangers to each other, and the number of animal generations in the pedigrees increased until no man could remember all of the foundation animals far back in the pedigrees. To supply this knowledge and to prevent (as far as other breeders could) unscrupulous traders from exporting grades or common stock as purebreds, herdbooks were formed. The latter motive was very important[4] and generally the herdbook was established soon after there began to be a considerable export demand.

The first herdbook was "An Introduction to the General Stud Book" for the Thoroughbred horse and appeared in 1791. In it were recorded the pedigrees of the horses winning important races. It was aimed, therefore, at recording the pedigrees of performers rather than of all members of the pure breed. The Shorthorn herdbook, which first appeared in 1822, was the next one formed and may be taken as an example of the modern type of herdbook which aims at including the pedigrees of all animals of the pure breed. The Shorthorn herdbook, however, like the one for the Thoroughbred horse, would accept for entry outstanding individuals or performers which would be called high grades in the United States. Later herdbooks were largely modeled after the more successful of the early ones. An English Hereford herdbook was published in 1846 and a Polled Herd Book (for Aberdeen-Angus) in 1862. The first swine herdbook in the world was that of the American Berkshire Association, which appeared in 1876. The Berkshire Society in England was established first in 1883. The first herdbooks in the continental countries of Europe appeared at a later date than in Britain. Studbooks for horses were founded in France in 1826, in Germany in 1827, and in Austria in 1847. The first cattle herdbook in France was established in 1855, the first German one in 1864, the first Dutch one in 1874 and the first Danish one in 1881.

The Shorthorn herdbook was undertaken as a private venture by George Coates, who had been a Shorthorn breeder in a small way. A number of the Shorthorn breeders helped finance him and each was to receive a copy of the book. The other copies were to be his personal property to sell for whatever profit he could. He was already acquainted with many breeders, and from each of them he secured such information as he could about the animals which that man regarded as genuine Shorthorns. No doubt there was plenty of dispute about that. That is, some breeders would say that certain animals should be included in the records while others would think that those animals were not sufficiently desirable to be included as genuine Shorthorns. Coates was criticized, of course, for many of these decisions; and it was even charged[5] (perhaps unjustly) that his favoritism to his personal friends went to the extent

[4] For example, many events in Bates' book (1871) illustrate the incentive which the American demand for Shorthorns gave to the formation of the Coates herdbook.

[5] *See* p. 38 in Bates' *History of Improved Shorthorn or Durham Cattle.*

of printing for their cattle false pedigrees which would make them sell well to the American trade, then becoming important. Where Coates included animals which the majority of the breeders thought were not really pure, the breeders themselves could remedy the situation by having nothing to do with those animals or their descendants—a course of action which is still open today and which is still used freely wherever falsification of pedigrees is suspected but evidence is not complete enough to justify canceling the registration. Wherever Coates omitted from the first volume of his herdbooks animals which the majority of breeders thought should have been included, such mistakes could be corrected by including these pedigrees in succeeding volumes of the herdbook—a process no longer available wherever herdbooks are entirely closed to all but the offspring of registered parents.

Some of the very early breeders objected to furnishing pedigrees of the animals they sold, believing that they would thus give away valuable trade secrets. The demand for full information about pedigrees, however, finally prevailed over the "trade-secret" idea; and it became accepted as a matter of course that anyone selling breeding stock should furnish full identification of their immediate ancestors.

Doubtless many of the contemporary breeders felt that this herdbook of Coates was only a hobby of his which would disappear with his death; but, as the breed became more popular and the number of breeders increased and the number of generations to be remembered in the pedigrees grew larger, the difficulties which first prompted Coates to the formation of the herdbook became greater. Eventually every breeder admitted the necessity of the herdbook, in view of the customer demand for pedigrees, and depended upon it in his purchases and sales of breeding stock. When this stage was reached, those who owned the herdbooks had the power to charge exorbitant fees for registration and transfer or to use their influence to favor the business of certain breeders and to harm that of others. While there was rarely any widespread complaint of this kind, yet it generally seemed wiser or even necessary for the breeders to organize breed associations in order to manage the herdbook, conduct breed promotion, and attend to any other matters which could be handled best by co-operative action.

The typical history of the formation of the British breeds (the breeding practices in the United States are patterned closely after the British ones) was about as follows: First, came the existence of a type which was more useful and desirable than the ordinary type, but which was not yet distinctly different in pedigree from the other animals in the community. Second, some of the best animals of that type were gathered into one or a few herds which then ceased to introduce much outside blood. Then followed some rather intense inbreeding among these animals and their descendants until the animals of those herds became distinct from the other animals in the community, not only in type but also in inheritance; that is, until they were really welded into a breed. Third, if this process had been moderately successful in producing a desirable kind of animal, the breed became more and more popular and more and more herds were established. Fourth, necessity for a central herdbook arose when the breed became so numerous and the breeders so many that no man could remember all the information needed for the proper use of pedigrees. Fifth, a breed society was formed to safeguard the purity of the breed, conduct the herdbook, and promote the general interests of the breeders. From the very beginning many of these breeders emphasized that the males they produced were especially valuable for crossing on other races or on common stock. An important function of these pure breeds was to produce sires for commercial use on unrelated stock, even for crossbreeding.

In not all breeds did the breed history develop in just exactly these steps. Sometimes there was a breed society before there was a herdbook. Thus, even in Bakewell's time a Dishley Society was founded, with the primary object of protecting the pure breeding of the animals descended from those bred by Bakewell and the commercial promotion of the interests of those who were breeding animals of the Bakewell strains. Often there was no intervening stage of private ownership of the herdbook, but the breed society established the first herdbook itself. In practically every case the breed was a well-established fact before any herdbook was considered. People did not say to each other: "Let's establish a breed." Rather they said: "Here we already have a useful and profitable breed. We should protect its purity and our own interests as possessors of this valuable breeding stock and the interests of the purchasers who want genuine animals of this breed."

In the continental countries of Europe pure breeding and registration were generally organized at a later date than in Britain. In Germany and adjoining lands (Engeler, 1936), extensive efforts at improvement developed first in sheep breeding,[6] then in horse breeding, and then in cattle breeding. In the period about 1800 it was common practice to cross extensively, even for producing seedstock, in accordance with the idea expressed by Buffon (1780) that perfection could be attained only through widespread crossing and mixing of all individuals which had any of the desired points, regardless of race or regional

[6] As long ago as 1779 Daubenton was measuring wool fineness with a micrometer and in 1802 Abilgaard wrote in detail about the reasons for marking sheep individually so that their production could be recorded and used as a basis for selections.

origin. Then for a half century the trend changed toward following the successful English example of pure breeding, that is, of improving a breed from within itself.[7] Some of the writers (but perhaps not many of the breeders themselves?) carried this to an extreme form in what became known as the theory of "racial constancy." This held that each animal transmitted according to its race and not according to its own characteristics. The latter were unimportant except as they indicated the animal's purity of race. Under the influence of that doctrine, herdbooks were only records of genealogy, and official attention was focused almost wholly on purity of breeding.

Sharp reaction to the theory of "racial constancy" developed about 1860, and the pendulum swung far the other way, at least among the writers. Thenceforth attention was devoted more to the individual. They sought more and more to make the herdbooks contain full information about each animal's characteristics, productivity, conformation, reproductive performance, longevity, etc. To collect this information the herdbook societies were organized around semi-official local records which might be either the private herdbook which the breeder was required to keep himself or the records kept by a local breeding association organized somewhat like a dairy herd improvement association in the United States. In either case the records to be kept were definitely prescribed and were inspected more or less regularly by officials of the herdbook society or of the government. From those local records the central herdbook society collected regularly the information thought useful there.

Because of this background, the continental breed associations make more use of formal scoring or other inspections or production requirements as a prerequisite to registry than is done in Britain, where the responsibility of deciding whether a purebred animal is good enough for registry is still left almost entirely with the individual breeder. In Britain it is thought that the reputation of his herd and the resultant prices which the cusomers will pay will more or less automatically reward or penalize the breeder if his efforts have been above or below average.

Often the continental associations have only tentative registry at birth; final registry in a printed herdbook is postponed until an animal is mature or even until it is dead and all of the data on its lifetime performance, prizes won, scores for type, etc., can be printed, too.

At the Strickhof agricultural school at Zurich, Switzerland, the pro-

[7] Thus Krünitz wrote in 1815 with surprise: "The English improve a race from within itself. They choose carefully the best individuals they can find within the same race and mate these together. In this way they keep the stock unmixed and produce a race in which the desired qualities are retained permanently."

duction of all cows has been recorded continuously since 1871. The first cow-testing association in the world was established in 1892 at Vejen in Denmark. For the last two decades about 40 per cent of the cows in Denmark have had their milk weighed and tested. Those thought to be best among these are admitted to registry each year. Often the continental associations were built around some such plan for recording production. At the beginning of 1938, 67 per cent of all cows in Germany were on test but the figure had only very recently risen that high.

In the lands of their origin the breeds usually continued for a long time to register what would be called high grades in the United States. A common rule—which still holds there for many breeds—was that females with four top-crosses of registered sires were eligible to registry themselves if they came up to certain standards of individual excellence. In importing lands, such as the United States and Argentina, the herdbooks have usually been closed from the very beginning, and fashions in pedigrees have often gone to greater extremes in waves of speculation than has been the case in the native lands of the breeds. The greater emphasis which the importing countries placed on strict purity of breeding is illustrated by the fact that for some breeds, e.g., Berkshire swine, Holstein-Friesian and Ayrshire cattle, and Hampshire sheep, herdbooks were established in the United States before they were in the native land of the breed.

Breeders of poultry have never attempted general registration of all eligible individuals. The short life and comparatively small value of the individual birds have made that uneconomical. There have, however, been a number of attempts to register individuals in connection with a scheme of advanced registry for outstanding producers, e.g., Lancashire (England) Poultry Society, Record of Performance in Canada, the Record of Performance in the United States.

As an illustration of the difficulties encountered in assembling the first herdbooks, we may take the pedigree of the first bull in the Coates herdbook and wonder how Mr. Coates collected and verified the information printed for it. That pedigree is as follows:

> "(1) Abelard, Calved in 1812, bred by Major Bower; got by Cecil (120), d. (Easby) by Mr. Booth's Lame Bull (359), g.d. by Mr. Booth's Old White Bull (89), gr. g.d. bought at Darlington."

For a more specific account of some of the kinds of mistakes later found in those first herdbooks, consult the second edition of "The Polled Herd Book" (Aberdeen-Angus in Scotland) and read the preface and the notes in brackets under the pedigrees of bulls numbers 1, 2, 3, 4, 12, 17, 29, 35, 49 and 51.

As an example of the controversies which arose over the purity or non-purity of certain animals may be cited the long controversy in early American Shorthorn history over the "seventeens"[8] which were imported in 1817 and hence were not recorded in the Coates Herdbook since it did not appear until five years later. Also, Bates makes many references to the long controversy over the "Galloway alloy," which one of the early Shorthorn breeders was thought by some of his fellows to have introduced into his Shorthorns.

REFERENCES

Anonymous. 1894. Jour. Royal Agr. Soc. of England, 55:1–31.

Allen, R. L. 1847. Domestic animals. 227 pp. New York: Orange Judd & Company.

American Shorthorn Herdbook, 2:33–69.

"The Major." 1920. Robert Bakewell's great work. The Shorthorn World and Farm Magazine, 5 (No. 7):3–4 and 63–65.

Bates, Thomas. 1871. History of improved Shorthorn cattle (from the notes of Thomas Bates, edited by his foreman, Thomas Bell). See especially pp. 11, 19, 37–38, 213–15 and 226.

Engeler, W. 1936. (In) Neue Forschungen in Tierzucht. Bern. Pp. 39–70.

Harrison, Fairfax. 1917. Roman farm management. New York: The Macmillan Company.

Sanders, A. H. 1915. At the sign of the Stockyard Inn. (Especially pp. 45–88.) Chicago: Sanders Publishing Company.

———. 1918. Shorthorn cattle. (Especially pp. 29–113.) Chicago: Sanders Publishing Company.

———. 1936. Red, White and Roan. (*See* especially pp. 1–24.) Chicago: American Shorthorn Breeders' Association.

Van Riper, Walker. 1932. Aesthetic notions in animal breeding. Quart. Rev. of Biology, 7:84–92.

Wilson, James. 1926. The history of stockbreeding and the formation of breeds. In Proc. of the Scottish Cattle Breeders Conference for 1925, pp. 17–25.

[8] *See* pp. 165–72 of "Shorthorn Cattle," by Alvin H. Sanders. Sanders Publishing Company. 1918.

CHAPTER 4

The History of Animal Breeding Methods in the United States

The first settlers in the New World brought with them such animals as they thought would be most useful. In most cases those came from the same communities as the immigrants themselves. Little detail is known about the animals they brought; but that is not surprising, since most of the pioneering period was ended in the region east of the Mississippi River before the period of herdbooks and pedigree breeding began in Britain. Then, too, during pioneering times the problems of defense against marauding men and wild animals and the problems of learning to raise the new crops and of adapting the old crops to the new climate overshadowed in importance any question of animal breeding methods.

Where the new conditions demanded a new type of animal, the pioneers or the first generations of settlers which followed them, seem to have been ready to produce that type. Thus, there were developed the Vermont Merino, the cornbelt breeds of swine, and horses like the Narragansett pacer, the Conestoga, the Morgan and the Standardbred, the American Saddle Horse, the Quarter horse, and many another race of less fame, each of which fitted some local need well enough to become known, but many of which never reached the stage of having an organized herdbook. Many of them have ceased to exist, either because they were engulfed in the flood of undiscriminating enthusiasm which came later for registered stock or because the economic and physical conditions for which they were adapted had ceased to exist.

It is an interesting but perhaps an idle speculation to wonder whether animal industry in the United States would have been more efficient today if in more cases the good local races, already well adapted to local conditions, had been preserved, either as pure breeds or as foundations for breeds combining some of the good traits of the local stock and of the imported breeds used for improving them. Examples of the latter process are such widely separated cases as the American Saddle Horse and Poland-China swine. Allen (*Domestic Animals,* pages 26 and 27), writing in 1847, says of cattle: "Every country and almost every district has its peculiar breeds, which by long association have become adapted to the food and circumstances of its position and, when found profitable, they should be exchanged for others, only after the most thorough trial of superior fitness for the particular location, in those proposed to be introduced."

Appreciation of the usefulness of improved breeding stock followed hard on the heels of the pioneering period. Cattle of the Shorthorn, Hereford, and Devon breeds were imported around 1800; although there were yet no herdbooks for those breeds. The new pedigree breeding methods of Britain were followed with interest. The individual pedigrees for the Herefords and for the Devons in those early importations were not permanently preserved, but some of the pedigrees in the early volumes of the American Shorthorn Herdbook trace in part to animals imported as long ago as 1817. Enthusiasm for pedigree breeding sometimes reached the stage of extreme speculation, as with some of the Merino sheep breeding early in the nineteenth century and with the Duchess Shorthorns in the 1870's.

General interest in pedigree breeding was mainly confined to the Shorthorn breed and to certain light horses and to the early sheep breeding in New England until the era of agricultural expansion which began soon after the Civil War. Then purebreeding became fashionable for all kinds of animals. The period from 1870 to around 1890 saw the founding and rapid expansion of breeding societies for almost all kinds of livestock, each with its herdbook, scorecard, etc. Most of the breeds were introduced from Britain; although there were a few from other places, such as the Holstein-Friesians from Holland and the Brown Swiss from Switzerland and draft horses from Belgium and France. Also, there were some native breeds like the cornbelt breeds of swine and the Standardbred and American Saddle Horses. The agricultural colleges, most of which were founded just before or early in this period, promoted the movement toward the general use of purebred sires as one of the quickest ways to improve the quality and efficiency of animal production. Experiments to find or demonstrate the value of the purebred sire in grading up common stock were conducted at some agricultural colleges until well into the 1920's.

The purebreds seemed obviously superior to the common stock in many ways, especially as the country was becoming urbanized and the available markets for animals were becoming similar to the markets for which the British stock had been bred for many animal generations. The initial gulf betwen the purebreds and the native stock was a wide one in most cases, since no large group of high grades with which to compare the purebreds had yet been produced. The most urgent need of the times in animal breeding seemed clearly to be a wider use of purebreds.

This expansion in the use of purebred sires made a generally expanding or rising market for the business of producing purebred sires. In turn, that favorable economic situation led to the establishment of still more herds and flocks of purebreds. Of course this rapid expansion in numbers of purebreds could not continue forever. For most breeds it came to a rather abrupt end with the economic depression which began for most agricultural enterprises about 1920. That economic crisis merely hastened the end of the remarkable expansion in purebred numbers which began about 1870. There is still room for promotion of the wider use of purebred sires, and doubtless some expansion in the total number of purebreds will yet be seen; but it will be at a slower percentage rate than was generally true for the half century ending in 1920. There are almost enough purebred flocks and herds to produce as many purebred sires as can be sold at a profitable price for use in commercial flocks and herds.

This end to the long period of expansion in numbers has meant a serious readjustment of the business of those who produce purebreds, and it has also had an effect on the way they regard current breeding problems. There may be nearly enough registered sires to supply all the demand which exists or can be aroused by good salesmanship, but there are not enough highly meritorious registered sires. Much of the emphasis which used to be given to breed expansion is being changed to breed improvement.

A contributory cause to this change of emphasis is that for most breeds large numbers of high grades have already been produced. Some of those are individually more meritorious in a practical way than the average of the purebreds, even though the average of the grades remains below that of the purebreds. Most of the experiments with the use of purebred sires for grading have shown that the averages of grades with more than two crosses of pure blood (that is, with more than 75 per cent pure blood) are very little below the average of the pure breed concerned. When there were no high grades for comparison, the differences between different purebreds seemed small and not worth much emphasis, compared with the gulf which often existed between the introduced purebred and the common native stock. Now the differences between members of the same pure breed are often large compared with the small average differences between purebred and high grade.

The idea of breed improvement is not new in kind—it is merely receiving more emphasis than formerly. Nearly all of the breed associations included improvement of the breed in their very first statements of the objectives of their association. An excellent example of that was the insistence by many of the early Holstein-Friesian breeders on the adoption of a system of official testing of production before they would merge their two breed associations, one of which lacked this feature. Nearly every breed association, very early in its existence, adopted a scorecard or some other form of official description of the ideal toward which they were working. Occasionally they even went to considerable trouble or expense trying out plans which were intended to improve the breed more rapidly. Often these did not work as expected and had later to be discarded. Examples are the bounties which the Holstein-

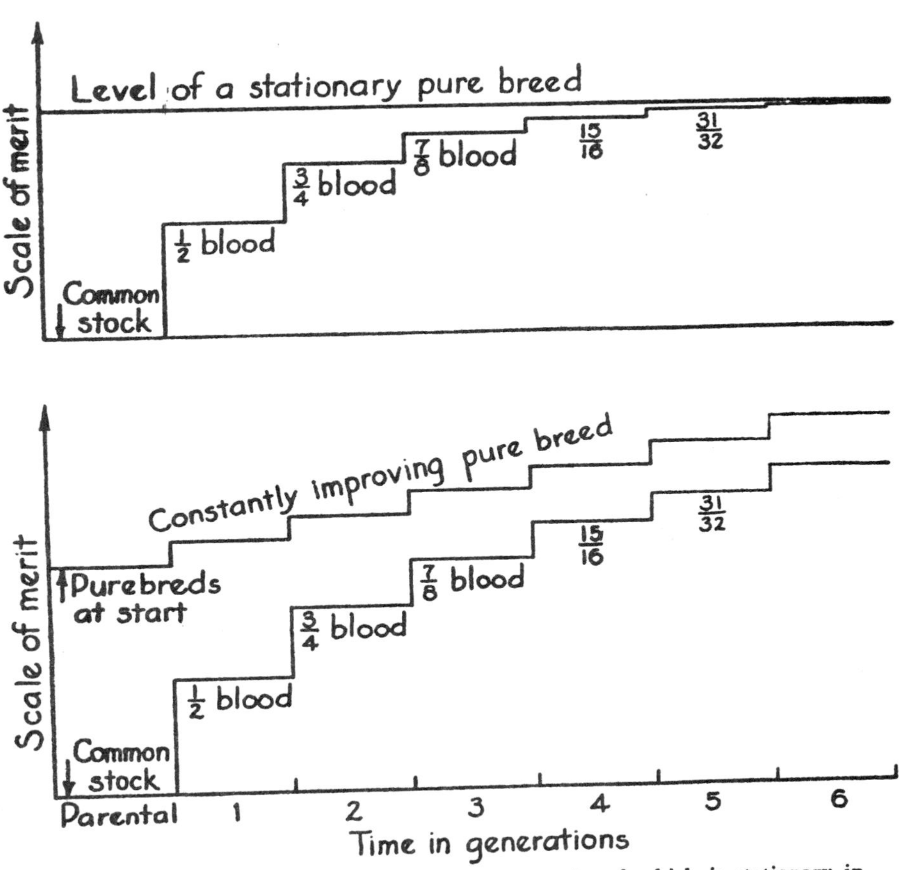

FIG. 2. The contrast between grading to a pure breed which is stationary in merit and grading to a pure breed which is improving its merit by a constant amount each generation.

Friesian association offered in 1889–1891 for butchering or castrating bull calves which would have been eligible for registry, and the ruling of the Hereford association (in effect during 1895–1897) that 10 per cent of all applications for registry of bulls would be refused. The Aberdeen-Angus Association in 1887 had a rule that of every ten bull calves eligible to registry, one must be castrated or two shall be dropped from record. Actually it appears that 70 were castrated and 8 were unrecorded while this was in force.

The very fact that the purebred has generally been found useful in grading up common stock carries with it the necessity for the continued improvement of the purebred itself if the distinction between pure breds and grades is to be maintained. If the pure breed did not continue to improve, the average merit of the grades would soon be so nearly equal to that of the purebreds that the distinction between them could not long be maintained. Figure 2 illustrates this on the assumption (which is true in general, although with many exceptions) that offspring average about half way between their parents.[1] The upper half of the diagram shows what would happen under continuous grading to a pure breed which was not itself being improved. The lower half deals in a similar way with a pure breed which is being improved each generation by an amount equal to one-tenth of its initial superiority to the common stock. The difference betwen upper and lower halves of the diagram is not important in the first generation or two but becomes pronounced after several generations of grading and shows how a continuously improving breed may maintain a commercially important margin of superiority over its grades as long as it can continue its own improvement. The diagram, of course, is simpler than the facts (as will be seen in more detail in the chapters on selection); but it demonstrates the necessity of continued breed improvement to maintain the superiority and popularity of the purebred sire.

Animal breeding history in the United States may be divided into four periods. First, there was a pioneer period when the kind of livestock was not of much importance. Second, there followed a period of developing local races and a little experimenting with introduced breeds. Third, came a period of extensive experiments with introduced breeds and widespread expansion of organized pure breeding characterized by rivalry between breeds and by breed promotion. In the present period the dominant interest is in breed improvement, not only for successful competition with the other breeds, but also for keeping the

[1] Allowance for regression toward the mean of the groups from which the parents were chosen needs to be made if the parents were selected individuals. Also *heterosis* sometimes causes the first-cross generation to be better than the diagram indicates.

purebreds distinctly ahead of the grades of the same breed in order to maintain a steady demand for purebred sires. These periods did not begin and end suddenly but merged gradually into each other, and some of the elements which characterize each period were present in lesser degree in all periods. At present it seems likely that plans for breed improvement will go on within the framework of the existing systems of purebreeding, but here and there occurs some experimenting with the combination of more or less blood from two or more breeds, just as was done by some of the breed founders in the days before strictly pure breeding and registration became the standard methods of improved breeding.

CHAPTER 5

The Relation of the Breed Association to Breed Improvement

The activities of the breed associations are intended to maintain the present merit of the breed, to improve the merit of the breed, and to promote the business interests of the members. Some activities serve all three of these purposes, and many serve two of them.

The primary object of breed associations, particularly in countries to which the breed is not native, is to *safeguard the purity* of the breed and to furnish accurate pedigrees to all breeders who desire them. In practically all cases the breed association took over the conduct of the herdbook early in the history of the association. Most of the clerical work in most associations is used for issuing and checking the accuracy of registrations and transfers. Most of the errors discovered in applications for registry or transfer are matters of carelessness or neglect, but close watch is kept for fraud, and occasionally a member is expelled for this reason or his registrations are canceled. Because of the possibility of legal complications and internal dissension, this is not usually done unless the proof is quite conclusive. The accuracy of a pedigree depends mainly upon the honesty and carefulness of the man who signs the application for registry. But the breed association's policy of rejecting or returning applications not accurate in detail and of investigating within reasonable limits cases where fraud is suspected prevents many errors which would come into pedigrees if there were no such supervision. Some associations publish the facts when a member is expelled or registrations are canceled because of fraud. Others keep it quiet to avoid scandal. Probably the first policy is generally the wiser. In either case the news usually spreads fast. Safeguarding the purity of a breed does nothing to *increase* its merit but does act to some extent as a ratchet mechanism to maintain whatever special merit the breed already has and to hold any future improvements which may be made in its average merit. Preserving the present merit of the breed is particularly important if the breed is an introduced one, few in numbers, and surrounded by animals of distinctly different origin.

A breed association tries to *improve* the merit of the breed by guiding the ideals which the breeders use when making their selections and by making official tests or ratings of the productiveness or conformation of individual animals. The ideals of the breeders are influenced by such activities as adopting a score card, or other verbal description of the breed ideal, or a series of "true type" pictures or models. A different approach to the same goal is through control of judging standards at the various fairs. Sometimes that amounts only to advising the fair management upon request whether the breed association considers a certain individual competent to judge that breed. In other cases the breed association prepares a fairly small list of men considered competent, and its contributions of prize money are contingent upon the fair management's selecting its judge from that list. Sometimes the association prints circulars or uses advertisements which serve the double purpose of promoting wider interest in the breed and showing pictures of animals which are considered nearly ideal for that breed. The association sometimes instructs its members as to what is considered important in pedigrees (See example in *Holstein-Friesian World,* 33:300, April 4, 1936).

The general purpose of all these ways of teaching the breeders the official ideals for that breed is that the breeders thus informed shall follow those standards when making their own selections and cullings and will thus move the breed average nearer to the breed ideal. The breed association's efforts end with presenting the lesson to the breeder. Unless he asks for further help from the fieldman or other officer, it is entirely up to him whether he accepts the official ideal and how much or how little he uses it when making his selections and cullings.

Official tests of the speed of individual horses were characteristic of the Thoroughbred and the Standardbred from the very beginning. In the case of the latter, a certain speed was necessary for registration—hence the name of the breed. As long ago as 1832 the preface to the list of Thoroughbred horses in Prussia contained the statement: "Unless herdbooks contain production tests, they will be useless and without interest, since they will contain only names, of which no one knows anything and which mean nothing." (Engeler, 1936). *Official* testing of dairy cattle began around 1880[1] with the Holstein-Friesian breed in the United States, largely as a result of the work of Solomon Hoxie. (Private production records were being kept at least as long ago as the time of Thomas Bates.) For a number of years there were serious doubts as to whether official testing would become popular enough to be retained.

[1] Pages 15 and 16 of *Holstein-Friesian History,* by Prescott, Price, Wing, and Prescott.

However, it eventually proved so useful, not only for breed improvement but also as an aid in advertising and selling, that no dairy breed association would now try to do without its department of official testing. But such official tests are not yet a prerequisite to registration. Whether the breeder will test or not is still entirely voluntary with him, except as the demands of a considerable portion of his customers put some economic compulsion on him to test.

In the United States, official inspection of whole herds, with a rating of each animal for type, was begun on a voluntary plan in 1929 by one of the dairy breeds and has since been adopted by most of the others. This plan was used by only a few breeders during the depression of the 1930's but now seems to be spreading more rapidly. Type inspection or scoring, sometimes on a compulsory basis, has been practiced much longer in some other countries; for example in The Netherlands or Switzerland.

Many of the swine associations have required the man who registers a litter to report the number of pigs farrowed and raised. Some have published in the herdbooks the number farrowed in each litter. Recently Records of Performance have been established by several of the swine breed associations, the Hampshire having been the first. These are based mostly on weight of litter weaned at 56 days. They are still voluntary and are just beginning to be tried on an extensive scale. They seem to be a sound step forward, but their wide adoption will probably depend on how insistently the breeder's customers ask for such information when they come to buy boars.

Tests for measuring the practical productiveness of beef cattle are being studied at experiment stations,[2] but these have not been adopted by any of the beef breed associations.

There have been some discussions of a Record of Performance for sheep, particularly for the fine-wool breeds. The shearing records made in New England a century ago were a kind of crude beginning in that direction. No definite plan is in actual operation today.

Although there has been some experimenting with endurance rides for cavalry horses and with pulling tests for draft horses, these have not yet been made an official part of breed association activities.

The associations in the United States do not usually keep official lists of the prizes won by individual animals at the important shows, although some of them do so in an unofficial way. In many cases the breed paper performs that service.

With a very few exceptions—such as the speed requirement for the

---

[2] See Montana Agr. Exp. Sta., Bul. 397; also Minnesota Agr. Exp. Sta., Tech. Bul. 94; also *Empire Jour. of Exp. Agriculture*, 8:259–68.

Standardbred horse, the type inspection given to Brahman cattle which are admitted as foundation stock, and the flock inspections which are a part of some plans of poultry improvement—the purely voluntary nature of the production-testing and type-rating done under the auspices of the breed associations in the United States is like that in Britain, but is in marked contrast to the compulsory inspections or testing in some breeds in continental Europe. Those are discussed in more detail in the chapter on selective registration. In considering how far it would be possible or wise for the American associations to go in that direction, there are broad general questions as to how far collective policies and efforts can or should replace or supplement individual freedom to follow whatever breeding policies or use whatever purebred animals one pleases, without regard to whether those would be approved by a majority of one's fellow breeders or by an official inspector of the association. Besides such questions of general principle, there are also some immediately practical questions of expense which might make impossible, in breeds as widely scattered as many of those in the United States, procedures which are feasible in lands where the breed is highly concentrated in one or a few localities. Most directors of the breed associations in the United States are reasonably eager to adopt any new practices or requirements which will increase the merit of their breed faster than at present. They usually demand that the new plan shall be tested in actual operation, however, before they commit their association definitely to it. They have sometimes tried a plausible scheme only to find that it did not work as well as they had felt sure it would, and traces of the confusion and discontent which resulted have remained for years to plague them.

Naturally, many activities of the breed association are directed mainly at promoting the present business interests of the members. Examples are the efforts to expand the breed numbers by getting new breeders to establish herds, the promotion or management of sales and the correspondence which the secretary's office has with would-be purchasers of breeding stock. Most associations prefer to give support and encouragement to sales efforts but not actually to manage the sale themselves lest the dissatisfactions which inevitably arise about some transactions should result in animosity toward the association itself.

The provision of prize money at the more important shows is intended to promote the breed by bringing out larger numbers, thus giving more advertisement to the breed. It is also intended to improve the breed by teaching more people the ideal for that breed.

Many of the larger breeds have one or more papers devoted mainly to promoting the interests of the breed. Most of these breed papers are privately owned and managed, but some are owned and operated by the association itself. The contact between the association and the breed paper is close and important to both parties, even when the paper is privately owned. Most of the activities of the breed paper are devoted to the immediate business interests of the members—advertisements and news of sales, merchandise for breeder's needs, etc.—but some of these papers carry articles and information helpful in improving the merit of the breed, but not otherwise a matter of financial profit to anyone. Besides the breed papers, most associations print leaflets, buy advertising space in other magazines which reach stockmen, and make occasional use of the radio as part of their regular efforts at breed promotion.

The activities of the fieldmen or breed extension service are in considerable part devoted to expanding the numbers of the breed and helping new breeders get started. The fieldmen work also with boys' and girls' clubs and help established breeders with their problems.

The forms of government of the breed associations vary widely. Usually the policies are determined by an unsalaried board of directors, preferably with overlapping terms and only one-third elected each year to prevent erratic changes in policies. The executive work of carrying out those policies is administered by a paid secretary. Sometimes there is a fixed number of shares of stock, as in most industrial corporations, and one can become a stockholder only by buying a share from someone else. More often the number of shares is not limited but any breeder approved by the board of directors may become a member by paying a fixed (usually small) sum for a non-transferable membership. Sometimes all members present at the annual meeting can vote. In other associations members not present can vote by proxy. Other associations, especially those with large financial reserves, have more or less elaborate systems of delegates or representatives chosen by districts, to ensure proportional representation and to avoid abuse of the proxy system. The form of government is important insofar as it may promote or endanger stability in conducting the association's activities, make it easy or difficult for the board of directors to see that the secretary carries out the general policies they wish, and make it easy or difficult for a minority to seize and hold control. Some of the cases where there are two or more associations for the same breed had their origin in intense dissatisfaction on the part of a group which was not in control, either because it was a minority or because control had been seized by another group which could not be ousted with the existing machinery for governing that association. At one time there were seven separate organizations recording Poland China pedigrees. The number was not finally reduced to one until 1946.

The current problems of the breed associations are numerous. Financial problems have been acute with many associations since 1920. Nearly all of the swine and sheep associations and some of the cattle and horse associations have suspended publication of their herdbooks. What the final substitute for the printed herdbook will be is not yet evident. In some breeds there are still two or more registration associations, with some duplication of operating expenses.

An innovation which is still in the experimental stage in the United States, although it was used by breeders of Thoroughbreds a century ago and advocated by Thomas Bates as a means of overcoming what he thought were defects in the conduct of the Coates Herdbook for Shorthorns, is the filing of birth certificates which act as tentative registrations. They keep the date of birth and the records of parentage straight but permit the owner to wait until the animal is mature to complete the registration. This is a common practice in those European associations which require inspection for conformation. Since such inspection must wait until the animal approaches maturity, records of parentage might become lost or incorrect in the interval if this precaution of filing a birth certificate were not taken. The increasing percentages which are incorrect when registration is delayed is illustrated by the following data[3] on 19,172 Holstein-Friesian applications received during a six-weeks period.

Age	Percentage Incorrect
Under 2 months	13
2– 6 months	21
6–12 months	25
12–18 months	33
18–24 months	44
Over 24 months	53

The Jersey association has recently experimented with selective registry and a pedigree rating system (the "star bull" plan) for bulls. Several dairy associations are trying plans for calling attention officially to unusually meritorious proved sires; for example, the Ayrshire approved sire plan, and the Holstein-Friesian publication of sire indexes.

Herdbooks, as printed in the past, permit the tracing of pedigrees in only one direction; that is, one can learn from the herdbook what an animal's ancestors were but cannot find what offspring it had. Often a full list of an animal's offspring is more valuable than all that could be learned by studying its pedigree. Most breed associations maintain office records which will permit making such lists (at least from

[3] *Holstein-Friesian World*, June 21, 1941, page 708.

females), but those are not published. The nearest approach to published lists of this kind is in the reports of official testing in the dairy breeds where the tested offspring of a given sire or dam may quickly be found. Usually little or nothing is published about the offspring which were not tested officially, although the Ayrshire association attempts to learn what became of each untested daughter of the bulls in its "approved sire plan."

In the United States the breed associations receive no direct financial support from the state or national governments. Correspondingly there is no governmental control or supervision of the associations or their activities, beyond whatever legal regulations apply in general to all nonprofit corporations or associations. However, the representatives of the United States Department of Agriculture and of the state agricultural colleges cooperate in many ways with the breed associations in activities which are expected to improve the practical merits of the breeds or to benefit the buyers of purebred sires. Examples are the supervision of official production tests for the dairy breeds, helping in the management and promotion of livestock shows, conducting purebred sire campaigns, promoting the use of proven sires, etc. In some countries the governments extend considerable financial aid and correspondingly exert some control over association policies. The details of such arrangements vary widely. Examples are Switzerland, The Netherlands, and (especially since 1936) Germany. In other countries, such as Canada and the Union of South Africa, the government cooperates in supervising registration and printing the herdbooks but does not participate in or control other activities of the associations. In yet other countries, such as Denmark and Argentina, the herdbooks are conducted by farmers cooperatives or by a "Rural Society," and the breed association either does not exist or is an advisory and promotional body.

Most of the breed *improvement* has to be done by the breeder himself. The association stands ready to help and advise him, but it does not select the animals he shall use or decide which he shall cull, except in the comparatively few cases where animals are barred from registry because they possess some undesired characteristic. Nor does the association decide which males shall be mated to which females. The actual selections and the choice of a mating system are left almost entirely to the individual breeder to do as he sees fit, provided his animals are purebred and the correctness of their pedigrees is unchallenged. Breed associations must remain more or less responsive to the opinions of the majority of the breeders and therefore cannot be expected to do much pioneering or testing of new and unpopular ideas. This will have to be done by venturesome breeders or by public research institutions. Practical experience indicates that breed associations are necessary for breed improvement, since practically every breed which has persisted long has soon developed an association to look after its interests. Yet the association's part is on the whole the conservative one of helping hold whatever average merit the breed has already attained and acquainting the beginner and the public with what is considered ideal by most breeders of that breed. Most of the aggressive positive work toward improving the merit of the breed to still higher levels will have to be done by individual breeders who are unusually able, energetic, persevering, or lucky.

REFERENCES

Anonymous. 1930. "Notice—Registration Cancelled." Holstein-Friesian World, 27 (No. 24):1177.

Advanced Registry Office. 1936. Pedigrees. Holstein-Friesian World, 33:300.

Engeler, W. 1931. Studies on the development and situation of pedigree registering in the cattle-breeding industry. 92 pp. Internat. Inst. of Agr., Rome.

———. 1936. Die Entwicklung des Herdebuchwesens unter dem Einfluss der Lehren von der Vererbung und Züchtung bei den landwirtschaftlichen Haustieren. (In) Neue Forschungen in Tierzucht. Bern.

Peterson, Guy A. 1919. Swedish herdbook registration. Hoard's Dairyman, 74:62.

Plumb, Charles S. 1930. Registry books on farm animals.

Van den Bosch, I.G.J. 1930. The Holstein industry in America. The scoring system in the Netherlands. Holstein-Friesian World, 27:283 and 337.

Wilson, C. V. 1925. A study of the breeding records of a group of Shorthorn cows. West Virginia Agr. Exp. Sta., Bul. 198.

Early volumes of the herdbooks of the breeds in which you are interested. Read especially the minutes of some of the early business meetings.

CHAPTER 6

The Mendelian Basis of Inheritance

THE PARTICULATE AND DUPLICATE NATURE OF INHERITANCE

The essence of Mendelism is that inheritance is by particles or units (called genes hereafter) and that these genes are present in pairs, one member of each pair having come from each parent. Each gene maintains its identity generation after generation instead of blending with the other genes to form a new kind or blend of hereditary substance, as was thought in pre-Mendelian days. When the individual reproduces, it transmits to each offspring one or the other, but not both, of the genes in each pair it possesses. Thus the parent gives to each offspring *only a sample half of its own inheritance.* The laws of chance govern this sampling, subject to the restriction that each sample must contain one gene of every pair. This sampling nature of the process of inheritance, scarcely suspected in pre-Mendelian days, allows a parent to transmit different inheritance to different offspring. More precisely, if we let *Aa* represent a pair of genes in a parent which has two offspring, there is one chance in four (the exact result in individual cases varying, of course, according to the laws of sampling) that both offspring will get *A*. There is one chance in four that both will get *a* and there are two chances in four that one will get *A* and the other will get *a*. Similar probabilities apply to every other pair of genes. Thus, about half of the genes which two offspring receive from the same parent (i.e., about one-fourth of all the genes they have) are exact duplicates; but the other genes the two get from that parent (another fourth of all the genes they have) were opposite members of the pairs in that parent. In those pairs of genes for which the parent was homozygous it will not matter which gene of the pair is transmitted, for the result will be the same with either. Most parents are heterozygous for many pairs of genes. Here lies the explanation of the fact that identical pedigrees do not mean identical inheritance, although they usually do mean a considerable degree of likeness.[1] That identity of pedigree (as of full broth-

[1] To illustrate the Mendelian basis for the fact that identity of pedigree generally means *similarity* but not *identity* of inheritance, let us consider the probable results of a particular mating in a breed heterozygous for many pairs of genes and mating at random with respect to each pair. If the contrasting alleles in each pair are equally frequent in the breed, then the most probable Mendelian formula of mates chosen at

ers) does not mean identical heredity was well known in pre-Mendelian days but either was regarded as one of the unexplained mysteries of heredity or interpreted to mean that a large amount of entirely new inheritance (mutations we would say today) had arisen in each individual.

Since half of the inheritance comes from each parent, except in the case of sex-linked genes, and since in each pair the gene received from the sire and gene received from the dam are equally likely to be transmitted to any one offspring, most of the facts of inheritance when expressed in quantitative form involve the fraction 1/2. It is only a small exaggeration to say that the mathematics of genetics is the algebra of 1/2!

Dominance is not an essential part of Mendelism, although Mendel himself noted it. It is a ready explanation of some cases of "reversion" or "atavism," but not the only explanation for those. The chief part played by dominance is to increase the variability of the population slightly and to make certain genotypes indistinguishable from each other. There is nothing in the mechanism of inheritance which would cause a dominant gene to increase in numbers at the expense of its recessive allel, or the reverse. If q is the proportion of A genes and $1 - q$ is the proportion of a genes in the population and p is the proportion of heterozygotes, then no matter what system of mating prevails, the zygotic ratio will be $(q - p/2)AA : p\ Aa : (1 - q - p/2)\ aa$. If these have equal opportunity to reproduce (that is, if no selection for or against either of the three genotypes prevails), then the proportion of A genes in the next generation will be $q - p/2$ from the AA *individuals plus* $p/2$ from the Aa individuals which equals q in the whole

---

random can be illustrated as below, the mates being alike in some genes and unlike in others. How many different kinds of full sibs could there be from this particular mating? How many different kinds of half sibs could come from this one sire (or dam) mated to all the different kinds of mates which exist in the breed? How many different kinds of individuals can exist in the entire breed? The Mendelian formulae of the parents and the answers to each of these three questions may be indicated as follows:

Formula of sire: *AABbccDdeeFfGGHhIiJjKKllMMNnooPp*
Formula of dam: *AABbCcDDeeffGgHhiiJjKkLLmmNNOoPp*
Kinds of full sibs: 1x3x2x2x1x2x2x3x2x3x2x1x1x2x2x3 = 20,736
Kinds of paternal half sibs: 2x3x2x3x2x3x2x3x3x3x2x2x2x3x2x3 = 1,679,616
Kinds in breed: 3x3x3x3x3x3x3x3x3x3x3x3x3x3x3x3 = 43,046,721

Since there are more than twenty thousand kinds of full sibs possible from this mating, it is unlikely that two full sibs even from among a large number would happen to be exactly alike. Yet less than one two-thousandth of the total number of kinds of individuals possible in the whole breed are possible at all in this particular set of full sibs. This sire could not possibly sire a twenty-fifth of the kinds possible in the whole breed, no matter what kind of mates he had. This example is schematic in two important respects: first, many more than 16 pairs of genes are doubtless heterozygous in all breeds; and, second, it would be a surprising coincidence if the unlike alleles were equally numerous in more than a few of those pairs.

population, the same as it was in the preceding generation. Without selection the Mendelian mechanism itself does not change gene ratios except, of course, that it permits sampling variations to occur in the segregation process at each generation. Those are slight except in very small populations. There they cause the phenomena of inbreeding.

POST-MENDELIAN ADDITIONS TO THE LAWS OF INHERITANCE

Mendel knew nothing of *chromosomes* or *linkage*. The achievements of genetics in the last third of a century in identifying the chromosomes as the carriers of the genes have not changed the laws which Mendel discovered, except to modify the law of independent assortment so that it is now known to apply only to genes which are on different chromosomes. Cytological investigations of the mammals and birds are unusually difficult because the number of chromosome pairs is large, the chromosomes are small, and the processes of killing, fixing, and staining are apt to cause the chromosomes to "clump" together, so that the observer cannot be sure how many there are.

Because of these difficulties, mammalian and avian chromosomes have not been so well investigated as those of most farm crops and of many lower animals. In several cases investigators do not yet agree in their counts. Some species have been studied by only one investigator. Most of the findings quoted in table 1 are still subject to confirmation.[2] Work much older than 1920 is quoted only where no subsequent work has been reported, or where this earlier work has been quoted widely. In general the later work is more apt to be correct. On account of the "clumping," the larger numbers are more likely to be correct wherever there is not yet substantial agreement.

Most farm animals have around 20 to 30 pairs of chromosomes; hence two genes chosen at random will nearly always be independent of each other. Yet if one is considering a trait affected by more than six or seven pairs of genes, there is likely to be linkage among some of them.[3] On account of linkage, genes which were transmitted to the parent together (i.e., which both came to it from the same one of its parents) will be transmitted together to the offspring more often than if they were independent. Yet in the population as a whole, if crossing-over occurs at all, the "repulsion" and "coupling" phases soon become equally frequent, thus causing linkage to hinder selection (in a hitherto unselected population) in about as many cases as it helps. Hence link-

[2] For a more complete list and references, see: Oguma, Kan, and Kakino, Sajiro. 1932. A revised check-list of the chromosome number in vertebrata. *Jour. of Genetics*, 26:239–54, and for birds: Miller, R. A. 1938. *Anatomical Record* 70:156–58.

[3] The chicken, the mouse, and the rat are the only farm animals for which the construction of linkage maps is yet well along. See *Jour. of Heredity* 31:232–35 and 36:271–73. Some mapping has begun with the silkworm and with cattle.

age does not offer the breeder much chance to get one gene by selecting another closely linked to it. The effects of linkage in causing two genes or characteristics to be together or apart more often than not are conspicuous only in the first few generations after a cross between two unusually homozygous races—a condition which rarely confronts the animal breeder under our present purebreeding systems. The existence

TABLE 1

Recent Reports of Chromosome Numbers in Mammals and Poultry

Animal	Number of Pairs of Chromosomes	Heterogametic Sex	Investigator and Date of Publication
Farm animals:			
Horse	30	Male	Painter, 1945
	19	Male	Wodsedalek, 1914
	19	Male	Masui, 1919
Ass	32	....	Sokolov, 1937
Mule	Unpaired, 51 in all	....	Wodsedalek, 1916
Camel	35	....	Novikov, 1940
Cattle	30	Male	Krallinger, 1931
	19	Male	Wodsedalek, 1920
	30	Male	Makino, 1944
Yak	30	....	Zuitin, 1938
Buffalo	30	....	Pchakadse, 1939
	24	....	Makino, 1944
Sheep	27	Male	Berry, 1941
	30	Male	Several workers, 1931 to 1940
Goat	30	Male	Krallinger, 1931
	30	Male	Warwick, 1935
Swine	19	Male	Krallinger, 1936
	19	Male	Crew and Koller, 1939
Dog	26 ±	Male	Painter, 1925
	39	Male	Ahmed, 1941
Cat	19	Male	Minouchi, 1928
	19	Male	Koller, 1941
Farm poultry:			
Chicken	33	Female	White, 1932
	39	Female	Yamashima, 1944
	40	Female	Miller, 1938
	16 to 37	Female	Various authors since 1923
Duck	38 to 39	Female	Werner, 1927
	39	Female	Sokolov, *et al.*, 1936
Turkey	23 to 39	Female	Various authors, 1929 to 1936

TABLE 1 (*Continued*)

Peacock	18 to 29	Female	Tiniakow, 1934
Pigeon	31	Female	Oguma, 1927
	25±	Female	Hance, 1933
	34±	Female	Painter and Cole, 1943
Various mammals:			
Man	24	Male	Many recent authors
Chimpanzee	24	Male ?	Yeager *et al.*, 1940
Old World monkey	27	Male	Painter, 1925
New World monkey	27	Male	Painter, 1925
Nyctereutes (raccoon dog)	21	Male	Minouchi, 1929
Red fox	21	Male	Wodsedalek, 1931
Fox: V. vulpes	17	....	Andres, 1938 and Wipf, 1942
V. lagopus	26	....	Andres, 1938
V. fulva	16	Male	Bishop, 1942
Rat	21	Male	Many recent authors
Mouse	20	Male	Many recent authors
Deer mice (Peromyscus)	24 to 29	....	Cross, 1938
Rabbit	22	Male	Painter, 1926
	22	Male	Minouchi and Ohta, 1932
Guinea pig	19	Male	Harman and Root, 1926
	31	Male	League, 1928
	33	Male	Mols, 1928
Mink	14	Male	Schackelford and Wipf, 1947
Hamster, golden	19	Male	Koller, 1938
" striped	7	Male	Pontecorvo, 1943
Twelve other species of rodents	22 to 43 (only 3 above 27)	....	Cross, 1931
Peccary	15	....	Krallinger, 1936
Armadillo	30	....	Painter, 1925
Bat	24	Male ?	Painter, 1925
Hedgehog	24	....	Painter, 1925
Opossum	11	Male	Hoy and George, 1929
	11	Male	Painter, 1925
Kangaroo	6	....	Painter, 1925
Eight other genera of marsupials	6 to 10	Male	Various authors

of linkage causes the genes to tend to segregate in large groups at any one cell division and hence causes the population to behave as if there were fewer genes than there actually are. But this effect is probably unimportant, since the cross-overs in different cells, even in the same individual, may be at various places on the chromosomes and since genes with opposite effects are as likely to be linked together as are genes with similar effects. Linkage is some hindrance to progress by selection (after the first generation) since it keeps desired and undesired genes from recombining into separate gametes as often as they would if not linked. This lessens slightly the variability among the immediate offspring of selected parents and consequently reduces the amount which can be accomplished by selecting among them.

Mendel did not know of sex-linkage which, in the heterogametic sex, is an exception to the rule that inheritance is in duplicate and comes equally from both parents. Only one pair of chromosomes carries the sex-linked genes. Presumably something like one-twentieth or one-thirtieth of all the genes are sex-linked, although that is only a rough approximation since the sex-chromosomes might carry more or less than their proportionate share of the genes. To ignore sex-linkage will generally lead to but little error; yet there doubtless is some sex-linkage in all farm animals, and probably there are occasional characteristics which are affected by a disproportionately large share of sex-linked genes. The general effect of sex-linkage is to make parent and offspring of opposite sex resemble each other slightly more than parent and offspring of the same sex do. Here and there it has some conspicuous special effect, such as making possible, in some matings, the identification of sex in very young poultry. Partial sex-linkage is known genetically in man and presumably exists in all other mammals in which portions of the X-chromosome cross over with portions of the Y-chromosome. The practical consequences are like those of sex-linkage but even less noticeable.

IS THERE ANY NON-MENDELIAN INHERITANCE?

Each year of genetic research brings added evidence that all inheritance is Mendelian in the broad sense of being in duplicate and particulate, with the particles maintaining their identity. The only well-established exceptions are plastid inheritance, known only from plants, and polyploidy, where inheritance is particulate but present in more than duplicate. Polyploidy seems to be very rare in animals, although it appears to have been important in the evolutionary history of the plant kingdom. On account of the sampling nature of inheritance, polyploidy must be a temporary condition lasting only a few generations until the polyploid individuals either die out or develop a new diploid division among their chromosomes. A few cases of inheritance through the maternal line only have been reported. Most of those have later been found to require some other explanation. Often it is found by further experiment that these were characteristics for which the unfertilized ovum was already organized and the genes which the sperm brought are merely delayed a generation in expressing their effects. Whether the shells of certain snails coil to the left or to the right is an example.

It may never be possible to prove rigorously that all inheritance is Mendelian in this sense, for so long as the inheritance of any characteristic is unknown, an obstinate skeptic might still say: "Perhaps the inheritance of that characteristic conforms to some other rule." It is a fact, however, that many cases of inheritance which were at one time thought not to behave in the Mendelian manner, have been shown by more thorough analysis to behave in that very way except that the number of factors is large or that the interactions between different factors are unusually complicated or that the egg was already so highly organized that the genes in the sperm cell could not show some of their effects until the next generation. Moreover, the ineffectiveness of selection within pure lines seems to be some positive evidence that there can be no appreciable amount of truly "blending" inheritance. (Cf. Fisher, pp. 17–18.)

THE NUMBER OF GENES

There are at least four kinds of evidence which show something about the total number of pairs of genes in certain species. They leave little doubt that the breeder of farm animals must contend with a genetic situation in which the number of different pairs of genes heterozygous in his herd or flock is at least many scores, probably many hundreds, and perhaps even a few thousands.

The first kind of evidence is the number of genes which have actually been found in the organisms studied most. In *Drosophila melanogaster* more than 500 different loci have already been located on the chromosome maps, and many more are known. In *D. pseudoobscura* there are in the third chromosome alone at least 289 loci which can mutate to lethal genes (*Genetics* 26:39). Baur and his co-workers found some 300 genes in the snapdragon, *Antirrhinum majus*. In corn (*Zea mays*) some 400 genes had been catalogued up to 1935 (Rhoades and McClintock). In the wasp, *Habrobracon juglandis,* about 100 mutations at separate loci are known. In *Datura* (the genus which includes the Jimson weed) about 500 genes were known, 77 of them located on the chromosome map, by 1941. The inheritance of more than 100 characters has been studied in barley (*Genetics* 28:419). In man more than 200 genes have been reported, some 45 of them on the X-chromosome, but the evidence for many of those is scanty because controlled experimental matings are not possible. The number of genes reported for most species of farm animals is only a few dozen,[4] and many of those are not very certainly established, but only a few thousand farm animals have been observed under such circumstances that genes would have been identified readily.

This kind of evidence has two limitations: First, only genes with effects conspicuous enough to permit their ready identification can be catalogued in the Mendelian manner. Genes with minor effects can only be lumped together in an indefinite background of "modifying factors." Second, the number already found provides scarcely any basis for guessing whether that number is only a tiny fraction or a large fraction of the total number which exists. Certainly the number reported is less than the actual number.

A second kind of evidence comes from cytology. Work (*Journal of Heredity* 33:403–7, 1942.) on the chromosomes in the salivary glands of Drosophila shows about 5,000 distinct bands; and the cytogenetics of deletions, inversions, etc., makes it seem plausible that each of these is a gene, although it remains possible that there may be several genes in some bands. Belling in studying the chromosomes of the lily was able to distinguish about 2,200 to 2,500 "chromomeres," or distinct segments of its chromosomes; but the genetics of the lily is not well enough known to show how closely the genes and the chromomeres correspond to each other. Each chromosome in the farm animals surely must carry many genes; but the cytology of mammals and birds is difficult and little except number is known of it so far.

A third kind of evidence comes from some indirect reasoning, based on the number of times certain mutations recurred. The first such estimate was a figure of about 1,800 loci, but it was recognized that the assumptions involved made this lower than the true figure. Gowen and Gay in 1933 reached an estimate of 14,280 loci in Drosophila. This estimate had a large sampling error but was thought to have no consistent bias either in the direction of largeness or of smallness. Muller and Prokofyeva in 1935 reached the conclusion that in Drosophila the total number of loci ". . . is of the order of a few (ca. 5–10) thousand."

A fourth kind of evidence comes from the experiments on quantita-

[4] For example, Ibsen in 1933 listed 17 pairs of genes and one multiple allelic series of genes affecting color in cattle. He mentions several other color characteristics, the mode of inheritance of which was not yet clarified. In the fowl 21 genes are already (1940) located in six linkage groups (*Jour. of Heredity* 31:231–35), and other genes are known but not located.

tive inheritance which require for their interpretation a minimum number of genes if those express their effects in the very simplest way (i.e., without dominance or other nonadditive interactions). Sumner's experiments with mice of the genus Peromyscus showed that many of the quantitative characteristics in which two varieties or local races differed were each affected by many genes for which neither population was homozygous. "Student" interpreted the Illinois Agricultural Experiment Station's work in selecting corn for high and for low oil content as showing that the oil percentage in the initial stock of corn ". . . was conditioned by the presence, or absence, of a number of genes, at least of the order 20-40, possibly of 200-400, and not at all likely to be of the order 5-10." The Illinois Station's work with the Bowlker herd, which was produced by crossing Guernseys and Holstein-Friesians, has been interpreted (page 123, Annual Report for 1928–29) as requiring more than 10 pairs of genes to explain the breed difference in milk yield and several more pairs of genes to explain the breed difference in the percentage of fat. The Tranekjaer experiments with Jersey and Red Danish cattle in Denmark indicate that at least seven pairs of genes were concerned with the difference in fat percentage between those breeds.[5] Jull concludes that in the fowl sexual maturity, rate of laying, and persistency of laying are each ". . . affected by a relatively large number of genes, some of which probably influence more than one character." Many other examples might be cited, each yielding figures of from 4 or 5 up to more than 20 as the minimum number of pairs of genes affecting a given quantitative characteristic.

The usual result of an experiment on the genetics of a quantitative characteristic is that the number of genes involved "cannot be less than" a certain number, but might be larger. Usually the longer the genetic investigation continues, the more genes are found. In such experiments the evidence which throws light on gene number usually involves differences between the parental means and between the variabilities of the parental groups, the F_1 or the F_2 or the back-crosses, or it concerns the change produced in the mean and in the variability by a given amount of selection. Often the numbers are small and the sampling errors are high. Those may make the answer obtained either too large or too small. Nearly all other sources of error make the answer smaller than it should be. Thus, this kind of evidence can only show that the number of genes is more than a certain small number which, however, is usually too large to leave any reasonable hope that even the

[5] Wriedt postulated that one pair of genes would account for the breed difference; but, as Skovsted pointed out, not enough of the parental types reappeared in the back-crosses or in the F_2 to justify that. The figure seven mentioned here is based on comparisons of parental means and F_1, F_2 and back-cross variabilities.

most thorough study will enable a breeder to know the Mendelian formula for all the important genes in any of his animals.

A fifth line of reasoning, which perhaps scarcely deserves to be called evidence until more is known about the physiology of how the genes produce their effects,[6] is that the development and functioning of each organ in the animal is so complex and is dependent upon such a delicate interplay of various tissues, hormones, fluids, etc., each acting at the proper time, that it is scarcely conceivable that a small number of genes can initiate and control all of this. The term "unit character" which was freely used in the early days of genetics tends to be avoided now, lest it confuse by implying that one gene by itself is enough to produce the whole characteristic. The gene, not the characteristic, is the unit of Mendelism. In a sense it may be legitimate (and it is often convenient) to refer to *the contrast* between two characteristics—for example, red eye and white eye in Drosophila—as a unit character since, in some matings at least, the difference between the two characteristics is caused by a difference in only one pair of genes. Yet it is confusing to speak of *red eye* as a unit character, since more than 40 genes have been found thus far which must all be present if the usual red eye of the wild fly is to develop. If one of them is absent, the eye may be "purple"; if another is absent, it may be "peach"; etc.; but the co-operation of them all (and doubtless of still unknown genes) is required to produce the normal eye. The cooperation of at least 100 genes is necessary to produce normal green chlorophyll color in corn. (*Amer. Nat.* 80:431).

The fact that many distinct abnormalities and defects are caused by a change in only one gene is to be expected if that gene interrupts, at some important stage for which there is no substitute, the long chain of physiologic processes by which the normal characteristic usually develops. For example, in the normal process of horn formation in cattle there may be several stages which, if interrupted, would alter or prevent all the later stages of development; but it is not easy to imagine that any one gene could guide the whole course of horn development, including the growth of the bony core, the blood vessels, nerves, etc. Thus, it may be legitimate to speak of a single gene for hornlessness: but it is not legitimate to infer that the allelic gene to that one is the gene which produces horns. The case is analogous to that of destroying a house by a single act, such as applying a match to it at any one of a number of places. But a house can be built only by the timely co-operation of an enormous number of individual acts. It is scarcely legitimate to speak of refraining from applying the match as an act which builds the house.

[6] See *Quart. Rev. of Biology* 13:140–68 and *Physiological Reviews* 21:487–527.

The eradication of single-gene defects, such as lethals and semi-lethals, may be an important part of the task of animal breeding; but its practical importance can scarcely approach that of changing the fertility, vitality, growth rates, proportions of conformation, milk production, speed, wool production, etc., which are most of the economic differences between ordinary or moderately inferior and distinctly superior animals. The genetic evidence indicates that these are complex physiological characteristics in which most of the hereditary differences are caused by a large number of genes, each with an individually small effect.

CONSEQUENCES OF THE LARGE NUMBER OF GENES

When only one pair of genes is concerned, two kinds of gametes and three kinds of genotypes are possible. If there are two pairs of genes, each possibility for the one may occur in combination with each possibility of the other, thus permitting four kinds of gametes and nine kinds of genotypes. Three pairs of genes permit 8 kinds of gametes and 27 kinds of genotypes. The general formulae are: the number of kinds of gametes or of homozygous genotypes possible with n pairs of genes is 2^n (which may be written $10^{.301n}$); the number of different kinds of genotypes possible is 3^n (which may be writen $10^{.477n}$). These numbers become big beyond human comprehension if n is very large. Even if there are only 100 pairs of genes, 31 digits will be needed for writing the number of kinds of gametes possible; and there will be 48 digits in the number of kinds of genotypes possible. The possibilities for hereditary differences under this system are enormous. They may be visualized by comparing them with the number of animals of each species actually alive in the whole world at any one time. During the years around 1926 to 1935 these were as follows for man and for some of the more important farm animals (figures from USDA Yearbooks):

	TENS OF MILLIONS
Human beings	200
Cattle	70
Horses	10
Mules and asses	3
Sheep	69 to 74
Swine	25 to 28
Chickens (in the United States only)	44

Hence, if the number of different genes heterozygous in each species is as large as 40 (and it may well be thousands), the number of different hereditary combinations possible in each species is millions on millions of times as large as the number of animals which can actually be alive at any one time. It would be a remarkable coincidence if any two living things were exactly alike in all their heredity, except for a few special cases such as identical twins, asexually reproduced organisms, and possibly members of a strain which had been very highly inbred for a long time. Gesell says (*Science* 88:227): "Even in the detailed studies of animal respiration, it has been found that no two dogs breathe exactly alike."

Another comparison to show vividly the enormous number of different kinds of individuals possible is furnished by the physicists' estimate that the number of electrons in the universe is about 10^{80}. In a species in which only 200 pairs of genes are heterozygous there could be 10^{95} different kinds of individuals. This is a million billion times as many as there are electrons in the universe!

In these calculations it was assumed that there are only two allelic genes in each series. In several cases it is already known that there are three or more different kinds of genes in an allelic series (called "multiple alleles" in genetics), and it is possible that all allelic series are potentially multiple.[7] This increases the number of kinds of gametes and genotypes possible. If m alleles are possible in each of n allelic series, the number of different kinds of gametes possible is m^n and the number of different kinds of genotypes possible is $\left[\frac{m(m+1)}{2}\right]^n$.

Linkage does not affect the number of kinds of gametes which may be produced but does affect their proportions and thereby increases the size of population necessary to permit all kinds of genotypes to be produced. Also, it increases the number of genotypes possible because the multiple heterozygotes will now be different genotypes according to whether the linkage is in the coupling or repulsion phase. That is, $\frac{AB}{ab}$ and $\frac{Ab}{aB}$ will be different genotypes if linkage exists, whereas both would have been the same genotype, *AaBb*, if there had been no linkage. A triple heterozygote, where all three genes are linked, can exist in four different genotypes, a quadruple heterozygote in eight genotypes, etc.

[7] Several series have already been found in which there are more than four alleles. The albino series in the rabbit is an example. The maximum number yet reported in any organism is 45 alleles for a self-sterility gene in one of the primroses, *Oenothera organensis*. *Genetics* 26:469. More than 22 alleles are known in the white eye series in *Drosophila melanogaster* and more than 40 at the locus for "bobbed." In man at least four alleles exist to determine the blood types. *A*, A^1, *B*, and *O*, and more than a half dozen alleles at the locus for the *rh* blood factor have been reported.

Both these additional complications—multiple alleles and linkage—increase the number of kinds of genotypes possible. Unless the number of heterozygous genes is very small, there is no escaping the conclusion that the number of genetically different kinds of individuals possible in a breed or species is practically infinite. Except in the rare case of identical twins, one can confidently expect to breed cattle, or any other species of farm animal, a lifetime without ever having a second animal exactly like one he produced earlier.

If each gene had a different kind of effect and there was no confusion by environmentally caused variations or by dominance, the number of kinds of animals different in appearance or performance would be the same as the number of kinds of genotypes. If all pairs of genes showed complete dominance, but each pair of genes had a different effect, the number of kinds of animals would be the same as the number of kinds of gametes. But if very many genes are involved, some will produce effects like those of others, some will produce effects only when certain others are present, and some will produce the same effects as variations in environment do. Therefore, if the genetic situation is at all complicated, the outwardly distinguishable kinds of animals grade into each other in an almost continuous series. When we classify a large group of animals on the basis of outward appearance or performance in any one characteristic, even with considerable precision, we are almost certain to include in each class many genetically different kinds of individuals.

THE GENETIC INTERPRETATION OF THE "PURITY" OF PURE BREEDS

In animal breeding usage, purity of breeding refers to ancestry and is not the same as the genetic term "homozygosity," although there is some slight relationship between the two. The purebred animal is one whose ancestors all belonged to that same breed for as far back as is required by the rules governing registration in that breed. Since all breeds are finitely limited in the number of animals alive at any one time, and many breeds were very small in numbers for a long time during their formative period, a certain amount of homozygosity was produced by the resultant inbreeding. This is usually a faint force in purebreeding as practiced today in breeds which have become large and successful, but occasionally was intense during the formative period when the breed was very small. The Shorthorn breed, which has the oldest pedigree record, probably lost through the inbreeding process about 25 or 30 per cent of its initial heterozygosity in the century and a third from its foundation to 1920. Most of this was lost in the formative stage while the ancestry of the future breed was largely included in the herds of the Colling brothers, who both shaped their breeding operations to an unusual degree around one bull, Favourite. In most breeds yet studied, the breed is now losing something like one-half of one per cent of its heterozygosis per generation. In breeds of cattle and sheep, where the average length of generation is around four or five years, this would mean that in a century the mere fact of absolute purity of breeding would cause a decrease of about 10 per cent in the amount of heterozygosity initially present. This would be partially offset by the occasional registration of a grade through fraud, accident, or official permission, and by the new mutations which might occur and survive during that century. In addition to these three processes of inbreeding, introduction of outside blood, and mutation, selection may have helped either to increase or to decrease the average homozygosity of the breed. Selection, however—in marked contrast to its effectiveness in changing average merit—is a very feeble tool for changing homozygosity, except under the very simplest genetic situations, as we shall see in chapter 11. It is not likely, therefore, that selection has made much change in the average *homozygosity* of the pure breeds since they were first separated from the general population, although it has certainly changed the breed averages distinctly in many cases.

It is sometimes argued that, while the total number of genes may perhaps run into the thousands, yet most breeds (or subgroups of a species in nature) will be homozygous for all but a few of those. This seems improbable, since no genetic mechanism is known by which that condition would be likely to be attained in the first place nor by which it could be maintained very long if it ever were reached. If a breed or species ever became entirely homozygous for a given pair of genes, mutations—even at a very low rate—would cause that homozygosis to be lost bit by bit. Selection is too feeble to restore *complete* homozygosity as rapidly as mutation destroys it, especially if the mutations are usually recessive, although selection is amply powerful to keep consistently undesirable mutant genes from becoming abundant. Inbreeding can be powerful enough to restore that homozygosity, but it is doubtful whether it often is intense enough in nature or in the breeding of large and popular breeds to achieve that end. Wright estimates[8] that a freely interbreeding species of one million individuals at equilibrium with one new mutation in each 1,000 individuals could support permanently some 30,000 unfixed loci, which is larger than any of the current estimates of total gene number. In other words, few genes in such a species would be entirely homozygous all through the species. Smaller species

[8] *Genetics,* 16:119–21. See also table 5 in Fisher's *Genetical Theory of Natural Selection.*

would not support so many[9] unfixed loci. The inbreeding which the pure breeds of livestock undergo comes mostly not from the smallness of the breed in absolute numbers but from the circumstance that many breeders are simultaneously using sons or grandsons of a few currently famous sires.[10] When the pure breeds finally reach equilibrium between the production of heterozygosis by mutations and the loss of heterozygosis because the effective number of animals in the breed is small, it is possible that the pure breed may support only a few scores of unfixed loci. But it is unlikely that the pure breeds have come at all close to that equilibrium point in the comparatively short time (in terms of animal generations) since they were organized. It seems entirely conservative to estimate that the average pure breed is still heterozygous for hundreds of pairs of genes, although, of course, no animal in it is heterozygous for even half of them and probably no one gene is heterozygous in even half of the members of a breed.

MUTATIONS

Mutation is a rare process. The mutation rates observed in the laboratory under otherwise natural conditions are generally around the magnitude of one mutation of each gene in 100,000 or 1,000,000 generations (Stadler, Gowen). The rate is not the same for all genes, however, and can be increased by such extreme environments as exposure to X-rays, radium, ultraviolet light or barely sublethal temperatures. Also a few genes which alter the mutation rates of other genes have been found. Dr. H. D. King observed among 45,000 Norway rats 6 different mutations affecting hair; but the number of genes affecting hair (i.e., the number exposed to mutation) is not known, nor can one be sure how many mutations with small effects occurred but were not observed. White and Ibsen estimate (*Jour. Genetics* 32:47) that in cattle the mutation rate from horned to polled is about one in 20,000. Haldane estimates that the mutation rate to hemophilia in man is about one in 30,000. More typical is the finding of Dobzhansky and Wright (*Genetics* 26:32) that in the third chromosome of *Drosophila pseudoobscura* the mutation rate to lethals must be less than three in 289,000.

If there are 5,000 genes in each individual, then only about one animal in every 20 or 200 would have even one gene which was a new

[9] The same formula gives nearly 2,560 as the number of unfixed loci in a species of 100,000 individuals, 210 in a species of 10,000 and 16 in a species of only 1,000. Our ignorance of whether the rate postulated for mutation is too high or too low and our ignorance of whether the effective number of breeding individuals is far smaller or only a little smaller than the census number prevent us from using these figures with much confidence.

[10] See Calder's study of the Clydesdale breed of horses. *Proc. Roy. Soc. Edinburgh* 47 Part 2, No. 8:118–40. 1927.

mutation. If the breeder were looking for a mutation in one certain gene, he could expect to find it in only one animal among something like 100,000 to 1,000,000 animals examined.[11] Even then, if this mutation produced only a small change in a characteristic also affected by environment and by other genes, the breeder looking for it would have only a small chance of recognizing it when he did see it.

Mutations are not only rare, but they are prevailingly harmful. The larger the change made by a mutation, the less likely is the mutation to be beneficial to the animal. The reasons for this hinge around the facts that mutations seem to be random changes in the genes and that any living animal is already a reasonably successful and highly complicated mechanism. Any random change in its machinery has only a small chance of making it a still more successful mechanism but is very apt to make it operate less well. The bigger the change, the less likely it is to improve the operation of the mechanism.[12] Hence the breeder does not yet have any reason to think that he can help his practical operations by increasing mutation rates.

Because mutations are rare and prevailingly harmful, the only significance they have for the breeder is that a tiny part of his efforts in selection must be spent in keeping these undesired newly mutated genes from becoming too numerous in the breed. Mutations do have great significance in evolution because they provide raw material which can be used by selection and inbreeding or other breeding systems to change the existing kinds of organisms. Even if only 1 in 1,000 new mutations were beneficial, geologic time is so long that such rare beneficial mutations may be important in evolutionary changes. Moreover, mutation serves an important evolutionary purpose by keeping (against the efforts of selection) a certain store of currently undesirable genes available in case the environment should change so as to reverse the direction of selection. For example, suppose a species well adapted to a life in a humid climate were by migration or change of climate forced to become better adapted to arid conditions. If mutation has kept in the species a few genes which make their possessors poorly adapted to a humid climate but better adapted to an arid climate, then

---

[11] When Warren Gammon wished to establish a polled variety of purebred Hereford cattle, he sent about 1,500 letters to breeders inquiring if they knew of such animals. From the replies he learned of 14 purebred Herefords which were polled, but some of these animals may have inherited their polledness from the same original mutation. We can only guess how many horned cattle had been observed by the men who reported these 14 polled ones, or how many of the 1,500 men who received these letters knew of polled cattle but neglected to reply.

[12] See Fisher's *The Genetical Theory of Natural Selection*, pp. 38–41, for more detailed reasoning on this point and for formulae relating the magnitude of a mutation's effect to the probability of its being beneficial.

when the conditions change, some of the newly desirable genes are already present. Selection may begin at once to increase their frequency. If mutation had not kept this store of formerly undesirable genes present, the species might have had to wait for the very slow process of mutation to produce them after the changed conditions arose. Waiting for the mutation might have taken so long that the species would have become extinct first. This consideration may be important in evolution but probably is rarely of any importance to the practical animal breeder, since he is concerned with so much shorter periods of time. Perhaps it might have some slight bearing on such situations as that which occurred in the American breeds of swine between 1910 and 1920, when there was a marked change in ideals and the direction of selection in many particulars was reversed.

GENE FREQUENCY

A gene may be much more abundant than its allel or it may be rarer. Gene frequency will be represented here by the letter q, which can have values between zero and 1.0. It is the fraction of the loci of that allelic series in the whole population which are occupied by the gene in question. Two examples may illustrate q and its variations. In a count of the colors reported for the 6,000 parents of 3,000 Shorthorns chosen at random from the British, Canadian, and American herdbooks, Wright found that 8.6 per cent were white, 43.8 per cent were roan and 47.6 per cent were red. Assuming that the roan is the heterozygote between the red and white (which fits the facts better than any other explanation yet advanced, although there may be some environmental or developmental overlapping between dark roans and reds), and letting q stand for the frequency of the gene for red, it is obvious that 47.6 per cent of the genes in this locus are genes for red in the red animals and that 21.9 per cent of the genes in the population are genes for red in the roan animals, while another 21.9 per cent of the genes in the population are genes for white in the roan animals. The final 8.6 per cent of the genes which occupy this locus are genes for white in the white animals. The frequency, q, of the gene for red in the whole population is, therefore, $.476 + .219 = .695$; while the frequency of the gene for white is $.219 + .086 = .305$. If the population had been mating truly at random, the proportions of red, roan, and white would have been the square of the ratio of the two kinds of genes, or: q^2 *reds* : $2q\ (1 - q)$ roans : $(1 - q)^2$ whites. The actual count shows a slight excess of roans and corresponding slight deficits of reds and whites, as follows:

Color	Actual Percentage	Expected Percentage	Excess of Actual
Red	47.6	48.3	— .7
Roan	43.8	42.4	+1.4
White	8.6	9.3	— .7

The discrepancy, although slight, appears to be significant statistically and is probably to be explained as a result of the practical breeder's preference for roan and his having observed long ago that the proportion of roans was higher from matings of white by red than from any other type of mating. The chief interest in the above example, aside from its illustrating what is meant by q, is that it shows how slight is the departure from random mating, even in a simple one-factor case where there is no dominance to confuse and where ideals are such as to lead to a rather strong effort toward mating unlikes.

Another example of gene frequency may be taken from black breeds of cattle,[13] such as the Holstein-Friesian or Aberdeen-Angus, in which something like one calf in every 100 to 200 purebreds is born red. The difference between the black and red, in most cases at least, is a single-factor one; and dominance is so nearly complete that no one has yet found how to distinguish the homozygous blacks from the heterozygotes, (i.e., both *BB* and *Bb* individuals are black but the *bb* individuals are red). If we let q represent the frequency of the gene for black, we cannot obtain its value simply by adding the proportion of the homozygotes to half of the proportion of the heterozygotes, as was done in the Shorthorn example, since we cannot identify the heterozygotes. By assuming random mating with respect to this gene (which is reasonable, since all parents are *BB* or *Bb*, which cannot be distinguished, and since not much inbreeding is practiced) we can, however, get an estimate of q if we can get a dependable count of the proportion of red calves born. That proportion should be $(1 - q)^2$ where q is the frequency of the gene for black. If the proportion of red calves born is 1 in 200, then the frequency of the gene for red is $\sqrt{\frac{1}{200}}$ or about 1 in 14 and that of the gene for black is about 13 in 14. Then about 1 in 7 or 8 among the calves born black is heterozygous and the others are homozygous.[14] The accuracy of this estimate depends upon the accuracy of the observation that 1 calf in 200 born is red; and this,

[13] Wisconsin Agr. Exp. Sta., Bul. 313.

[14] The ratio of homozygous dominants to heterozygotes in a random breeding population will be $\frac{q}{2(1-q)}$.

for obvious reasons, is not very dependable. If the proportion of purebred calves born red is about 1 in 100, then q is about .9 and about one in five or six of the purebred blacks is heterozygous for red. Also the heterozygotes will not be uniformly distributed all through the breed but will be more abundant in those herds where heterozygous sires have been used recently.

Gene frequency will be low for genes against which selection has already been directed for many generations, as is the case with most lethals. There is no *a priori* way of estimating whether it will be high or low for genes which have been the object of selection for only a few generations or for genes affecting the magnitude of a characteristic for which the ideal is genetically an intermediate. In populations which have been very small for a long time or are otherwise intensely inbred, more genes will be fixed, or nearly so (i.e., they will have values at or near zero or 1.0); and fewer genes will have frequencies near one-half than will be the case in large random-bred populations under otherwise similar circumstances.

In the case of multiple alleles, it is usually sufficient to let q represent the frequency of the most desirable gene of the series, grouping all the other alleles together as less desirable and having a total frequency of $1 - q$.

THE BINOMIAL DISTRIBUTION OF ZYGOTES

If mating is random, the proportions in which the zygotes occur will be the square of the gametic ratio, as shown in Table 2. The typical Mendelian F_2 or unselected F_3 ratio is merely a special case of the binomial distribution where q happens to be exactly .5. Even small variations in q affect rather strongly the proportions of AA and of aa. However, the variations in the proportions of AA and aa cancel each other to some extent so that the percentage of heterozygosis changes only a little with variations in q, particularly when q is anywhere near one-half. For instance, it may be seen from Table 2 and Figure 3 that the percentage of heterozygosis varies only from .32 to .50 while q ranges between .2 and .8 which includes 60 per cent of the values which q may have. It is only when q is extremely high or extremely low that changes in it produce much change in the percentage of heterozygosis.[15]

The frequency with which each kind of mating occurs depends much on q. Thus, if there are $q^2 AA$ males and $q^2 AA$ females and mat-

[15] If there are multiple alleles, the percentage of heterozygosis will really be a little larger than that shown, since some of those designated here as among the $(1-q)^2$ aa individuals will really be heterozygous for two of the undesired alleles, i.e., will be $A'a$, $A'A''$, $A''a$, etc. Those are homozygous in the sense that neither of the genes is the "desired" one.

TABLE 2

VARIATIONS IN GENE FREQUENCY AS THEY AFFECT THE PROPORTIONS OF THE ZYGOTES IN A POPULATION MATING AT RANDOM

	AA	*Aa*	*aa*
q	q^2	$2q(1-q)$	$(1-q)^2$
.99	.9801	.0198	.0001
.95	.9025	.0950	.0025
.9	.81	.18	.01
.8	.64	.32	.04
.7	.49	.42	.09
.6	.36	.48	.16
.5	.25	.50	.25
.4	.16	.48	.36
.3	.09	.42	.49
.2	.04	.32	.64
.1	.01	.18	.81
.05	.0025	.0950	.9025
.01	.0001	.0198	.9801

ing is at random, the most probable proportion of matings of the type *AA* x *AA* is $q^2 \times q^2$, *or* q^4. Extending this to the other five types of matings possible, where there are only two alleles, gives the proportions shown in Figure 4. Matings of the kinds *AA* x *AA* or *aa* x *aa* can constitute anything from none to all of the matings in the population

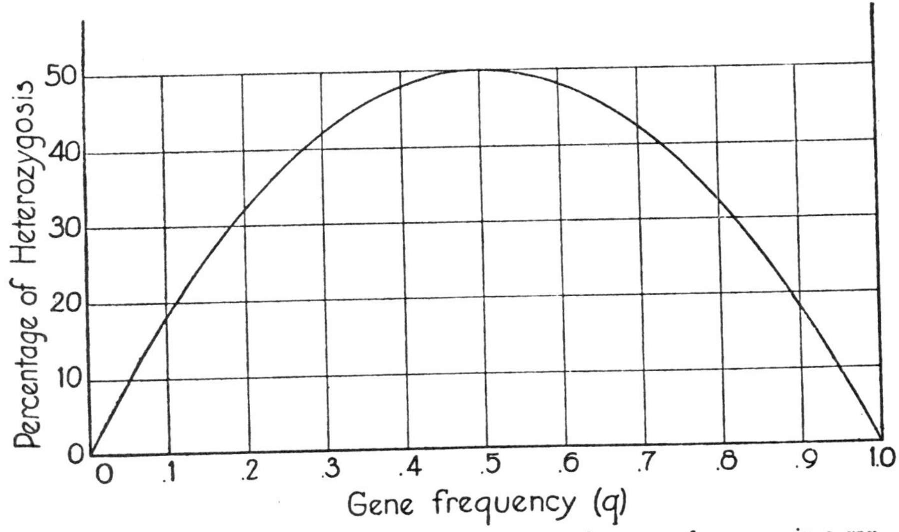

FIG. 3. The percentage of heterozygosis as related to gene frequency in a random breeding population.

according to the value of q. Matings of the kinds *AA* x *Aa* and *AA* x *aa* can constitute from none to 42 per cent of the matings. The maximum figure is reached for the *AA* x *Aa* mating when $q = .75$ and for the

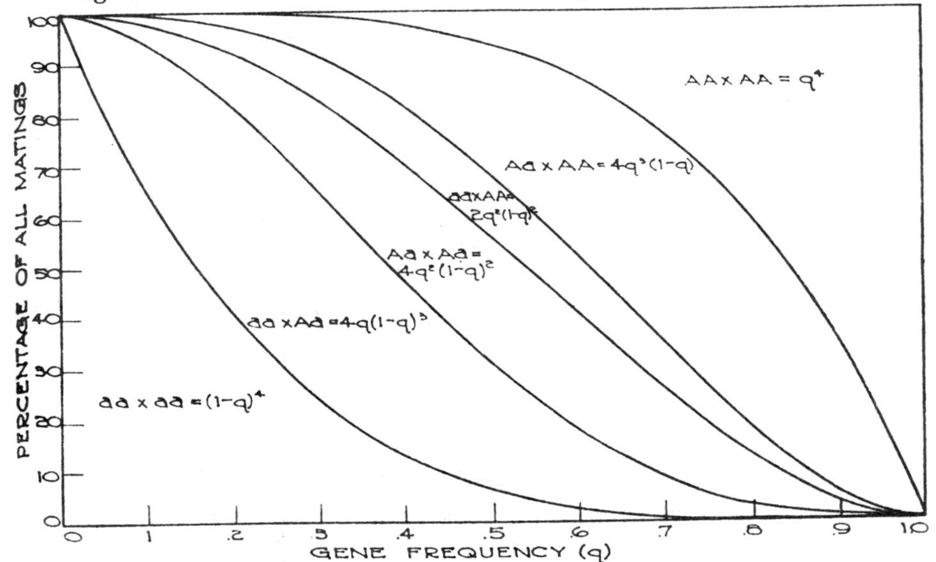

FIG. 4. Showing how the abundance or scarcity of each kind of mating changes with gene frequency in a population mating at random. Vertical distances are in proportion to the frequency of each kind of mating.

Aa x *aa* mating when $q = .25$. Regardless of q, matings of the type *Aa* x *Aa* are always just twice as frequent as matings of the type *AA* x *aa* under random mating. The maximum proportion of the matings they can constitute is when $q = .5$, when the *Aa* x *Aa* matings make up one-fourth and the *AA* x *aa* matings one-eighth of all matings.

The rule that in random breeding populations the zygotic ratio is the square of the gametic ratio can be extended to include nonallelic genes. That is shown geometrically for two pairs of genes in Figure 5, where it is assumed for illustration that the frequency of gene A is .4 and that the frequency of gene B is .7. Hence .16 of the population are *AA*, .48 are *Aa*, and .36 are *aa*; while .49 are *BB*, .42 are *Bb*, and .09 are *bb*. The proportion of individuals which are *AABB* is .16 of .49 =.0784. The other proportions are obtained similarly.

In the array of gametes the nonallelic genes will be combined with each other independently except under three circumstances. First, if the population is the result of a recent cross, the coupling and the repulsion phases of linked genes may not yet have had time to become equally abundant. Second, if the parents were produced by assortive

	q_A^2 AA	$2q_A(1-q_A)$ Aa	$(1-q_A)^2$ aa
q_B^2 BB	.0784 AABB	.2352 AaBB	.1764 aaBB
$2q_B(1-q_B)$ Bb	.0672 AABb	.2016 AaBb	.1512 aaBb
$(1-q_B)^2$ bb	.0144 AAbb	.0432 Aabb	.0324 aabb

FIG. 5. Showing how the rarity or abundance of the various genotypes in a random breeding population depends on the frequency of the genes in each pair.

mating (See Chapter 27), genes which produce similar effects will be together in the same gametes more frequently than otherwise. Third, if the parents are a selected group (rather than a random sample or a typical sample of their generation), the gametes they produce will contain slightly fewer extreme combinations and more intermediate combinations than if the same genes were combined entirely at random.

The two-factor F_2 ratio, used in genetics texts to introduce the subject of inheritance where more than one pair of genes is involved, is only the special case in which the frequency of both genes is exactly .5. The zygotic ratio when *n* genes combine at random can be had by multiplying all the zygotic ratios for each pair of genes together thus:

$$[q_A A + (1 - q_A)\,a]^2[q_B B + (1 - q_B)\,b]^2[q_C C + (1 - q_C)\,c]^2 \;.\;.\;.\; [q_N N + (1 - q_N)\,n]^2.$$

If the frequency of the desired gene is the same in each of the n pairs, the formula can be written in a simpler form: $[q_A + (1 - q_A)\,]^{2n}$. This shows why the search for breeding animals homozygous in *all* desired genes has no prospect for immediate success unless the number of genes desired is very small and the desired gene in each pair already has a high frequency. Table 3 shows the expectation for various values of q and n, figures being shown only where at least 1 in 1,000 is expected to exist.

TABLE 3

PORTION OF RANDOM BRED POPULATION WHICH WILL BE HOMOZYGOUS FOR n DESIRED GENES $= q_A^2\, q_B^2\, q_C^2 \;.\;.\;.\;.\; q_N^2$

Let $q_A = q_B = q_C = \;.\;.\;.\;.\;.\;.\;.\; = q_N$. Then portion equals q^{2n}

q	n			
	5	10	20	50
.50	.001	....	....	....
.60	.006	....	....	....
.70	.028	.001	....	....
.80	.107	.012	....	....
.90	.349	.122	.015	....
.95	.599	.358	.128	.006
.98	.817	.668	.446	.133
.99	.904	.818	.669	.366

Animals having at least one desirable gene in each pair but not necessarily homozygous for all those pairs are much more frequent. Thus, if q^2 are AA and $2q\,(1 - q)$ are Aa, those which carry A either in the homozygous or in the heterozygous condition are $q^2 + 2q(1 - q)$, which may be written $q\,(2 - q)$. Extending this to n pairs of genes gives the figures shown in Table 4. Animals possessing at least one desired gene in each pair are more apt to exist in large enough proportions that it will be finitely possible to find them and to breed from them, discarding all which do not come up to this standard, than is the case with animals homozygous for the desired genes.

In any actual population some of the animals will come nearer than others to having all the desired genes, even though none of them perhaps comes close to that ideal. The practical breeder's simplest reasonable hope is that by selection he can steadily increase the frequency of the desirable genes until sometime—perhaps not until after many gen-

TABLE 4

PORTION OF RANDOM BRED POPULATION WHICH WILL POSSESS AT LEAST ONE DESIRED GENE IN EACH PAIR $= q_A(2 - q_A)q_B(2 - q_B)q_C(2 - q_C) \quad . \quad . \quad . \quad . \quad q_N(2 - q_N)$.
If $q_A = q_B = q_C = \quad . \quad . \quad . \quad . \quad = q_N$, then portion equals $[q(2 - q)]^n$

q	n			
	5	10	20	50
.30	.035	.001	....	....
.50	.237	.056	.003	....
.70	.624	.389	.152	.009
.80	.815	.665	.442	.130
.90	.951	.904	.818	.605
.95	.988	.975	.951	.882
.99	.999	.999	.998	.995

erations—they may reach such a high frequency that perfect breeding animals may be born in his herd. But whether or not that goal is actually attained in his lifetime, the increasing frequency of the desired genes will carry the average merit of the population with it, so that he can reap the reward for his efforts in each generation in which there actually is any increase in the frequency of the desirable genes.

In most circumstances the practical breeder is much more interested in average genetic merit than in homozygosity itself. For example, if 10 pairs of genes affect a trait and a breeder is choosing between one animal which is homozygous for five of the desired genes but homozygous for the undesired gene in the other five pairs and another animal which is heterozygous for all 10 pairs, there will be little difference in their breeding usefulness to him. Each has 10 desired genes. On the average each will transmit five desired genes to an offspring, the former to every offspring without exception and the latter sometimes more and sometimes less than five. Probably the wiser choice would be for the more heterozygous animal since the greater variability of its offspring offers more chance for rapid progress by selection among them.

DEVIATIONS FROM RANDOM MATING AS THEY AFFECT THE BINOMIAL DISTRIBUTION OF ZYGOTES

Mating is random when mates are no more and no less like each other than if they were paired by drawing numbers out of a hat. Random mating does not mean promiscuity or carelessness. Deviations from random mating may be of four genetically different kinds: (1) inbreeding, (2) outbreeding, (3) mating like to like on the basis of somatic resemblance, and (4) mating individuals which are somatically unlike each other. There can, of course, be breeding systems which involve combinations or alternations of two or more of these. Selection, which is deciding that certain individuals shall have many offspring while others shall have few or none, may be practiced along with random mating among those selected to be parents. An illustration is the breeding practiced on large ranches where sires and dams may be highly selected to conform to the owner's standards, but after the selections are made, several sires and many females are turned loose in the same pasture and the matings within that pasture are random, so far as concerns any human control over them. This is selection plus random mating within the group selected to be parents. If the ideals toward which the selections are made differ on different ranches, mating may be random within the limits of each ranch but will not be random with respect to the whole breed, since animals of like types will tend to be on the same ranch and therefore will mate together more often than would be the case if all the sires and dams on all the ranches were in one pasture.

Each of these systems of breeding which is not random will be the subject of a later chapter. Inbreeding tends to make the array of zygotes more like that which would exist if the array of gametes were changed into zygotes by doubling all the genes in each gamete. Inbreeding promotes the formation of families of all kinds, thus adding to the diversity of the population. Outbreeding, which prevails when mates are less closely related to each other than if they were mating at random, is the reverse of inbreeding and cancels many of its effects. It is especially potent in destroying family differences which have been produced by inbreeding, and at producing hybrid vigor or "heterosis." Unless distinct families already exist, outbreeding cannot be carried far. The general effect of mating like to like, within the whole group of parents selected to produce the next generation of the breed, is to make the population more variable by providing more than a random chance for gametes from an extreme individual to meet with gametes from an individual which is extreme in the same direction. The mating of unlikes together makes the population more uniform by making it more likely that gametes from an extreme individual will unite with gametes from an individual extreme in the opposite direction than would be the case under strictly random mating. It is the most potent breeding system for producing immediate uniformity in a population but produces nearly all its effects in the first generation practiced. It is practiced much where the ideal is intermediate and the breeder seeks to mate them so that each will compensate for the defects of its mate.

REFERENCES

Belling, John. 1931. Chromomeres of liliaceous plants. Univ. Calif. Pub. Bot., 16: 153–70.

Cole, L. J., and Jones, Sarah V. H. 1920. The occurrence of red calves in black breeds of cattle. Wisconsin Agr. Exp. Sta., Bul. 313.
Fisher, R. A. 1930. The genetical theory of natural selection. (See especially pp. 17–18 and 38–41.)
Gowen, John W., and Gay, E. H. 1933. Gene number, kind, and size in Drosophila. Genetics, 18:1–31.
Ibsen, H. L. 1933. Cattle inheritance. I. Color. Genetics, 18:441–80.
Jull, M. A. 1934. The inheritance of sexual maturity, rate, and persistency of laying in the domestic fowl. Poultry Science, 13:286–89.
King, H. D. 1932. Mutations in a strain of captive gray Norway rats. Proc. Sixth Internat. Cong. of Genetics, 2:108–10.
Muller, H. J., and Prokofyeva, A. A. 1935. The individual gene in relation to the chromomere and the chromosome. Proc. Nat. Acad. Sci., 21:16–26.
Rhoades, M. M., and McClintock, Barbara. 1935. The cytogenetics of maize. Botanical Review, 1:292–325.
Skovsted, A. 1932. Nordisk Jordbrugsforskning, 5:201–5.
Stadler, L. J. 1932. On the genetic nature of induced mutations in plants. Proc. Sixth Internat. Cong. of Genetics, 1:274–94.
"Student." 1934. A calculation of the minimum number of genes in Winter's selection experiment. Annals of Eugenics, 6(Part 1):77–82.
Sumner, F. B. 1932. Genetic distributional and evolutionary studies of the subspecies of deer-mice (Peromyscus). Bibliographia Genetica, 9:1–106.
Wriedt, Chr. 1929. Den mendelske spaltning av fettprosenten ved kryssning av rødt dansk og jerseyfe. Nordisk Jordbrugsforskning, 103–21.

CHAPTER 7

The Genetic Basis of Variation

Variation—differences between individuals—is the raw material on which the breeder works. It is not necessary that the animals vary widely enough that the breeder can at the very start find some perfect ones to select, but merely that some of them will be closer to his ideal than others are.

The causes of variation are differences in the heredity with which individuals started life, and differences in the environments, internal and external, known and unknown, to which they were exposed during their development. Except in the case of identical twins, two individuals rarely if ever start with all their genes identical.[1] No two individuals ever develop under absolutely identical environments. Hence in practice an observed difference between two individuals must always be considered as the net result of some differences in their heredity and some differences in their environment. Hereditary and environmental differences may have been far from equal in the size of the effects they produced, but almost always both will have been present. They may have opposed each other or both may have worked in the same direction.

Besides these two main divisions of variation into hereditary and environmental portions, a third portion (necessary for logical completeness) comes from joint effects of heredity and environment which cannot fairly be ascribed to either one alone. Such joint effects may occur either if heredity and environment are correlated or if they interact in some nonadditive way so that the effect of a particular variation in heredity may be larger in one environment than in another or, con-

[1] In organisms which can reproduce asexually, (as many plants can by cuttings, budding, etc.,) large groups of individuals with identical heredity can occur. These are called "clones." A highly inbred line either of plants or animals, also approaches the condition in which all individuals in it have the same heredity, but this approach is asymptotic and it is rarely if ever possible to be sure that complete identity of heredity has been reached. The broad term, "isogenic line," which includes identical twins, clones, and completely inbred lines, is convenient where it is desired only to mean that all members of the group have identical heredity, regardless of how that group was produced.

versely, a certain change in environment may make a large change in individuals of some genotypes but only a small change in other individuals whose genotypes make them less labile.

Positive correlation between heredity and environment makes the whole population more variable by preventing the plus effects of variations in heredity from being canceled in individual animals by the minus effects of variations in environment, or vice versa, as often as would be the case if the two were uncorrelated. Such a correlation is an ever-present possibility in data which are collected from a variety of farms, since it often happens that the man who tries hardest to give his animals the best environment also tries hardest to select the animals with the best heredity and has some degree of success in both efforts. Correlation between heredity and environment is also likely to exist in data concerning the mental and social traits of man, since inherited aptitudes on the part of the parents tend to cause them to create in their own homes environments which favor the development of those same special abilities in their children. These children will also inherit some of the same genes which made those parents have those aptitudes in the first place.

It seems likely that the nonadditive combination effects of heredity and environment are generally small in amount,[2] but some interactions of this kind do occur.

MODES OF GENE EXPRESSION

The simplest way in which the effects of various genes are combined is that the substitution of a gene for its allel produces a certain plus or minus shift in the measurement of the characteristic affected and that this change—the "effect" of that gene substitution—is the same, regardless of what other genes are present. As a physical example, consider how adding or subtracting one more brick makes exactly the same increase or decrease in the weight of a brick pile, regardless of the number or kind of bricks the pile already contains. Some genes may combine their effects exactly in this simple way and many seem to do so to some extent, yet many genes are known to interact with each other so that the outward result of substituting a gene for its allel is larger in some genotypes and smaller or zero or even reversed in other genotypes. Thus the actual effect of the gene substitution in each separate individual may depend partly on what other genes are present.

A simple example is dominance. If dominance exists, the outward effect of substituting gene *A* for *a* is larger when the substitution is

[2] For some extreme examples, consult Chapter 5 of Hogben's *Nature and Nurture*.

made in an individual which is *aa* than when it is made in one which is *Aa,* although the effect on breeding value of the individual is the same; namely, that it now transmits *A* to one-half of its offspring which would otherwise have received *a* from it. Dominance is nonadditive combination of the effects of genes which are in the same allelic series. When the effect of making two such gene substitutions in an *aa* individual is not simply twice as large as the effect of one, some degree of dominance exists.

Genes which are not allelic may also modify the magnitude or even the direction of each other's effects. A classic example is Bateson's case of white and purple flowers in sweet peas. He found that two different pairs of genes, both showing dominance, were necessary for the production of the purple color. Plants which were *cc* had white flowers, no matter whether they were *RR, Rr,* or *rr;* and plants which were *rr* had white flowers, no matter whether they were *CC, Cc,* or *cc.* But plants which were either *CC* or *Cc* and were also *RR* or *Rr* had purple flowers. It is as if *R* produced an enzyme necessary for developing color and *C* produced the substrate on which the enzyme could work. Whether the substitution of *C* for *c* will produce a change from white to purple depends on whether *R* is also present, as well as on whether the substitution is made in a *cc* or in a *Cc* individual. The difference between purple and white is a *joint effect* the credit or blame for which cannot wholly be divided fairly, part to one gene and the rest to the other. An example in which the direction of the effect depends on other genes is the case of the *E* gene in guinea pigs, which darkens certain colors in the presence of the *P* gene but lightens them in *pp* individuals. Many other kinds of nonadditive combinations of the effects of genes are known. Some common examples are: inhibiting genes, threshold effects, and the general class of cases in which the outward extreme is genetically an intermediate. The latter may be very common among physiologically complex characteristics where the degree of expression of the characteristic depends on the harmonious interplay of a number of different organs and processes.

SUBDIVISION OF HEREDITARY VARIATION

In an actual population there will be genes acting in all these ways, and the number of genes and possible kinds of interactions between them is so enormous that there is no possibility of learning exactly what each gene does in every combination. The simplest way to think of this tangled situation is to imagine that one could average the effects (some of them large, some small, some positive, some negative, etc.)

which a gene substitution actually does have in that particular population and then proceed as if this *average effect* were the *actual effect* of that gene substitution in all genotypes and under all environmental circumstances which occur in that population. In effect this is what we do when we speak of a gene as "good" or "bad," or as "a gene for high production."

By adding these average effects of all the genes which an animal has we can obtain an "expected" value, or measurement of the appearance or individual performance of this animal. The expected and the actual characteristics may not be exactly the same, if the genes interact in non-additive ways. The expected value of an individual corresponds more closely to its breeding value than its own appearance or performance does.

The variation of the expected values from each other is the additively genetic portion of the actual variation. Differences between the expected and the actual values are deviations from the simple additive scheme. It is convenient to divide these nonadditive deviations into two groups, the first being the deviations caused by dominance and the second being the nonadditive interactions of genes which are not allelic to each other. For brevity these latter are called "epistatic" deviations in this book, although this is a broader use of epistatic than Bateson intended when he introduced the word.

To understand the principles of what we do when we separate the additively genetic variation from that due to dominance deviations, consider the two cases shown in Figure 6. Polled is considered a simple dominant over horns in cattle.[3] The frequencies of the three genotypes in this example were assumed to be: *PP*, .01; *Pp*, .18; and *pp*, .81. These are not far from the present frequencies of the three types in the Hereford breed as a whole in the United States. Probably there actually are slightly more *PP* and fewer *Pp* individuals than this.

The actual or phenotypic values are indicated by Y's and the expected or genetic or breeding values are indicated by G's. The G values come nearer to agreeing with the Y values than could any other three values which lie on a straight line.[4] In technical statistical terms,

[3] This fits most of the facts as far as yet known, but there are a few sets of data in which the situation seems more complicated. Also dominance is not always complete, scurs being some indication of heterozygosis.

[4] The *G* values must lie on a straight line since they are made to conform to the assumption of no dominance; i.e., the phenotypic change expected from substituting *P* for *p* is to be the same when the substitution is made in *Pp* individuals as when it is made in *pp* individuals. The *G* values are completely determined by the requirement that the sum of the squares of their deviations from the *Y* values shall be the least possible for any set of three values which are on a straight line. This is the "least squares" method for fitting a straight line as closely as possible to observations which do not actually lie on a straight line.

the line connecting the G values shows the regression of phenotypic values on breeding values. If there were no dominance Y_{Pp} would lie on a straight line connecting Y_{pp} and Y_{PP} and the G values would fit the Y values exactly. Hence the discrepancies between each G and the corresponding Y (on the vertical scale) are called dominance deviations. Vertical differences between the G values are the additively genetic deviations. If we let the difference between horned and polled be one unit on the vertical scale, we can compare on a quantitative basis the additively genetic variation and the dominance deviations and find how important both are. We get the following values:

Genotype	*pp*	*Pp*	*PP*
Frequency	.81	.18	.01
Actual phenotype, Y	y	y + 1.00	y + 1.00
G value	y + .01	y + .91	y + 1.81
Y − G	−.01	.09	−.81

When we summarize the variation in terms of "variance" (See page 79) we find that 18/19 of the actual variation is included in the variation of the G values, while only 1/19 of it has to be charged against the domi-

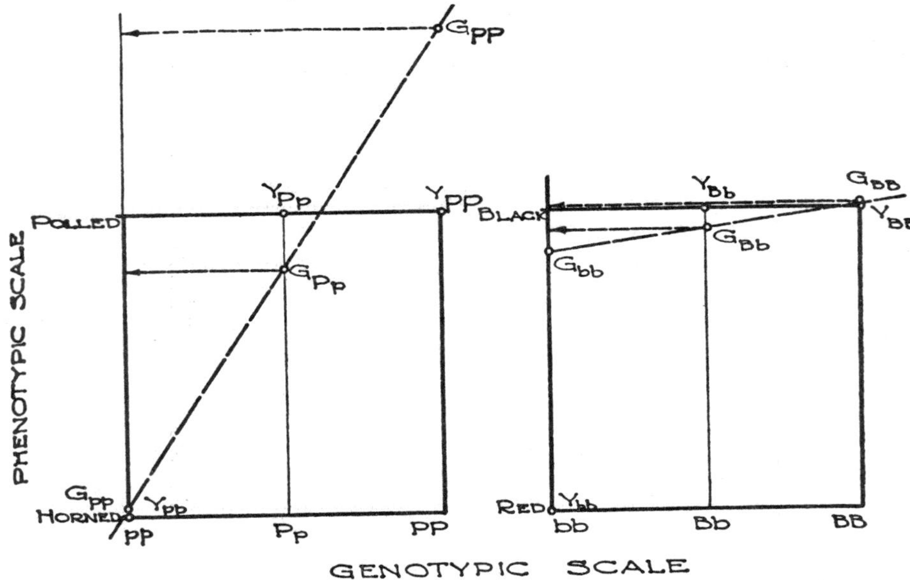

FIG. 6. Diagram of regression of phenotype on genotype where only additively genetic variance and dominance deviations are involved. Left: A case where the dominant is rare, the G-values are far apart, and most of the variance is additive. Right: A case where the recessive is rare and the dominance deviations cause much more variance than the differences between G-values.

nance deviations. In this case the additive scheme comes near to telling all the truth, even though dominance is complete. The G values are far apart and the variation between them is large. The discrepancy between G and Y is very small for the *pp* genotype which includes most of the individuals. Nearly all the rest of the individuals are in the *Pp* group where the dominance deviation is also rather small. The dominance deviation is large for the *PP* group, but the *PP* animals are rare and hence contribute only a few deviations to the total and do not have much influence.

The average effect of substituting *P* for *p* may be computed as follows: For every 100 individuals in the population there will be 180 *p* genes and 20 *P* genes. Of the *p* genes 18 are in *Pp* individuals where changing them to *P* would produce no phenotypic effect. The other 162 are in *pp* individuals and for any one of these a change to *P* would change the phenotype of its possessor one full unit. Hence the average effect of substituting *P* for *p* in this particular population would be $\frac{162}{180} = .9$ unit.[5]

Obviously, the size of the average effect and the comparative importance of additive and dominance variations will depend on the frequency of the genotypes, as well as on the degree of dominance. These are part of the description of this particular population.

The right side of Figure 6 shows for comparison how different the situation is when the recessive is very rare. The *Bb* pair of genes determines the contrast between black and red in cattle, black being completely dominant. The frequencies assumed for the genotypes are about what they would be in black breeds of cattle in which one calf in each 200 is born red. The actual variation is small and the variation between the *G* values accounts for only $\frac{14}{107}$ of it. Dominance deviations account for the rest, $\frac{93}{107}$. The contrast between the left and the right sides of Figure 6 illustrates the general fact (of which more in Chapter 11) that dominance is an *important* source of confusion or hindrance to progress by selection only when the undesired recessive is already rare. Even when dominance is complete, the variance among the G values will account for $\frac{2(1-q)}{2-q}$ of the actual variance in random breeding

[5] This simple arithmetical way of computing the average effect illustrates its meaning and will be correct when the heterozygotes are present in the same proportion as they would be under random mating. When they are more abundant or less abundant than that, the average effect should be computed by the more technical procedure which is called the least squares method of fitting a straight line.

populations, q being the frequency of the dominant gene.

For a numerical example of epistatic variance, let us take again the case of purple and white flowers in sweet peas. Its genetic basis is definite and well-known, and it is generally considered to be a rather extreme case of epistasis—although that idea may need revision when we learn more about the usual results of making several gene substitutions at one time. If the sweet peas were breeding at random,[6] and if the frequencies of the C gene and of the R gene were each .5, then the various genotypes would occur in the proportions shown in column 1 of Table 5.

If we let the difference between purple and white be one unit on the scale on which we measure color, then in this population the average effect of substituting C for c is 3/8 of a unit. That may be computed as

TABLE 5

ILLUSTRATION OF THE BASIS FOR SEPARATING ADDITIVE GENETIC VARIATIONS, FROM DEVIATIONS CAUSED BY DOMINANCE AND EPISTASIS, USING BATESON'S CASE OF PURPLE AND WHITE COLOR IN SWEET PEAS

Frequency With Which the Various Genotypes Would Occur	Fraction of the *c* Genes of the Whole Population Which Are in This Genotype	Values on a Scale on Which White = 0 and Purple = 1		
		Actual	"Expected"	Deviations Due to Dominance and Epistasis
1 ccrr	1/8	0	− 3/16	+3/16
2 Ccrr	1/8	0	+ 3/16	−3/16
2 ccRr	2/8	0	+ 3/16	−3/16
1 CCrr	none	0	+ 9/16	−9/16
1 ccRR	1/8	0	+ 9/16	−9/16
4 CcRr	2/8	1	+ 9/16	+7/16
2 CcRR	1/8	1	+15/16	+1/16
2 CCRr	none	1	+15/16	+1/16
1 CCRR	none	1	+21/16	−5/16

follows: One-half of all the c genes are in Cc individuals (the second, sixth, and seventh lines in Table 5), where, on account of dominance, the substitution would make no change in the color. One-eighth of the c genes are in $ccrr$ individuals (the first line in Table 5), where the substitution would produce no outward effect because the R gene, which is also necessary for the production of purple, is not present. The remaining three-eighth of the c genes are in $ccRR$ or $ccRr$ individuals, where the substitution of C for c would produce the full change from white to purple. One unit of change in three-eighths of the cases plus no change in five-eighths of the cases makes an average effect of three-eighths of a

[6] The sweet pea does not really fulfill this condition, since it is largely self-fertilizing.

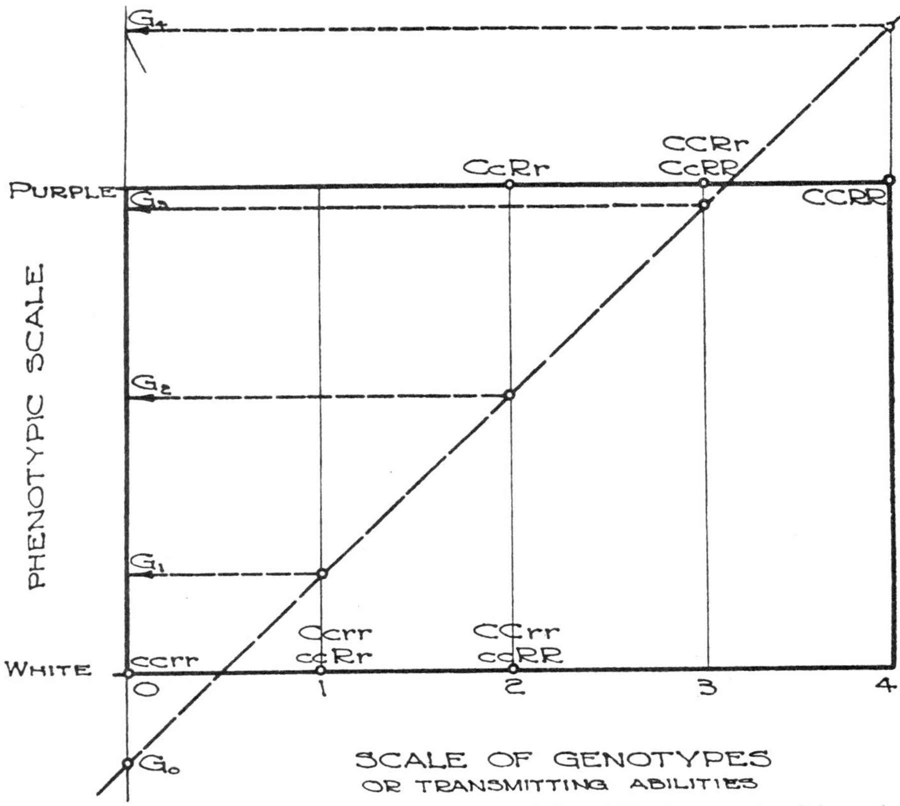

FIG. 7. Regression of phenotypes on transmitting ability in a case involving epistatic deviations in addition to dominance deviations and additive differences. G_0 - - - G_4 are the expected values or transmitting abilities of individuals which have 0, 1, - - 4; respectively, of the *C* and *R* genes. Differences (on the vertical scale) between the *G* values are the additive variation. Differences (on the vertical scale) between each *G* value and the corresponding actual value are deviations caused by dominance and epistasis together.

unit for substituting *C* for *c* in that population, although the actual change would not be exactly three-eighths of a unit in any one plant.[7] The expected values shown in Table 5 were found by the additional requirement that the average of the expected values must be the same as the average of the actual values. Figure 7 shows graphically how near the actual and the "expected" values are to each other. The differences between them (the last column in Table 5) are the deviations caused jointly by dominance and epistasis.

Variance due to dominance and variance due to epistasis can be

[7] For a more detailed discussion of this conception of the average effect of a gene substitution, see the first few pages of Wright's article beginning on p. 243 of volume 30 of the *Jour. of Genetics,* or pages 53–56 in volume 11 of *Annals of Eugenics,* 1941.

separated by setting up another series of values expected, on the hypothesis that dominance is complete but there is no epistasis. The final result in this example is that 4/7 of the actual variation can be gathered into and described by the simple additive hypothesis, 2/7 must be charged to dominance deviations, and only 1/7 to epistatic deviations. It will be seen that the expected values are partly determined by the frequencies with which the genotypes occur. They are partial descriptions of a particular population and may vary from one population to another, even where the same genes are involved. The relative importance of additive genetic variance, dominance deviations, and epistatic deviations will change also. Thus, in the same example but with different gene frequencies the ratio of additive to dominance to epistatic variance becomes 12:9:2 when q_R and q_C are both .6, and 24:8:9 when both are .4.

These illustrations are intended to make clear what is meant by dividing hereditary variation into these three portions. In actual practice the breeder or investigator will not know exactly what genes are present, nor the frequency of each, nor all of the ways in which they interact with each other. Rough estimates of the additive variance can be had from observing the results of selection or the likeness between different kinds of relatives. The additive variation is the part which contributes in the simplest and most direct way to the likeness between relatives. Knowledge of its size is important for estimating the probable results of any proposed breeding plan. For many practical purposes it may not be worth while to separate the dominance and the epistatic portions from each other.

The additive genetic variation caused by a gene can become zero only when the average effect of that gene is zero; that is, when the sum of all the plus changes which it causes is exactly equal to the sum of all the minus changes which it causes in other genotypes in that same population. Most of the variation which a gene causes will be included in the additive portion if its average effect is large. For most of its variation to be epistatic requires that its average effect be near zero but that it produce large plus effects in some genotypes and correspondingly large minus effects in other genotypes. As yet there are only a few actual data to indicate whether epistatic variations are abundant and important, or so rare and small that ignoring them in practice would not cause many errors.

THE MEASUREMENT OF VARIATION

The methods of measuring variation are inconveniently technical for those not trained in statistical methods. Moreover, there are several of them, and each has advantages for certain purposes. For reasons which do not concern us here, the importance of various causes in producing the variability of a population is most conveniently expressed in terms of the "variance" (σ^2) of that population. The variance may be defined as the average of the squared deviations of the individuals from the population average.[8] An equivalent definition is that the variance is one-half of the average squared difference between pairs of individuals chosen at random. The square root of the variance is called the "standard deviation" (σ), which, since it is expressed in the same terms as the original measurements, is often more convenient for expressing the variations of individual items than is the variance, which is expressed in squares of the original measurements. In a "normally distributed" population about two-thirds of the individuals differ from the average by less than the standard deviation, while about one-sixth of the individuals will be more than one standard deviation above the average. The remaining sixth will be below the average by more than one standard deviation. Only about one-fortieth of the individuals will be more than twice the standard deviation above the average and another fortieth will be more than twice the standard deviation below the average. In small populations the standard deviation is usually about one-fourth to one-sixth of the difference between the largest and the smallest individuals (the "range"); but that rule is not very accurate, since the range depends on only two individuals and is subject to large sampling errors. The arithmetic average of the deviations, neglecting signs, is about .8 as large as σ, but for various reasons is not as dependable and is almost never used.

Not all populations are "normally" distributed, although most of those encountered in breeding practice are nearly enough so that the statistics of the normal curve may be used with little error for practical purposes. The normal curve is frequently called the "Gaussian curve" after the mathematician who first studied it in detail, or the "curve of error" because its first application, which is still its principal application in some sciences, was in making allowance for unavoidable but random errors of observation. It is symmetrical and bell-shaped, as will be seen in Figure 8.

The statistical cornerstone for the genetics of populations is the "binomial distribution," which is obtained by expanding the expression $(a + b)^n$. This is just the mathematical description of what results

[8] Since the known average of the n individuals in the sample studied may not be exactly the same as the unknown average of the much larger population from which the sample comes, the sum of the n squared deviations of individuals from the average of the sample is divided by $n - 1$ instead of n to obtain the best estimate of the variance of the population.

naturally from the duplicateness of inheritance and the "one-or-otherness" of gene transmission; that is, of Mendel's laws of segregation and recombination. The Mendelian mechanism guarantees that in a random-mating population the zygotes will be distributed according to the square of the gametic ratio. The genotypes for characteristics determined by n pairs of genes with equal frequencies and equal effects will have the binomial distribution corresponding to $[qA + (1 - q)a]^{2n}$.

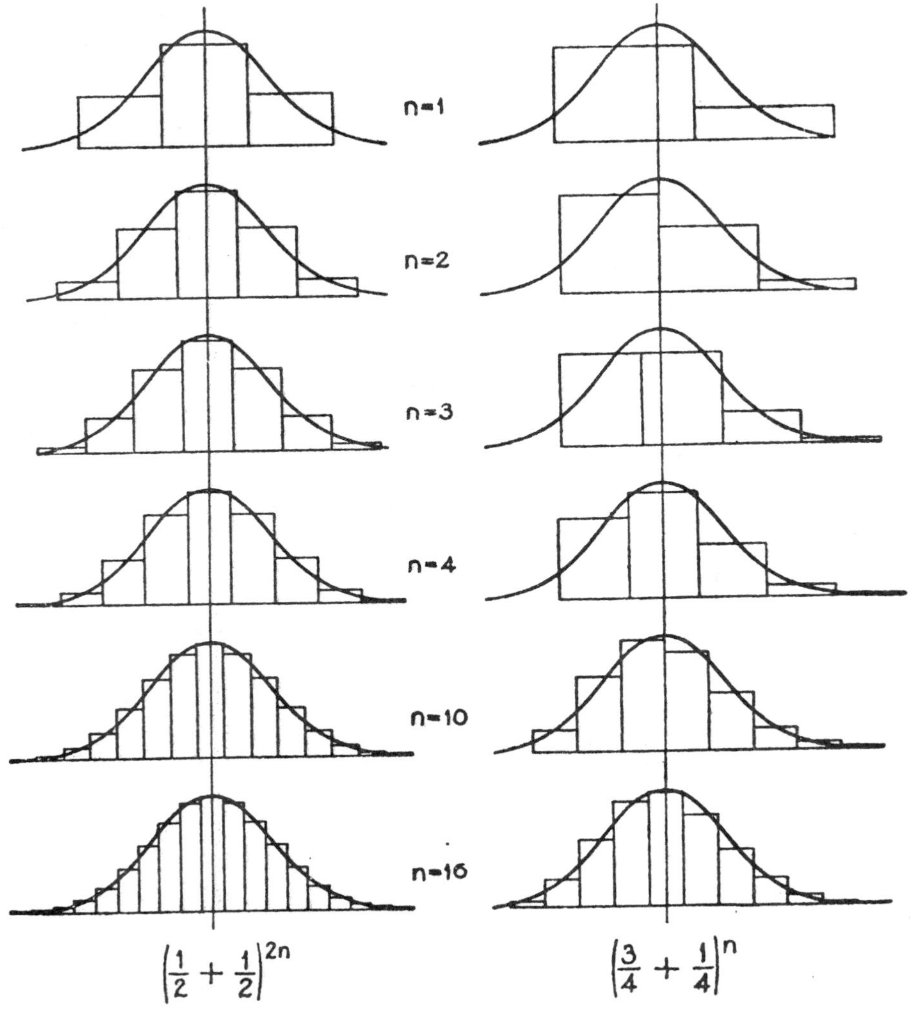

FIG. 8. Binomial distributions for n pairs of genes with equal effects and superposed normal curves with equal mean, equal area, and equal variance. Left: No dominance. Right: Complete dominance in the same direction in all pairs of genes.

If the gene frequencies are not the same for all pairs of genes, the distribution will be somewhat less variable than if they were all equal but had the same average. If the different pairs of genes do not have equal effects, the distribution of the genotypes is more variable than if the same number of genes each played equal parts in producing the same total effect. When $a = b$ (or $q = 1 - q$ in the genetic formula) and n is large, the binomial distribution approaches the normal distribution so closely that for practical purposes they may be treated alike, especially in small populations. The binomial curve can be distinctly skewed to one side if all of the genes which operate in the plus direction happen to be much more frequent than their alleles, or the reverse. It can also be skewed if the genes multiply each other's effects instead of combining additively, or if there are threshold or other epistatic interactions. Figure 8 shows on the left side several symmetrical binomial curves with normal curves superposed on them. With some additional environmental modifications to blur the accuracy of classification, the binomial curve, even with very small values of n, could not be surely distinguished from the normal curve in small populations. The right side of Figure 8 shows how the skewness produced by dominance, even when the dominance is all in one direction, is extreme when n is small but would be difficult to detect in small populations if there were as many as 10 pairs of genes involved, especially if there were also many variations from environmental causes.

One of the most important consequences of the Mendelian mechanism of inheritance is that it maintains the variability of an interbreeding population at a nearly constant level over long periods of time, thus maintaining a supply of variability available for selection or other breeding practices. The importance of this may be shown most clearly by contrasting it with what would be expected under the blending theory of inheritance. Under that theory, if the sire deviated x and the dam deviated y from the mean of the race, *every* offspring from that mating would be expected to deviate $\frac{x + y}{2}$. The average squared deviation (the variance) of the parents would be $\frac{x^2 + y^2}{2}$. The average squared deviation of the offspring would be $\frac{x^2 + 2xy + y^2}{4}$. If the parents mated at random, sires with positive values of x having no especial tendency to mate with dams which had positive values of y, the term $2xy$ would be zero (since the negative terms would cancel the positive ones); and the average squared deviation of each generation would be *only half as large as that of the preceding one.* The group would thus approach perfect uniformity at a tremendous rate. Even a pronounced tendency for like to mate with like would delay this approach only a little (by causing $2xy$ to have a positive value) unless the tendency of like to mate with like were perfect. Figure 9 shows how rapidly the

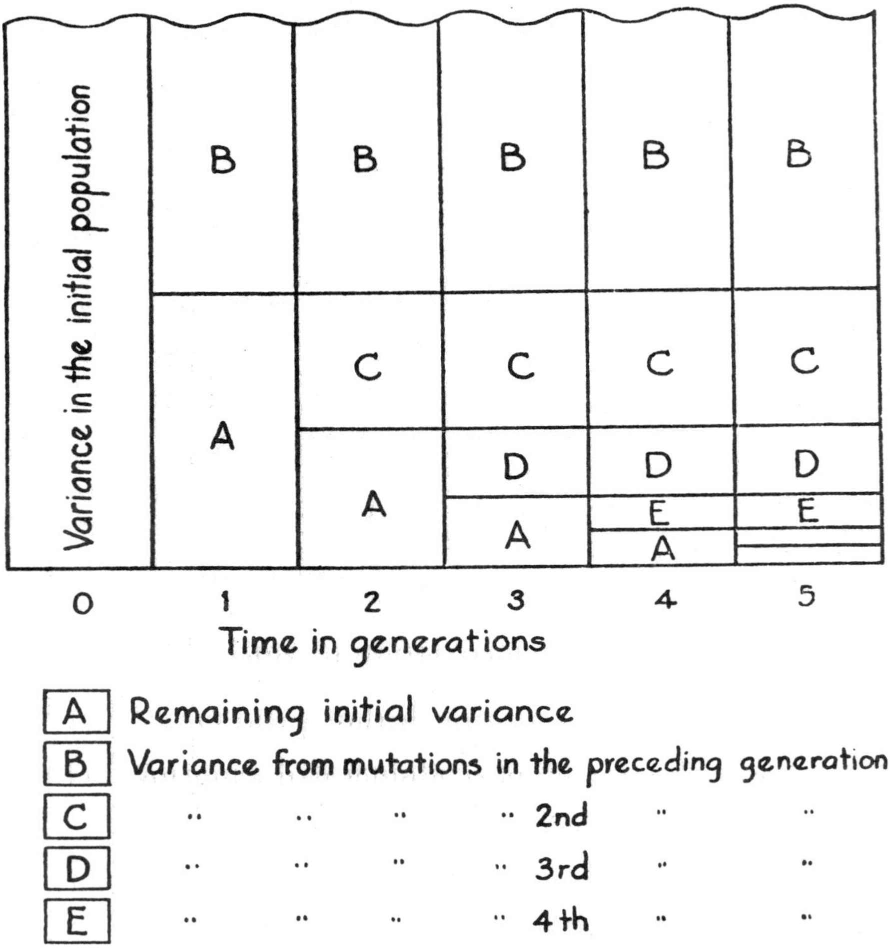

FIG. 9. Rate at which initial hereditary variance was supposed to be lost and the supposed recent origin of the hereditary variations existing at any one time, according to the former theory that inheritance really blended.

hereditary variability existing at any one time would be "swamped" if inheritance really were blending, and how much of the hereditary variability existing at any one moment must have come from mutations which had just occurred in the last few generations, if the variability of a population were to remain the same from generation to generation. Much of the skepticism about Darwin's theory of evolution by natural selection had its roots in the tacit assumption that inheritance is "blending" and the consequent belief that selection would have to be almost instant and perfect in its seizure of new variations if these were to be incorporated into the species before they were lost. In the older view, heredity was a conserving force and variation was contradictory to it. That is now obsolete as it is seen that the mechanism of heredity conserves individual variation also. Knowledge of Mendelism and of the hereditary variation to be expected between full brothers has freed us from the supposed necessity of thinking that mutations are frequent or important in practical breeding problems. It has also relieved us from the necessity of believing that selection must act almost at once if it is to utilize variations before they are "swamped" or lost. In many respects Mendelism has rounded out the Darwinian theory of the power of natural selection by showing that some of its most serious supposed weaknesses do not exist.[9]

Mendelism gives us a picture of a stable population composed of changing individuals, or of endless individual variations which added together result in an almost constant population. A Mendelian population may be compared to a group of bees around a hive. Almost every bee is constantly in motion and yet the average position of the swarm may remain almost the same hour after hour. The individuals are dynamic—the population is almost static. Changing a population by selection may be compared roughly to driving a herd of many hundred steers. At any one time there are steers moving in all possible directions, yet the drivers, by constantly discouraging those which attempt to move toward the rear and leaving the road open to those which go in the desired direction, can succeed in moving the herd a few miles each day.

The *correlation coefficient* is a measure of how closely two things tend to vary in the same direction. Examples are the tendency in human data for father and son both to be tall or both to be short and the tendency for tall men to weigh more than short men. In both cases the tendency is pronounced enough that it is a matter of common knowledge, even to those who have never heard it expressed quantitatively. In both cases there are frequent and striking exceptions. In the former case the measurements are the same kind (i.e., stature), and they are paired together on account of the genetic relationship of their possessors. In the latter case the measurements (i.e., stature and weight) are different in kind and are paired together because they apply to the same

[9] See pp. 1–12 of Fisher's *The Genetical Theory of Natural Selection.*

individual. The general idea of correlation is simple and universally understood, but the coefficient for measuring degrees of correlation is technical enough that considerable practice in computing it on various kinds of data is usually necessary for proficiency in understanding it. The coefficient of correlation is expressed on a scale running from + 1.0, where two characteristics vary in perfect step with each other, through zero, where there is no correspondence at all, to — 1.0, where there is a perfect tendency to vary in exactly opposite directions from each other.

Regression is the general statistical term for expressing how much one variable may be expected to change per unit change in some other variable. As a concrete example, in dairy cattle regression of daughter's fat yield on dam's fat yield is the average amount of increase in the fat production of the daughter which we may expect for each extra unit of fat the dam produced. Historically regression gets its name from a certain aspect of correlation wherein it was observed that the offspring of extreme parents are usually nearer to the average of the population than their parents were; i.e., they regress toward the mean of the race. This idea of regression of the offspring from the parents toward the mean of the race was later extended to all kinds of equations for predicting one variable from another, where the relation between the two is not perfect. Regression is now used in many cases which do not involve questions of heredity at all and even in cases where the idea of the correlation coefficient would be artificial, as in time trends. The customary symbol for the correlation coefficient, *r*, was originally chosen because it was the first letter of the word regression. Thus, it is a reminder of the once intimate relation between the two ideas!

SUMMARY

Variation is the raw material with which the breeder works. For purposes of subdivision into its constituent parts, variation is best measured in terms of "variance," which is the average squared deviation of the individuals from the population average. According to its causes, variance may be divided into three main parts: that due to variations in environment (σ_E^2), that due to differences in heredity (σ_H^2), and that due to joint effects of variations in heredity and environment (σ_{HE}^2) which are nonadditive or otherwise intertangled so that they cannot fairly be ascribed to heredity or to environment alone. The hereditary variance may be further subdivided into three portions: the additive genetic variance, which includes all that can be described by assuming that the effects of the whole combination of genes in the individual equal the sum of the average effects of those genes (σ_G^2); the variance caused by dominance deviations from the additive scheme (σ_D^2); and the variance caused by epistatic deviations from the additive scheme (σ_I^2).

For expressing the variation of individuals, or for expressing differences between expected and actual values, variation is most conveniently expressed in the form of the standard deviation (σ), which is the square root of the variance. In most populations about two-thirds of the individuals differ from the population average by less than the standard deviation.

The tendency for two different characteristics of the same individual, or for the same characteristic in pairs of individuals related in a certain way, to vary in the same direction is measured by the "coefficient of correlation." The equation for predicting the value of one characteristic which will most probably correspond with a given value of another characteristic is called a "regression" equation.

CHAPTER 8

Heredity and Environment

In the strictest sense of the word, the question of whether a characteristic is hereditary or environmental has no meaning. Every characteristic is both hereditary *and* environmental, since it is the end result of a long chain of interactions of the genes with each other, with the environment and with the intermediate products at each stage of development. The genes cannot develop the characteristic unless they have the proper environment, and no amount of attention to the environment will cause the characteristic to develop unless the necessary genes are present. If either the genes or the environment are changed, the characteristic which results from their interactions may be changed.

Nevertheless, it is often convenient to speak of a characteristic as "hereditary" or "highly hereditary" when we wish to emphasize that most of the differences we usually see between individuals in that characteristic are caused by differences in the genes they have, and only a few of the differences between individuals are caused by differences in the environments under which they developed. The difference between black and red coat color in cattle is such an example of a highly hereditary characteristic. Environmental circumstances, such as exposure to sunshine, may cause the black to vary from a jet black to a rusty or brownish black: but that is a tiny variation compared with the large difference between black and red which is caused by differences in genes.

With equal logic it is often convenient to call a characteristic "environmental" or "only slightly hereditary" when most of the differences ordinarily found between individuals in that population are caused by differences in the environments under which they developed and only a small part of those differences between individuals are caused by differences in the genes they have. Examples of such largely environmental characteristics are degrees of fatness and of lameness. In most populations variations in those are much more apt to have resulted from previous management, feeding, accidents, or condition of general health than from differences in heredity. Yet it is certain that some individuals have genes which make them fatten more readily than others, or have genes causing structural weaknesses which predispose them to lameness.

The whole matter of whether a characteristic is hereditary or environmental, if we find it convenient to state it in that way, is a question of how much of the variation in that characteristic in that population is caused by differences in heredity and how much is caused by differences in environment.

The question of whether heredity or environment is the more important can be phrased precisely for a particular trait in a particular population and, if the data are available, can be answered. It does not have a single answer true for all traits in one population nor for the same trait in all populations. Let $\sigma_O^2 =$ the actually observed variance, $\sigma_H^2 =$ that part of the variance caused by differences in the heredity which different individuals have, and $\sigma_E^2 =$ that part of the variance caused by differences in the environments under which different individuals developed. Then[1] $\sigma_H^2 + \sigma_E^2 = \sigma_O^2$ and $\frac{\sigma_H^2}{\sigma_H^2 + \sigma_E^2} = \frac{\sigma_H^2}{\sigma_O^2} =$ the portion of the observed variance for which differences in heredity are responsible. When this fraction is large, we say that the characteristic is highly hereditary; when this fraction is small, we call the characteristic slightly hereditary or largely environmental.

The value of $\frac{\sigma_H^2}{\sigma_H^2 + \sigma_E^2}$ can be altered by changing either σ_H^2 or σ_E^2. If we try to make the environment exactly the same for all individuals, as is usually attempted in genetic experiments, we may go far in that direction although we can hardly hope to control the environment perfectly. So far as we succeed in making σ_E^2 smaller than it is in the general population, we make the variations in that characteristic more highly hereditary in our material than they were in the general population. If we also enlarge σ_H^2 in our material by mating like to like while selecting for opposite extremes or by inbreeding in several separate lines, we increase σ_H^2 and make variations in that characteristic in our material still more highly hereditary. Unless we are quite aware of what we have done in partially controlling the environment (and thus making σ_E^2 tend toward zero) and in selecting or breeding to increase the genetic diversity of our laboratory material, we are apt to get an exag-

[1] In order not to confuse the argument, the nonadditive interactions of heredity and environment are neglected here. There is reason to think that those joint effects will generally be small. This definition includes as "hereditary" the dominance and epistatic deviations since they result from differences between whole genotypes, although they will not contribute so much to the likeness between relatives as the additive differences do.

gerated idea of the importance of heredity in causing variations in that characteristic in the general population. On the other hand, if we are experimenting with the effects of some procedure in nutrition or in management, we will probably try to make σ_H^2 as small as possible by using a uniform stock, discarding at the beginning of the experiment any which appear to deviate much in either direction from the population average, and minimizing the hereditary differences between lots by putting litter mates in different lots, etc. Then we will also make our experimental (environmental) treatments so contrasting that we will be reasonably sure to find differences in their results. Again, unless we are quite aware of what we did in neutralizing hereditary differences and in magnifying environmental differences in our laboratory material, we are apt to get an exaggerated idea of how important our environmental differences are in causing the variations in the general population. If this were clearly understood, a considerable amount of fruitless controversy would be avoided.

A quantitative statement of the relative importance of heredity and environment is a partial description of the causes of the variation in a particular characteristic in the specified population. It is useful in estimating the probable results of certain breeding systems in the next generation or two, but it tells nothing about the ultimate limits of the changes which might be made in that population either by breeding or by altering its environment.

METHODS OF ESTIMATING HERITABILITY

All methods of estimating heritability rest on measuring how much more closely animals with similar genotypes resemble each other than less closely related animals do. The techniques suitable for doing this vary with the material and according to whether environmental correlations between relatives and the peculiarities of the mating system, if it was other than random, can be measured and discounted by other means.

Variation within isogenic lines is wholly environmental. Comparing this with the variation in an otherwise similar random breeding population may give an estimate of heritability. This method is of little use in farm animals because among them are no isogenic lines except occasional pairs of identical twins. These are difficult to identify but if sought diligently and studied intensively might finally give reliable information. Identical twins are to be compared with ordinary twins, rather than with pairs of individuals unrelated to each other, lest the similarities in the environments of twins might lead to errors in the interpretation. The method of isogenic lines is the only method likely to measure all of the epistatic and dominance variations as well as the additive ones but, because of the rarity of identical twins in farm animals, is not promising as a source of information fairly free from sampling errors.

In selection experiments, if we can measure the amount by which those selected to be the parents exceeded the average of their generation, we can divide that into the amount by which the average of the offspring exceeded the average of the generation in which their parents were born. This gives a measure of the additive portion of the variance plus a portion (somewhat less than half) of the epistatic variance. In order not to be misled by unnoticed environmental changes, it is usually necessary that selection be practiced in opposite directions at the same time, so that the interpretation will be based on differences between the high and the low lines, rather than on the absolute values of the averages. Sometimes this method can be followed in experiments especially designed for this purpose at some research institution, but it is not often available to the breeder, since he can rarely afford to select in the undesired direction just to get information on heritability.

The resemblance between parent and offspring is the most widely useful method, but is likely to include some environmental correlation between parent and offspring. Also it will include something from the resemblance of the offspring to the other parent, if mating were not random. In using this method the procedure is as follows: (1) observe the correlation between parent and offspring, (2) subtract from that the environmental contribution, (3) double the remainder, and (4) divide by one plus the correlation between mates.[2] The second step is always likely to be difficult, and the fourth will be unless the deviations from random mating are known more exactly than is usually the case.

A useful dodge which makes steps two and four unnecessary is to divide the mates of each sire into a high and a low half on their own performance, combine the data for all sires, divide the difference between the daughters of the high and the low halves by the difference between the two groups of dams and double the results. This measures the additive portion and a bit of the epistatic portion of the differences between such dams as were mated to the same sire. It leaves unanalyzed the differences between the groups of cows which get mated to different sires. A similar division of the offspring of the high and low sires mated to the same dam would answer as well in principle but, because of the

[2] This should be the genetic correlation (coefficient of relationship) if the departures from random mating were of the inbreeding kind, but should be the actually observed correlation if the mating choices were based on each animal's own individuality for the characteristic being studied. One will rarely be certain about this.

usually small number of offspring per dam, is rarely possible with farm animals.

One of the first clearly analyzed cases of the relative importance of heredity and environment was Wright's study of the amount of white spotting in a stock of guinea pigs. It will illustrate the principles and the use of nearly isogenic lines and of correlation between relatives. Besides the control stock, in which even second cousin matings had been avoided, there was a stock which came from the same foundation but which had been inbred full brother and sister for more than 10 years (probably about 20 to 25 generations in most branches of the family, nearly all of which came from a single mating in the twelfth generation), so that it must have been almost entirely homozygous and could have retained but little genetic variability. By measuring the average likeness between parents, between parents and offspring, and between litter mates, Wright was able to separate the variance into a portion due to heredity, a portion due to environment common to litter mates, and a remainder due to environment or embryological accidents which were not alike even for litter mates. Table 6 shows the findings. The

TABLE 6

PIEBALD SPOTTING IN GUINEA PIGS. PORTIONS OF THE VARIANCE ACCORDING TO CAUSES (AFTER WRIGHT)

Causes of Variance	σ^2 in the Inbred Stock		σ^2 in the Control Stock	
	Actual Units	Percentage of Total	Actual Units	Percentage of Total
Heredity	.010	2.8	.271	42.2
Environment common to litter mates	.020	5.5	.002	.3
Environment not common to litter mates	.334	91.7	.370	57.5
Total	.364	100.0	.643	100.0

most illuminating fact for our present purpose is that the variance due to environment was almost the same in actual units of measurements (.354 and .372) in both stocks but was 97.2 per cent of the variance in inbred stock and only 57.8 per cent of the variance in the control stock. Here is a case where the same characteristic in two separate stocks derived from the same colony by different breeding methods is very slightly hereditary (2.8 per cent) within the inbred stock and nearly half (42.2 per cent) hereditary within the control stock. All that really happened was that in one stock the inbreeding had caused nearly all of the initial hereditary variability to be lost and thereby had altered greatly the proportion of hereditary to environmental variability.

The following are some examples of the kind of analysis which can be made by comparing correlations between relatives. Gowen studied Jersey Register of Merit data on milk yield and fat percentage, assuming that there was no correlation between the environments of daughter and dam. He came to the conclusion that about 50 to 70 per cent of the variance in milk production and about 75 to 85 per cent of the variance in fat percentage came from variations in the heredity of the individual cows. But if there was as much as .10 to .20 of environmental correlation betwen daughter and dam, as seems probable from the usually observed correlation between the records of herd mates, these figures are too high by .20 to .40. Plum's analysis of the records of cows in Iowa Cow Testing Associations led him to the figures shown in Table 7. Studies of intrasire regression of daughter on dam have generally given values of around .15 to .30 for heritability of differences in fat production between cows in the same herd where each cow was represented by only one record.

TABLE 7

RELATIVE IMPORTANCE OF CAUSES OF VARIATION IN BUTTERFAT PRODUCTION

Causes of Variation	Percentage of Total Variance	
Breed		2
Herd		
Feeding policy of herd	12	
Other causes (genetic or environmental)	21	33
Cow (mostly genetic)		26
Residual (year to year variations)		
Feeding variations within the herd	6	
Other year to year differences	1	
Length of dry period	1	
Season of calving	3	
Other factors	28	39
Total		100

Lush, Hetzer, and Culbertson, studying the birth weights of pigs born during 15 years at the Iowa Agricultural Experiment Station, came to the figures shown in Table 8. Part of the 29 per cent due to "other" environment common to litter mates may really have been hereditary as the result of hereditary differences among the dams, although it was environmental as far as the pigs themselves were concerned. There is much evidence that the dam's own size or other characteristics have much influence on the birth weights of her offspring.

The figures in Tables 7 and 8 illustrate how the actual data may permit subdividing the environmental variance into portions caused by certain tangible factors or groups of factors. Nearly always a considerable part of the variance will remain unidentified as to causes. These are most naturally inferred to have been individual unobserved (or at least unrecorded) variations in environment, but in some cases may really have been errors in observations or (in some methods of analysis) will also have included those portions of the epistatic or dominance variations which did not contribute to the likeness of the relatives studied. For other examples similar to Tables 6 to 8, yet each showing some special features of its own, see: *Genetics* 19:535; 21:360; 22:468; 26:217; 31:503; and *Onderstepoort Jour. Vet. Sci. and An. Husb.* 5:580.

While it is true that the animal at birth contains all the heredity it is going to have but has not yet been affected by many of the environmental circumstances which will affect it, yet the presence or absence of differences at birth is not a good criterion of whether those differences

TABLE 8

RELATIVE IMPORTANCE OF CAUSES OF VARIANCE IN BIRTH WEIGHTS OF PIGS

Causes of Variance	Percentage	
Heredity of the pigs		
Breed differences	2	
Sex	1	
General	3	
		6
Environment common to litter mates		
Litter size	7	
Year	5	
Ration	4	
Gestation length	2	
Other	29	
		47
Environment not common to litter mates		47
Total		100

are hereditary. Some of the differences found at birth are the result of previous differences in intra-uterine environment, or of what for lack of a better term may be called embryological accidents. On the other hand, many genes in which individuals may differ do not produce their effects until the individual reaches a certain stage of development. Examples are the genes which affect early maturity, milk production, shape and quality of teeth, and in man such specific things as baldness, prematurely gray hair, and Huntington's chorea.

PRACTICAL APPLICATIONS

Much individual variation is left even when either the heredity or environment is perfectly controlled. For example, if half of the variance in a characteristic is hereditary and half is environmental, perfect control of heredity would still leave the standard deviation 71 per cent (the square root of one-half) as large as before. If all environmental variations in a characteristic which is 80 per cent hereditary were eliminated, the standard deviation would still be nearly 90 per cent (the square root of 80 per cent) as large as before. If the hereditary variations were entirely eliminated, the standard deviation would still be about 45 per cent as large as before. Even if a characteristic were 99 per cent hereditary, complete elimination of all hereditary variability would still leave the standard deviation 10 per cent as large as it was originally.

Only those variations which are caused by differences in heredity are themselves inherited. Variations caused by environment can be large and very important economically, but they do not change the inheritance of the animal and are not transmitted to its offspring but must be produced afresh in those offspring by repeating the environmental treatments which produced them in the parent. There is not space here to repeat the proofs for the noninheritance of environmental effects; and, as with other negative concepts, it may be impossible to prove this one rigorously. But the many experiments carefully planned to test whether the effects of environmental treatments are inherited in such a way that the offspring inherit some degree of the modifications originally produced in their parents by the environmental treatment have all given negative or doubtful results. Even one who deliberately wishes to believe that environment does affect heredity in this way must admit that the effects are so slight that they are not practically important in any one generation.[3] Perhaps they do not occur at all. Both in improving the heredity and in improving the environment of his animals the breeder is likely to encounter the law of diminishing returns. Yet in improving their heredity there is the possibility that if he can achieve enough—for example, get his herd widely known as one of the three or four best sources of breeding stock in the whole breed—he may come again to a zone of increasing returns because of the high prices he will receive. The competition to get into that zone is usually very strong, however.

Improvements in heredity are permanent[4] and each generation

[3] Certain extreme environments, such as exposure to X-rays, or to barely sublethal temperatures, or to radium, do increase the mutation rate.

[4] Except for gains in the epistatic effects. Those tend to disappear as the genes recombine. One must keep on selecting to hold them.

stands on the shoulders of the preceding one, whereas improvements in the environment produce almost their full effect on the animals for which they are first made. Each new generation must again receive the improved environment or the gain will be lost. Hence, in the long run it may be profitable to spend considerable effort to make small improvements in heredity, since the expense of making such improvements in one generation may yield dividends for many generations. The expense of making improvements in heredity (so far as those are additive) is a capital investment; the expense of making improvements in environment is an operating expense. Naturally the breeder will wish to do both so far as they are profitable.

Besides the economic value of its direct effects on the animals, environment needs the practical breeder's attention in two ways; First, the animals should be kept in an environment which will permit them to show readily which of them come nearest to having all the genes which have effects the breeder wishes. Second, the breeder should observe carefully the environment which applies to each animal so that he can allow for this when making his selections. It is usually impossible to disentangle the effects of environment completely from the effects of heredity in individual cases, but some effort spent in trying to do that will often do much to make selections more accurate and progress more rapid.

Breeding animals should be kept under environment like that for which their offspring are being bred. If animals are being bred for resistance to unfavorable conditions, they should be kept under unfavorable conditions so that the breeder will have a chance to learn which ones have the genes that will make them most nearly what he wants.[5] If cows are being bred for specialized and intensive dairying, they should be well fed and milked three times a day, provided those are the conditions under which commercial cows are to be kept in the specialized dairying of the future. To feed them poorly or milk them only twice a day would prevent some of the genes, useful under the more specialized

[5] This reaches its extreme form in breeding for disease resistance, a practice to which is now devoted a large portion of the efforts in plant breeding, but which is still in the laboratory stage in animal breeding except where (as, for example, in the tropics or in the breeds of sheep native to marshy regions) some effective natural selection has been practiced automatically. For fewest mistakes when breeding for disease resistance, the exposure or inoculation dose should be severe enough that somewhere between 30 and 70 per cent of the animals would contract the disease. That may not be practical for diseases which have a high mortality. Those which cause only a low mortality may not be important enough to warrant much effort in breeding for disease resistance. Animal breeders now look first of all for prevention, vaccines, or medicaments as a more economical way out. But if those fail and we are driven to breed for disease resistance, the above considerations show how far we will have to depart from the (at present) more orthodox practices of sanitation and efforts to prevent exposure.

conditions, from showing their presence. On the other hand, to force the cows by extravagant feeding and by such extreme practices as milking four times a day would magnify the differences between their production records and to that extent would be a help in selecting animals adapted to these conditions; but for most practical purposes this gain would probably be more than offset by the fact that some of the cows would respond more than others to those forced conditions, without there being a corresponding increase in what they would produce under the usual conditions for which they are being bred. The breeder of dual-purpose cattle will make fewer mistakes in his selections if his breeding cows are tested under twice-a-day milking and other management like that under which he expects their descendants to be used.

The question of testing under forced conditions or under ordinary conditions can perhaps be clarified by an analogy. If an athletic coach were to examine all the men in a college to find the best runners for his track team and were to test them in a race of only one length, such as a two-mile race, it is true that most of those who did well in this long race would also do better than average in the short races; yet there would certainly be some who had not the endurance for the long race but could do well in the 100-yard dash. There would be others who could win in the long race but have not the bursts of speed necessary to make them good performers in the short races. In brief, each man has a certain amount of general ability to run, and that will manifest itself in races of all lengths; but each individual also has certain special abilities or disabilities which will help him or hinder him in certain lengths of races but not in others. In order to find the very best runners for each kind of race the coach will need to test them in that kind of race. Yet if the correlation between their abilities at one length of race and at others is high, perhaps he might conveniently eliminate half or more of the whole group from further consideration, by trying them on one kind of race, without much risk of losing a runner who would be really outstanding at some of the other lengths of races. Likewise with farm animals there is doubtless much correlation between an animal's ability to do well under many different environments, but that correlation is far from perfect. The genes which enable it to do well or poorly in a certain environment will manifest themselves most clearly only when the animal is kept under that environment. In the writings about race horses there is much mention of "stayers" and "sprinters," indicating that many horses are good in one of these respects but not in the other. Also they sometimes speak of "mudders" which can do well on a wet track but are outclassed by many others when the track is dry.

If a breeder can foresee that general conditions of management are going to change in a certain way in the future, it is of course to his advantage to change his conditions of testing now, so that, when the general change comes, his animals will already have been selected for a generation or two toward adaptability to those conditions. In doing so, of course, he runs the risk that his prediction of the coming change may be wrong. If it is, his stock may be changed farther away from the real goal of the future than if he had made no attempt to foresee a change.

Also, a high record made under forced testing has considerable advertising value. Not all of the potential customers will discount this high record as much as they should for the environmental circumstances under which it was made. While this may even be a hindrance to breed improvement, yet in some cases it has a commercial value to the individual breeder which he cannot afford to ignore.

Mistaking the effects of the environment for the effects of genes dulls the keenness of selections, makes the breeder sometimes save animals he would cull if he knew what genes they had and sometimes cull animals which have better genes than most of those he saves. This source of error is very important for such things as growth rate, fertility, health, vigor in general, size of fleshy parts, ability to fatten, etc., which are economically important and physiologically complex and often much modified by environment. The most important practical consequence is a regression of the offspring toward the mean of the race. That is, the offspring of parents which are extreme in either direction will not usually average as extreme as their parents. The simplest quantitative expression for this is that for each unit which the selected parents average above the mean of their race, their offspring will most probably average about $\frac{\sigma_H^2}{\sigma_H^2 + \sigma_E^2}$ as far above. This would be literally true if the genes all combined their effects additively. In actual fact there will be some dominance and some nonadditive gene interactions which will produce more regression toward the mean of the race than this formula shows.[6] Mistaking the effects of environment for the effects of genes, next to matters of health and fertility in some species, is usually the biggest obstacle to the breeder's rapid progress toward his chosen goal.

The remedy for confusing the effects of enviroment with those of heredity is either to control the environment physically by eliminating variations in it, testing all animals under standard conditions and thus reducing σ_E^2 toward zero, or to control the enviroment statistically by using correction factors to allow as best one can for unusual individual

[6] The formula would be more nearly correct if the numerator were only the additive genetic portion of the variance; but that is a slight understatement of the case, since a portion of the epistatic variance also belongs in the numerator.

circumstances when judging what each animal would have been under standard environmental conditions or what the difference between two individuals would have been if they had been under the same environments.

Physical standardization of the enviroment can never be perfect. Partial control may be too expensive to carry far. Statistical control may actually introduce errors through use of the wrong correction factors. Yet so far as one makes allowances or corrections which are more often right than they are wrong, he will eliminate more of σ_E^2 than of σ_H^2. Therefore a larger fraction of the variance in his corrected or adjusted estimates will be hereditary than was true of the variance in the original observations. Such allowances for differences in environmental conditions may vary from the vaguest kind of a mental allowance to intricate correction factors which sometimes approach the limit where the increases they make in the accuracy of selections do not pay for the labor of making the corrections. The man who sees his animals every day has an important advantage over the man who does not work with them himself and sees them only at rare intervals or knows them only through the report of his herdsman, since the former knows the environmental differences better and can make fairer allowance for them. The man who works with his animals daily is, however, more likely to make too much allowance for his favorites without being aware that he is doing so.

That the offspring of extreme parents generally are nearer to the average of the breed than their parents were (especially as concerns characteristics of low heritability), does not automatically make the breed become more uniform as time passes. The offspring from each parent will vary among themselves and environment will shove some of them far up and others far down. The extreme ones thus produced will replenish the supply of extreme individuals in the next generation.

In medical writings a distinction used sometimes to be made between "hereditary" and "familial," the former referring to cases where the offspring was obviously like one parent, and the latter to traits (such as recessives or those with low "penetrance") which "run in families" but in which the individual might not resemble either parent. Progress in genetic knowledge has now made that distinction obsolete.

"Hereditary" in the broad sense of the word has nothing to do with abundance or scarcity of a characteristic or with dominance, although some methods of estimating heritability do not gather up the variance due to dominance deviations. Black is no more and no less hereditary than the much rarer red in black breeds of cattle; rather the question of heritability concerns the cause of the contrast, black versus red.

SUMMARY

1. All characteristics are both hereditary and environmental in the strictest sense of those words.

2. Characteristics called hereditary for convenience are those for which most of the usual differences between individuals are caused by differences in the genes those individuals have.

3. Characteristics called environmental or nonhereditary are those for which most of the differences between individuals result from differences in the environments to which the individuals were exposed.

4. The effects of environment are not inherited except as extreme environments (like heavy X-ray radiation) produce mutations, and those are not *adaptively* related to the environment which produced them.

5. The breeder should keep his animals under the environments in which they and their descendants are intended to be used so that the desired genes may have a chance to express their effects and be recognized for selection.

6. The breeder will often mistake the effects of environment for those of genes and will thus make mistakes in his selections. Such mistakes are usually the most important cause of the fact that the offspring of selected extreme parents average nearer than their parents to the mean of their race.

REFERENCES

Gowen, John W. 1934. The influence of inheritance and environment on the milk production and butterfat percentage of Jersey cattle. Jour. Agr. Res. 49: 433–65.

Jennings, H. S. 1930. The biological basis of human nature. 384 pp. New York: W. W. Norton & Co. (See especially pp. 127–37, 203–17 and 218–21.)

Lenz, F. 1939. Was bedeutet "Erblich" und "Nicht-erblich" beim Menschen? Proc. Seventh Intern. Cong. of Genet. pp. 187–90. Cambridge.

Lush, Jay L. 1940. Intra-sire correlations or regressions of offspring on dam as a method of estimating heritability of characteristics. Proc. Amer. Soc. An. Prod. 293–301.

———; Hetzer, H. O.; and Culbertson, C. C. 1934. Factors affecting birth weights of swine. Genetics, 19:329–43.

Plum, Mogens. 1935. Causes of differences in butterfat production of cows in Iowa cow testing associations. Jour. of Dairy Sci., 18:811–25.

Wright, Sewall. 1920. The relative importance of heredity and environment in determining the piebald pattern of guinea pigs. Proc. of the Nat. Acad. of Sci., 6:320–32.

CHAPTER 9

The Nature of Differences Between Breeds (or Races or Species or Other Groups)

Averages vary less than the individual items on which they are based. This happens because the items averaged together are not all alike and their variations partly cancel each other. The average of a sample of *n* individuals will differ from the average of the population from which it came by one *n*th of the sum of the individual deviations which did not happen to be canceled. If the sample is selected at random, that difference will be small, especially if the sample is large. The variance of the averages of such random samples is one *n*th as large as the variance of individuals. But, if the sample is selected by some method which tended more often than not to choose individuals with plus (or with minus) deviations, those individual deviations will be prevailingly in the same direction and will not cancel each other completely. Instead, the sample average will tend to be different from the population average by an amount determined by the method of selection. In such a case the variation within the sample will be less than if it had been selected at random.

As a numerical example we may take annual butterfat production, for which the standard deviation among cows in Iowa cow testing associations is not far from 100 pounds. We are not much surprised if we choose two cows at random and find that their records differ by as much as 100 pounds. In fact, nearly half of all such pairs would differ by that much or more. But we would be surprised if the average of two groups of 10 cows, each selected entirely at random, differed by as much as 100 pounds. We would not expect that to happen oftener than once in about 20 or 25 such comparisons. If it did happen, we would wonder whether the two groups really had come from the same population or whether they had been selected in some biased way which tended to bring higher records into one group and lower ones into the other. If the one set had all been selected from one herd and the other set had all come from another herd, we would not be so surprised by that big a difference, since good or poor management or breeding in either herd would tend to shove all the records from that herd up or all down together. In choosing only two herds at random it might easily happen that we would get one herd with poor management and another with good management; while, if each record came from a different herd, it is not at all likely that the 10 in one set would all come from well-managed or well-bred herds and the 10 in the other set would all come from poor herds, unless there was some difference in the method of choosing the records for each set.

Because averages are less variable, it is often possible to be sure that there is a real difference between two groups which have averages not very different and in which the individuals vary so widely that the two groups overlap in much of their range. This is the general situation for most breed differences, especially for differences in economically impor-

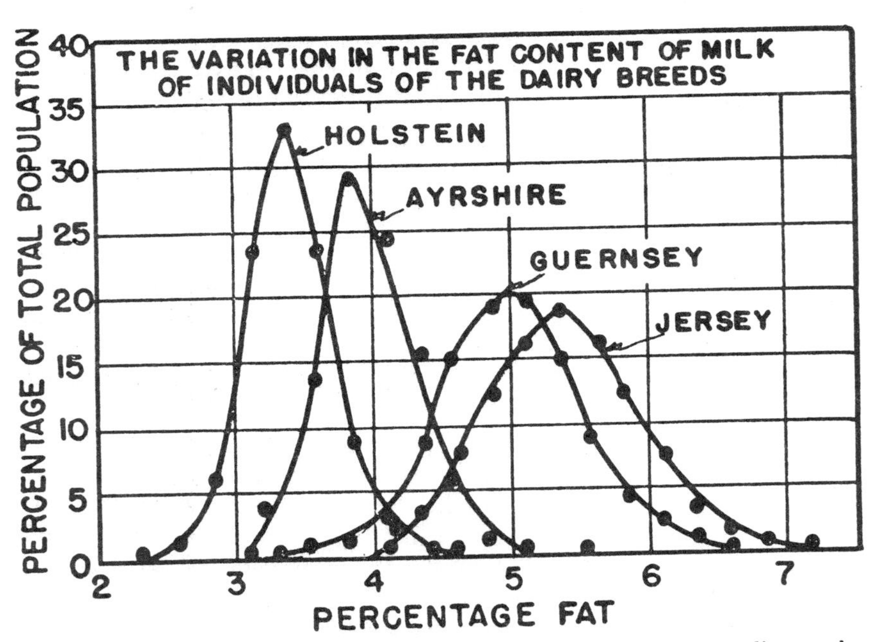

FIG. 10. Distributions of individual cows in four dairy breeds according to the percentage of butterfat in their milk. The breeds overlap, but the differences between their averages are real. (Adapted from Bul. 365 of the Mo. Agr. Exp. Station.)

tant and physiologically complex characteristics like milk production, fertility, size or shape of muscular parts, etc. Figure 10 shows this for percentage of fat in the milk of four breeds of dairy cattle. For some characteristics, notably color or details of bone dimensions or conformation, there may be no overlapping at all between breeds. Figure 11 illustrates graphically the relation between individual differences and group differences. The fairly common saying, "There is more difference within breeds than there is between breeds," is often true in the sense that the difference between the breed averages is small compared with many of the differences between individuals which belong to the same breed. This saying is quite misleading, however, if it is interpreted to mean that the differences between breeds are not real after all. Group differences of the same kind as are illustrated in Figure 11 often prevail between races and may prevail between families or other groups within a breed.

It is to be expected that breeds which have been kept separate from each other in their ancestry for many generations will usually have drifted apart in many of their characteristics on account of the sampling variations of Mendelian inheritance in small populations, or will have been drawn apart by selection which has not been equally successful in

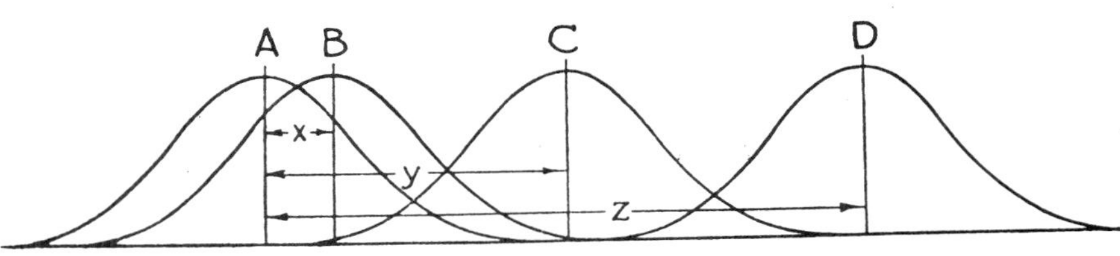

FIG. 11. Overlapping frequency curves showing the nature of differences between population averages and differences between individuals. x is the difference between the averages of populations A and B, which overlap in most of their ranges. y is the difference between the averages of populations A and C, which overlap only a little. z is the difference between the averages of populations A and D, which do not overlap at all. If two individuals are chosen at random from one population, the difference between those two will be larger than x in far more than half of such pairs and in a few pairs will be larger than y.

both breeds or has not been directed toward exactly the same ideals. Breed differences may often be so small that they are economically unimportant, especially if the breeds have been selected toward almost the same ideals; yet it would be a remarkable coincidence if the breed averages were exactly the same for any characteristic. Even where two breeds have been selected toward the same ideal and their phenotypic averages are almost the same, the sets of genes by which that phenotype is produced are likely to have become qualitatively different if the two breeds have been kept entirely from any crossing with each other for tens of generations.

The genetic basis of differences between breeds may be of two kinds. In the first place, one breed may be homozygous for one gene and another breed may be homozygous for an allel of that gene. If that were true for all genes, we could write the Mendelian formulas of two breeds as follows:

Breed No. 1	*AABBccddEE* . . . . *NN*
Breed No. 2	*aabbCCddEE* . . . . . *nn*

This would indicate that the two breeds are homozygous for different genes in the series for *A*, for *B*, for *C* and for *N*, but are alike in the series for *d* and for *E*. This conception of breed differences appears to have been rather widely held by those who discussed the possible practical applications of Mendelism in the early years of genetics, but it is expressed less frequently now, as the evidence accumulates that few genes are homozygous in all members of the breed. The other kind of genetic situation which may be the basis for distinct breed differences is that a pair of genes is not entirely homozygous in either breed but the proportion of one gene to its allel may be widely different in the two breeds. The Mendelian formulas for the two breeds could be written something like this:

Breed No. 1:

$$[q_A A + (1 - q_A)\,a]^2[q_B B + (1 - q_B)\,b]^2 \;.\;.\;.\;.\; [q_N N + (1 - q_N)\,n]^2$$

Breed No. 2:

$$[q'_A A + (1 - q'_A)\,a]^2[q'_B B + (1 - q'_B)\,b]^2 .\;.\;.\; [q'_N N + (1 - q'_N)\,n]^2$$

The blood groups in the Guernsey and Holstein-Friesian breeds of cattle are a good example of this very situation (*Jour. Ani. Sci.* 3:315–321. 1944). In an extreme case of this sort, such as is illustrated in Figure 12, two breeds might be alike in the sense that every kind of gene which exists in the one also exists in the other, and yet be distinctly different outwardly and in average genotype. Genetic differences of both kinds probably exist between most breeds. Really, the first kind of difference is only an extreme limit of the second where $q = 1.0$ in one breed and is zero in the other.

A complete description of a breed involves not only a statement of the genes for which it is homozygous and different from other breeds but also a statement of the frequencies in it of genes for which it is not homozygous and which other breeds may also possess in larger or smaller frequencies. Thus, a complete description of a black breed of cattle may contain, besides the statement that the "type" (the most frequent kind) is black, the statement that in this breed the frequency of gene *B* is .93 and of *b* is .07. Perhaps in another black breed the frequency of *B* may be .97 or perhaps only .80, while in a breed like the Shorthorn, where solid black is unknown, the frequency of *B* is .00. Perhaps other alleles in this series may yet be found. For example, the genetic explanation of the black pigment present in fawn or brindle breeds or of such dark pigment as occasionally appears around the eyes or muzzles of Shorthorns is still uncertain. If such other alleles are found, a complete description of some other breed may include the statement that in it the frequency of B is .40, of B_1 is .45, and of b is .15. Naturally it will be difficult to get such complete information except for a few genes which individually have conspicuous effects. For a long time to come it is likely that the practical description of a breed will consist mainly of its averages, or "type" in traits which do not lend themselves readily to

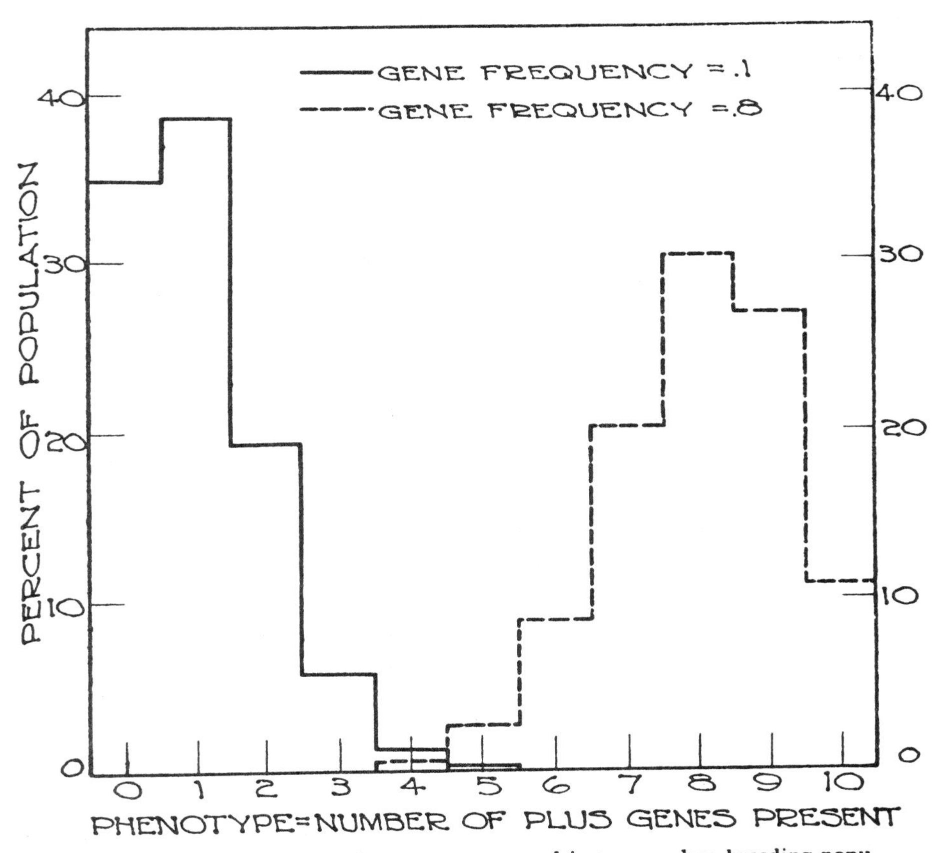

FIG. 12. The distribution of genotypes expected in two random breeding populations each of which is heterozygous for five pairs of genes with equal effects. The genes lack dominance and combine their effects additively. In one population the frequency of the plus gene in each of the five pairs is .1, while in the other it is .8. This will illustrate how two breeds could both have exactly the same ***kinds*** of genes and yet overlap so little that there would be practically no mistakes in classifying individuals.

numerical averaging, with a few sketchy semi-quantitative comments about its variability in those features.

Two breeds may overlap in every observable characteristic, and yet it may be possible to identify with certainty the breed to which every individual belongs. If 90 per cent of the individuals in breed No. 1 but only 10 per cent of those in breed No. 2 are spotted, we cannot with much assurance clasisfy the breed of an individual by examining it for spotting alone. There are too many exceptions to the "type." But if 90 per cent of the individuals in breed No. 1 are black in their colored areas, while only 10 per cent of the individuals in breed No. 2 are black, then we would have considerable assurance in classifying a black spotted individual as belonging to breed No. 1, since 81 per cent of the members of breed No. 1 but only 1 per cent of the members of breed No. 2 are of that type. If we are not satisfied with that degree of assurance, or if we are puzzled about such individuals as those which are spotted but not black, and therefore conform to the type of one breed in one characteristic but not in the other, we have only to look at other characteristics in which the most frequent type in the two breeds is distinctly different. Thus, if 90 per cent of breed No. 1 but only 10 per cent of breed No. 2 have dished faces, then a spotted black individual with a dished face would be the commonest type in breed No. 1 (about 73 per cent of all individuals being of that phenotype with about 24 per cent more deviating in only one of the three characteristics), while such an individual would occur only once in a thousand times in breed No. 2. By extending the number of characteristics observed, we can reach any degree of assurance we wish.

This is the process widely used for classification in such sciences as anthropology, where it is often difficult to find any differences which are 100 per cent true of all individuals in the races being compared although average racial differences certainly exist. This process is also used to some extent in taxonomy, although in dealing with differences between species there is more chance of finding one or two criteria which are so nearly true of all individuals in the species that they can be used alone for identification. The taxonomist may rarely need more than a half dozen criteria in order to be sure of his classification, while the anthropologist, because of the more extensive overlapping of his groups, may need a score or more to reach the same degree of assurance. Table 9 shows in some detail the average number of deviations from "type" which may be expected per individual in a population where n independent characteristics are being examined and in each of those characteristics t is the fraction of individuals which deviate from type. The average number of deviations expected in an individual is nt. The figures which follow the $\pm$ signs are standard deviations computed by the formula, $\sigma = \sqrt{nt(1-t)}$, which must be interpreted with reservation on account of the distinct skewness of the distribution where t is far from .5. The standard deviations show that, when the average num-

TABLE 9

AVERAGE NUMBER OF CHARACTERISTICS PER INDIVIDUAL EXPECTED TO DEVIATE FROM THE "TYPE" OF THE GROUP WHERE n CHARACTERISTICS ARE OBSERVED AND t IS THE FRACTION OF INDIVIDUALS WHICH DEVIATE FROM TYPE IN EACH RESPECT

t	n		
	5	10	20
.01	.05 ± .2	.1 ± .3	.2 ± .4
.05	.25 ± .5	.5 ± .7	1.0 ± 1.0
.20	1.0 ± .9	2.0 ± 1.3	4.0 ± 1.8
.40	2.0 ± 1.1	4.0 ± 1.5	8.0 ± 2.2

ber of deviations from type is large, there may be no individuals which are exactly like the "type." For example, when $n = 20$ and $t = .40$, the average individual will deviate from "type" in 8 respects, only about one-eighth of them will deviate from type in as few as 5 respects, only 1 in 20 in as few as 4 respects. Less than 1 in 27,000 will conform exactly to type in all 20 respects. In this sense it may be true that in finite populations there is *no such thing as an average individual,* but that does not impair the usefulness of the average for describing the group. The description of the group will be more complete if something is also stated about the variation to be expected in each characteristic.

The same principles used in classifying individuals may be extended to the classification of groups wherever it appears likely that the groups are random samples or systematic samples selected fairly. One will often see a Jersey cow which is broken-colored and in that respect is more like the Guernsey breed than the Jersey; but one will almost never see a herd of Jerseys which are all broken-colored, although that is the usual description of a herd of Guernseys. A broken-colored Jersey without any black pigment, although rare, sometimes occurs and might be mistaken for a Guernsey if color alone were considered. One would never see a herd of Jerseys which were all broken-colored and free from black pigment.

DIFFERENCES BETWEEN SPECIES

A species is a more or less continuous and interbreeding group as actually found in nature. To practical taxonomists *discontinuity* between two groups is the criterion of whether or not they are really different species. This leaves room in some cases for disagreement as to whether the continuity and freeness of interbreeding are sufficient to justify calling the population one species. Two groups of similar individuals which are not connected by any intermediates are universally recognized as valid, or "good," species. On the other hand, two groups of individuals may be widely different in the averages of many of their traits and yet, if they are connected by a complete series of intermediate individuals, most practical taxonomists would regard them as really one species with a tendency to exist in several varied forms or subspecies. This situation is much the same as might be encountered in deciding in a mountainous region whether one were describing two distinct mountain ranges or one group of mountains. If there were a broad and deep valley separating the mountains, everyone would agree that they were two distinct ranges. If the intervening space were occupied by a group of large hills or low mountains, irregular in outline, some might still wish to call them two mountain ranges, but others would think the whole group should be called by one name. The fact that there is sometimes room for argument about whether a group is one species with two subspecies or is two separate species does not contradict the fact that there are such things as species groupings in nature, any more than the reality of mountains is contradicted by a dispute as to whether some particular elevation is one mountain with two peaks or is two separate mountains.

Besides being a matter of practical convenience, the taxonomic system attempts to describe the general fact that the existing plants and animals are not scattered uniformly all through the infinite field of possible combinations of form and function and appearance and behavior, but are definitely and irregularly clumped around certain types, most of which are separated from the nearest similar types by a considerable void of conceivable intergrades which do not occur. This is like the distribution of matter in astronomical space, which is highly discontinuous and irregular with definite clumps, like planets, stars, etc., which are themselves clustered in irregular groups like solar systems, constellations, nebulae, etc. The interesting cases about which there is dispute as to species classification are those where a few of those conceivable intergrades actually do occur and form a nearly continuous connection between two nearby clumps, or where the clump itself seems to have two or more separate centers of density around which the existing forms are clustered more closely than elsewhere in the clump. Quite conceivably these may be centers of incipient formation of new species, but in particular cases they may be still so close together and connected by so many intergrades that there can be no reasonable argument for elevating them to specific rank.

The early taxonomists often had an exaggerated idea of the supposed uniformity of wild species. Traces of that idea still linger in biological literature. Much of this doubtless comes from a more or less unconscious deference to the opinions of Linnaeus (Carl von Linné, 1707–1778), the Swedish naturalist who devised the present binomial system of naming plant and animal species.[1] Today it is recognized with increasing clarity that the more one studies a wild species, the more differences he finds among the individuals composing it. If he makes several different collections of individuals of the same species, each collection having been trapped in a different locality, he will usually find each collection different enough from the others that if he were to see another *collection* made from one of those same localities he would be able to identify the locality from which it came, although he would not be able surely to identify the locality of each animal presented to him singly. In short, even the "good" species differ from locality to locality, particularly among animals which do not travel far.[2]

The definition of a species has a peculiar relation to the idea of special creation. Linnaeus believed firmly in the special creation of each individual species, as did most other people of his time. The belief in special creation necessitated a belief that there was a real nature-made difference between species. Hence, there would be some fundamental valid distinction between species which, if one could only discover it, would serve as an accurate touchstone for deciding in accordance with natural laws what was and what was not a true species. Linnaeus, as well as all contemporary and succeeding naturalists, recognized that orders, genera, etc., were divisions made by man, as a matter of convenience, for classifying together organisms which had certain general resemblances. They regarded subspecies, varieties, breeds and such divisions of a species as matters of convenience or as the result of man's handiwork and therefore outside the scheme of nature. This insistence that the difference between species is a fundamental one on an entirely different basis from other differences in classification explains the intensity of many arguments about what constituted a real species in particular cases. The acceptance of the idea of organic evolution carries with it

[1] For example, on page 128 of *Biology and Its Makers,* by Locy, we read: "Ray had spoken of the variability of species but Linnaeus in his earlier publications declared that they were constant and invariable"; and on page 129: "While Linnaeus first pronounced upon the fixity of species, it is interesting to note that his extended observations upon nature led him to see that variation among animals and plants is common and extensive, and accordingly in later editions of his *Systema Naturae* we find him receding from the position that species are fixed and constant. Nevertheless, it was owing to his influence more than to that of any other writer of the period, that the dogma of fixity of species was established."

[2] Sumner, Francis B. 1934. "Taxonomic Distinctions Viewed in the Light of Genetics." Amer. Nat., 68:137–149.

the consequence that species differences are no less and no more man-made than differences between genera or orders. The difference between species is taken out of a special category and becomes only a matter of degree in the general system of classification.

The modern genetic idea of species differences is that they are similar in kind to the breed differences explained above but are much more extreme, so that discontinuity is an essential part of the species definition. Also, in most cases, the two sets of genes are so different that they will not work together harmoniously in producing their physiological effects. Hence, crosses between species are usually (but not always) impossible or, if possible, are usually sterile. Since inheritance is in duplicate, it will sometimes happen that the first cross hybrids which have a complete but single set of the genes from each parent can function all right in their own physiology (as is the case with the mule) but cannot reproduce because the genes from the different kinds of parents are so unlike that they will not pair properly enough for the reduction division to take place. Even if reduction does occur, the resulting sample of genes will rarely have all that are necessary for harmonious functioning. It is somewhat as if one were to try to make a workable automobile from two different kinds by fastening together cylinders from one, pistons from the other, a fuel pump from the first, a carburetor from the second, cam shaft from one, distributor from the other, etc. If the two makes of automobiles were very different, it would be the rarest kind of a coincidence if such an automobile would run at all.

Discontinuity of ancestry between breeds is a fairly recent thing and does not go back much farther than the beginnings of the herdbooks in some cases. Separateness of ancestry between species is a vastly older thing, going far back into geologic time in most cases. Whenever two subgroups of a single group cease to interbreed, their genetic averages tend to become different, either through the random drift of gene frequencies in the sampling which takes place each generation during the reduction division, or because the direct effects of selection may be different if the subgroups live in different regions or otherwise occupy different ecological niches. Both processes can be supplemented by mutations which may not be the same in both groups. Once discontinuity of interbreeding is established, it is easy to see how two subgroups of a species not only might but must in geologic time drift apart so far that they could not cross.

The most puzzling problem in the origin of species is how the discontinuity of interbreeding first arose in each case of a dividing species. Natural selection differently directed in different regions might have made portions of a species become unlike; but, unless they were so distinctly separated that there was almost no interbreeding between the two groups, it seems likely that crosses between them would usually prevent that from going far enough to form two separate species. Discontinuity of interbreeding might have been brought about by geographic isolation; but, as nearly as we can interpret the geologic record, changes of sufficient magnitude to bring that about seem to have been too slow to provide enough isolation to account for the facts of evolution. Irregularities in the chromosome mechanism, such as polyploidy or frequent and extensive inversions, might have brought about discontinuity of interbreeding even without geographic isolation. These probably did play a considerable part in the evolution of plants, but it is generally thought that their effect on animals must have been much less important, both because the chromosome numbers among so many of the modern mammals are so similar and because self-fertilization and asexual reproduction are not possible in higher animals. One with a chromosome abnormality probably could not find a mate like itself, and it or its offspring from normal mates would usually be sterile. Assortive mating may have played a part, although most students of the subject now think the importance of that was overestimated by Darwin in his writings on sexual selection. Inbreeding is a powerful force for differentiation of groups, but there is uncertainty about how much of that actually occurs in nature.[3] Perhaps all these processes played a part, varying in importance in different cases. The subject is full of interest to the philosophy of biology but contributes little to the routine practice of the animal breeder since he can establish discontinuity in his own breeding operations in whatever way he wishes.

SUMMARY

1. Because of the lessened variability of averages it is often possible to distinguish breeds (or other groups) whose averages are not very far apart even though the individuals within each breed vary widely and the breeds overlap.

2. Breeds may differ in one's having a gene which is entirely absent from the other, but more commonly they differ in the frequencies of genes which are present in both.

3. By considering a sufficient number of characteristics in which the averages of two breeds are different, it is possible to identify with any desired degree of certainty the members of each, even though both breeds overlap each other's range in all those characteristics.

[3] For a general survey of the possibilities in that, see: Wright, Sewall. 1940. "Breeding Structure of Populations in Relation to Speciation." *Amer. Nat.* 74:232–48. See also the chapter on isolating mechanisms in T. Dobzhansky's *Genetics and the Origin of Species.* 1941. Columbia University Press.

4. Differences between species are thought to have the same genetic basis as the differences between breeds but are so much more extreme that crosses between species are usually sterile, if possible at all.

5. A species is a more or less continuous and interbreeding group as actually found in nature. To the practical taxonomist the discontinuity rather than the magnitude of the differences between species is the most satisfactory criterion of whether they are really two species or a single variable one.

6. The most puzzling problem in the origin of species is how the discontinuity of interbreeding first arose in each case. Several explanations are possible, but it is not certain which of them has been the main explanation in most cases.

CHAPTER 10

The Means Available for Controlling Animal Inheritance

Man can do only a few kinds of things to change the heredity of his animals. First of all he has some power to decide which of them shall have many offspring, which shall have few and which shall have none. That is selection. Selection has always been practiced by animal breeders, and among many of them it is almost the only breeding method used. There can be various degrees of it. It can be based on individuality, on ancestry, on progeny, or on combinations of those in different degrees.

In the second place, since those chosen to be parents will not be exactly alike, either in pedigree or in their own somatic appearance and performance, there are many different ways in which the breeder may decide which of the chosen males are to be mated to which of the chosen females. But in their genetic nature and practical consequences all these systems of mating, if they deviate at all from random mating within the group of chosen parents, may be classified as the mating of like to like or as the mating of unlikes. Likeness or unlikeness may be based either on blood relationship or on individual appearance and performance.

Mating systems, wherein the mates have a closer blood relationship to each other than if mating were at random, are *inbreeding* in the broad sense of that word, although most animal breeders reserve that term for the closest degrees of inbreeding. Mating systems wherein the mates are less closely related to each other by blood than they would be under random mating are *outbreeding*. The consequences and uses of inbreeding and outbreeding were but vaguely known in pre-Mendelian days. Inbreeding and outbreeding are still used only a little by most breeders of purebreds; but now that the reasons for their results are understood and measures of their intensity have been found, considerable increase in their use seems likely in the future. The results of mating like to like or of mating unlikes on the basis of their own individual characteristics, regardless of pedigree, are very different from the results of inbreeding and outbreeding. Since selection and these four general systems of mating are the only tools with which man can change the inheritance of his animals, it is important for the practical breeder to know what kind of change each is apt to produce, what things each will do well and what each will do poorly or not at all, what are the chief dangers or difficulties in each, and what are the most useful means of overcoming those dangers and difficulties.

Any of these four kinds of mating systems can be practiced in combination with or in alternation with any of the others. They are almost always accompanied by some degree of selection. That makes possible an almost infinite number of specific breeding plans. The probable consequences of each of those may be predicted in a general way, but the chance involved in Mendelian segregation and recombination will leave room for surprising results in individual cases. Moreover, the combination of one mating system with another will sometimes give results which are not simply the sum of what each would accomplish if practiced separately—an epistatic effect among the breeding systems themselves, so to speak! There is no immediate prospect that reliable predictions of the outcome of breeding plans can become so detailed and accurate that they will remove from the business of livestock breeding the sporting element of hope and uncertainty which has been one of its great attractions and has led many wealthy men to take it up as a hobby.

Man's knowledge of how the mechanism of inheritance operates is fairly complete, but that knowledge has not yet given him any ability to interfere with some of the processes so as to change their outcome in the direction he wishes. Thus, no way has yet been found to control the segregation of genes so as to produce from heterozygous parents gametes which contain more than a random proportion of the desired genes. Nor is there any prospect that such control over assortment at segregation will ever be achieved. All that man can yet do in this respect is to select from among those animals available for parents the ones which suit him best and then accept whatever gametes they produce. But even after the gametes are produced, he cannot select those which most nearly have the genes[1] he prefers or promote the union of those which are most like each other or least like each other. All he can do is to let the array of gametes from the chosen sire unite at random with whatever

[1] P. C. Mangelsdorf has shown (1931, *Proc. Nat. Acad. Sci.*, 17:698–700) that it is physically possible by the use of certain mechanical sieving methods to separate corn pollen grains which carry the gene for "sugary" from those pollen grains which carry the allelic gene for "starchy," but the method has not found practical application. Several attempts to separate male-producing from female-producing spermatozoa by physical or chemical methods have been tried without success.

ova the chosen dam has produced. There is extensive evidence from plants that selective fertilization exists in nature,[2] but the general importance of this is in some doubt. In the mildest forms of selective fertilization in plants, pollen tubes containing genes like those in the tissue of the plant being fertilized grow down the style toward the ovule a little faster (or a little more slowly) than pollen tubes which carry unlike genes. This gives some kinds of genes an advantage over others in reaching the ovule and fertilizing it, although the handicapped genes are perfectly capable of doing so if they have no competition from the favored genes. Whether there is anything to correspond to this in the higher animals is uncertain, although the processes of animal courtship may possibly indicate that there is considerable assortive mating in nature.[3] The most extreme forms of selective fertilization in plants are cross-sterility or self-sterility, which are phenomena well-known among certain horticultural crops such as some varieties of apples. Definite genes for self-sterility, often long series of multiple alleles, are well known in some cases (*Genetics* 27:333-38. 1942). Occasionally it happens with animals that a female is bred several times to one male without conceiving but conceives promptly when bred to another male, although the first male was fertile in matings with other females. Yet such cases rarely furnish any very plausible indication of selective fertility since one cannot know whether that female would have conceived if she had been bred again that last time to the first sire. There are almost never enough such cases at any one time for a statistical investigation to be decisive.

Neither can the breeder change the laws of Mendelism nor the number of genes nor their linkage relations. He cannot change their mutual physiological interactions, such as dominance, except as he can find and increase the frequency of other genes which modify in the desired way the physiological effects of the first genes. To a very limited extent genes can be changed into other kinds by such violent treatments as exposure to X-rays, radium, etc.; but such treatments usually result in a high degree of sterility in the treated animals, and the mutations produced are so nearly all undesirable[4] that the production of mutations offers no help to the practical breeder.

This leaves as the breeder's only practical means of controlling the heredity of his animals his partial freedom to decide how many offspring each animal shall have and his freedom to choose, within the

[2] Jones, D. F. 1928. *Selective Fertilization*. 163 pp. University of Chicago Press.

[3] For a summary of knowledge concerning that in Drosophila, see *Biological Symposia* 6:277–79. 1942.

[4] Gowen and Gay found in their material (*Genetics* 18:1–31) that 92.2 per cent of all mutations were actually lethal and many of the rest obviously caused low viability.

group selected for breeding, which shall be mated with which. These opportunities the breeder possessed and used before Mendel's work was known. But he used them with many mistakes, and he neglected many opportunities to use mating systems which could have forwarded his work. Full use of the genetic knowledge available today should make the mistakes in selection fewer, although it cannot prevent them all, and should enable freer use of inbreeding, outbreeding, and the crossing of types than breeders would have dared before the principles underlying those practices were understood. Perhaps the analogy is not too fanciful if we compare pre-Mendelian animal breeding with the animal breeding now possible in the same way we would compare the

Systems of mating which may be combined with various kinds of selection	Mating like to like	*By pedigree:* Inbreeding, including linebreeding, staying within one family, etc.
		As individuals: Mating big with big, little with little, medium with medium, compact with compact, rangy with rangy, active with active, sluggish with sluggish, etc.
	Random mating	Mates no more and no less alike, either as individuals or in pedigree, than if they had been mated by drawing lots from within the group selected to be parents.
	Mating unlikes together	*By pedigree:* Outbreeding, ranging from species crossing through crossbreeding, to crossing strains within the breed.
		As individuals: Compensating for defects, crossing extremes to produce intermediates, mating large with small, coarse with refined, active with sluggish, etc.

The above classification may make clearer the kinds and definitions of breeding systems.

common practice of making soap on farms less than a century ago with the processes now used in modern soap factories. The fundamental chemistry of soapmaking has not changed; but the product has changed tremendously in variety, usefulness, adaptability and dependability, as a result of accumulated refinements in the purity of the ingredients, closer control of temperatures and concentrations and the use of small amounts of certain ingredients whose effects were formerly but dimly understood.

CHAPTER 11

How Selection Changes the Genetic Composition of a Population

Causing or permitting some kinds of individuals to produce more offspring than other kinds do is selection. It is the number raised and added to the breeding herd rather than the number born which matters, since those which are born but get no chance to reproduce cannot affect the composition of the future population. Under some circumstances selection may quickly cause large and permanent changes in the population. Under other circumstances it may cause marked changes, but the moment selection is relaxed the population returns to its original condition. Under still other circumstances selection may be almost powerless to produce any change unless it is combined with some mating system like inbreeding.

The changes which selection produces in the underlying genetic composition of a population can rarely if ever be seen or measured directly, since the observer will not know what genes are present, nor the frequencies of each, nor the frequencies of their various combinations, nor the amount of change which selection makes in those.

Selection creates no new genes. It merely causes the possessors of some genes or of some combinations of genes to have more offspring than those which lack those genes or combinations. Its primary genetic effect is to change gene frequency and the frequency of gametes carrying certain gene combinations. All its other effects are consequences of that, and their magnitude depends on how much the gene and gamete frequencies were changed. Changes in gene frequency are permanent even if selection ceases, unless counter-selection in the opposite direction begins and is effective. Changes in gamete frequency, other than those which result from changes in gene frequency, are temporary because the genes recombine when segregation takes place in forming the gametes for the next generation. Because of this segregation and recombination, the gains from selecting for epistatic differences are temporary, and selection must be continued merely to hold those, although the gains from selecting for additive differences are permanent and remain even when selection is relaxed and abandoned.

Selection can be creative only in the sense that new types can be produced when selection has moved the average of the population far from the original position, as is shown in Figure 15. For example, suppose there are five desirable genes, each having a frequency of .1. If mating is random, only one individual in ten billion will be homozygous for all five of the desired genes. For practical purposes this is nonexistent. But if selection in favor of the individuals with the larger number of these genes were practiced long enough to raise the frequency of each gene to .7, then about 28 individuals in each thousand would be homozygous for all five genes. This is frequent enough that some of them would be found. In that sense selection may be said to have created something new, somewhat as an architect can create a building of an original design, although all the materials were already in existence before he began.

ONE PAIR OF GENES

If from a population containing the three genotypes, *AA*, *Aa*, and *aa*, only the *AA* individuals are allowed to reproduce, the next generation will be homozygous for *A* which will then have a frequency of 1.0 in that population. Selection will in one generation have done all it could if it were to be continued for many generations. Similarly, if only the *aa* individuals had offspring, the whole population in the next generation would be homozygous for *a*, the frequency of *A* would have fallen to zero and change by selection would come to an end, as far as that pair of genes is concerned. In actual practice, selection can practically never be that accurate and extreme. Instead, some of the undesired genotypes are kept and some of the desired ones are culled, either because there are not enough of the desired ones to permit culling all the others, because the breeder is careless, or because dominance and environmental variation mislead him. The net result is that selection increases the frequency of the favored gene by at least a little each generation and thus leads to some change in the genetic composition of the population.

There may not be enough individuals to permit discarding at once all of the undesired ones. If all Shorthorn breeders were to decide suddenly that they wished their breed to be white, there are probably not enough white Shorthorn bulls alive that every breeder could secure a white bull to head his herd, even if no attention were paid to anything but color. Some would have to use roan bulls for at least a generation until the number of white bulls had increased. As for cows, they would not only have to use all the whites but probably all of the roans and even some of the reds. In the next generation they might have enough whites and roans that they could cull all of the reds, but it would probably be several generations before they could discard all of the roan cows.

The animal is the smallest unit which the breeder can select or reject, but in the animal the genes come in pairs rather than singly. This makes progress by selection slower than if selection could be gene by gene. For example, suppose 25 per cent of the animals in a Shorthorn population were white, 50 per cent were roan, and 25 per cent were red, and the breeders should suddenly decide that they wanted the breed red, but could only afford to cull half of each generation. The best they could do in the first generation would be to keep all of the red animals and half of the roans, discarding the whites and the other half of the roans. There were enough genes for red in the population that the breeders could have discarded all the genes for white and have changed the population completely in one generation if they could have selected gene by gene, but instead they must select or reject the genes a pair at a time. Selection between zygotes thus changes gene frequency more slowly than selection between gametes would. It seems unlikely that the breeder will ever be able to select between gametes, although some natural selection at that stage does take place in plants and perhaps some also in animals.

The effects of environment may duplicate or hide the effects of genes, thus causing the breeder to discard some animals which he would keep and to keep others which he would discard if he knew what their genotypes really were. Dominance may do the same thing by preventing him from knowing which individuals are *AA* and which are *Aa*. Naturally every mistake of this kind means that the undesired genes are transmitted by more individuals and the desired genes by fewer than would have been the case if these mistakes had been avoided. Such mistakes lower the rate at which selection increases the frequency of the desired gene. They do not stop the process but merely cause more time to be required to produce the same amount of change.

RATE OF CHANGE IN GENE FREQUENCY

The amount which gene frequency is changed by one generation of selection could be computed if we knew the rates of reproduction for each of the three genotypes and the frequency of each genotype. If the numbers of offspring produced by the same number of *AA*, *Aa*, and *aa* individuals are in the ratio: $1 : 1 - hs : 1 - s$, we can consider s as a measure of the intensity of selection against the *aa* individuals and hs as measuring the intensity of selection against the heterozygote. For example, if for every 100 offspring which *AA* individuals produced, the same number of *Aa* individuals would on the average produce 95 offspring and an equal number of *aa* individuals would produce only 80 offspring, then s would be .2, hs would be .05, and hence h would be .25. In a random breeding population the change (Δq) produced in the frequency (q) of gene A by one generation of selection is a tiny bit larger than $sq(1 - q)[1 - q + h(2q - 1)]$. The height of the three curved lines in Figure 13 shows how large Δq would be at each value of q and for each of three conditions of dominance. The values of h are, respectively 0, .5, and 1. Selection is most effective, (Δq is largest) when gene frequency is somewhere near the middle of its possible range, and is least when q is near zero or 1.0.

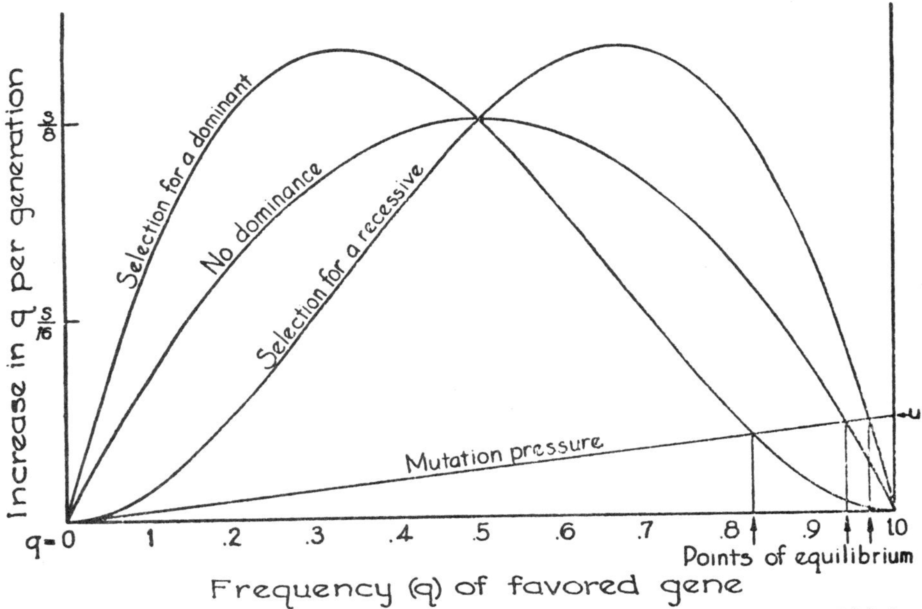

FIG. 13. Rate of change in gene frequency under constant selection, s, which is opposed by a constant mutation rate, u. Drawn with u equal to .03 s, which is rather weak selection. The height of the curved lines indicates the rates at which selection would change the frequency, q, of the desired gene under the three conditions specified for dominance. The height of the straight line, "mutation pressure," indicates the rate at which mutation would change gene frequency in the absence of selection. The difference between the height of the curved lines and the height of the "mutation pressure" line indicates the net rate at which gene frequency is changed by selection and opposing mutation. Arrows indicate gene frequencies at which selection pressure and opposing mutation pressure are equal. (After Wright in *Genetics*, 16:104.)

Dominance of the favored gene is a help to selection when that gene is rare, but a hindrance when the favored gene is more abundant than the undesired recessive. Thus, those who are breeding Hereford cattle for polledness are now fortunate that the gene they want is dominant, because that enables them to distinguish between the heterozygotes and the homozygous recessives. The gene for polled is still rare enough in the Hereford breed that its frequency can be increased by using heterozygous animals for breeding purposes and there are not yet enough homozygous polled animals to permit discarding all of the heterozygotes. If the trend toward polledness continues long enough, the time will come when the favored gene will be more abundant than its recessive allel. Then the breeders will want to discriminate more strongly against the heterozygous polled animals. When that time comes, they will wish that the gene for polledness were recessive. If breeders of Aberdeen-Angus cattle were suddenly to decide that they wanted their breed red, very few of them could find red bulls at first and it would be several generations before there were many red cows. Something like 12 to 16 per cent of the animals in the breed at present are heterozygous for red but, because red is recessive, those cannot be identified directly. These heterozygotes would be useful to any breeder wishing to breed for red, but because black is dominant, he cannot locate them readily. If red were dominant, as polledness is, it would advertise its presence in the heterozygous animals and they could be used extensively until the homozygotes were so abundant that it was no longer necessary nor advantageous to use heterozygotes. This will illustrate why progress is so slow when first selecting for a very rare recessive gene. Under those circumstances one can find and favor only a tiny fraction of the desired recessive genes which are in the population. Dominance aids selection when the desired gene is dominant and rare, but is a hindrance when the desired gene is dominant and has a frequency above .5.

Figure 13 will show why progress in making a simple recessive defect rare, merely by culling all of the defectives, is so slow. If the population is breeding at random, $(1 - q)^2$ of the genes will be undesired genes in defective individuals, where they will be exposed to selection. But $q\ (1 - q)$ of the genes will be undesired genes hidden in the heterozygous individuals. Thus q of the undesired genes will be in heterozygous individuals where dominance shields them against selection. This becomes a larger and larger fraction as the undesired recessive becomes rarer. Consequently, although discarding the defectives is always to be recommended if the defect is serious, and will decrease the proportion of defectives rapidly when they are abundant, it produces only slight changes when the defectives are already rare. Selection is abundantly able to make an undesired gene rare but is almost powerless to elminate it entirely from the population.[1]

[1] Among the nonrandom mating systems, only inbreeding alters this situation much. Under it the heterozygotes shield from selection only $q\ (1 - F)$ of the undesired genes, F being the inbreeding coefficient (Chap. 21) and ranging from 0 to +1.0.

It is sometimes said that selection makes more rapid progress at first and that further progress per generation becomes slower and slower. Inspection of Figure 13 will show that this need not be so. If the desired gene is very rare, the increase in gene frequency made by selection would be small at first, simply because there is not enough genetic variability in the population. As the gene frequency rises toward the values near the middle of its range, progress would become faster and faster until it reached a maximum. After that it would decrease.

In actual practice s cannot be large against many genes unless each is rare, as lethals are. If 10 per cent of the population is *aa,* 10 per cent is *bb,* 10 per cent is *cc,* etc. each of these being undesirable; and if many such traits are to be considered, it will be impossible to find animals which are free from all of these defects. In any population which is constant in numbers, at least enough offspring must reach breeding age to replace their parents. Some animals which have a few defects must be saved for breeding because they are better than average in other respects. Any mistakes caused by dominance or by the confusing effects of environment will also decrease s. Since many genes affect the net desirability of each animal it is reasonable to use a general value something like .01 for s in these formulas, although of course s will vary widely from one gene to another. It will be as high as 1.0 for lethals and doubtless will approach zero for many genes. In actual practice s is likely to change as selection changes the population. Then more intense selection for some genes may become possible, and less intense selection than formerly may be needed for others.

We can compute how many generations will be required to change gene frequency from one value to another if s is known and remains constant and if the frequencies of the different genotypes are known. The figures necessary for such computations[2] are shown in Table 10 for a random mating population. These figures will show how much time selection may need for increasing the frequency of a gene by a large amount. Other than for demonstrating this principle Table 10 is not of much practical use since one will rarely know the frequency of any of the genes in his herd. Still more rarely will he know how intense his selection for each gene actually is. The following example will show how Table 10 may be used. The time required to change q from .01 to .05 when selecting for a complete dominant is $\frac{1.69}{s}$ generations, which equals 169 generations if $s = .01$, but only 1.69 if $s = 1.0$. In either case

[2] They are derived from integrating the equations for Δq under those three special conditions of dominance. The correction factors in the last column and in the bottom row allow for a denominator which is not quite 1.0.

there is a correction, 1.61, to be subtracted. The final answer is a little more than 167 generations for the mild selection, and only .08 generations for the intense selection. For the same problem, except that selection is for a complete recessive, the final answers are 8,083 generations for the mild selection and .04 generations for the intense selection. The

TABLE 10

APPROXIMATE TIME REQUIRED FOR SELECTION TO INCREASE THE FREQUENCY (q) OF A FAVORED GENE BY VARIOUS AMOUNTS

q to be Changed From q_1 to q_2		Time, Expressed in $1/s$ Generations			
q_1	q_2	Selection for a Complete Dominant ($h = 0$)	Selection When There Is No Dominance ($h = .5$)	Selection for a Complete Recessive ($h = 1.0$)	Correction Factor x
.01	.05	1.69	3.30	81.65	1.61
.05	.10	.81	1.49	10.75	.69
.10	.20	.95	1.62	5.81	.69
.20	.30	.72	1.08	2.21	.41
.30	.40	.68	.88	1.28	.29
.40	.50	.74	.81	.91	.22
.50	.60	.91	.81	.74	.18
.60	.70	1.28	.88	.68	.15
.70	.80	2.21	1.08	.72	.13
.80	.90	5.81	1.62	.95	.12
.90	.95	10.75	1.49	.81	.05
.95	.98	30.95	1.89	.98	.03
.98	.99	50.70	1.41	.71	.01
.99	.995	100.70	1.40	.70	.00
From answer *in generations* subtract:		x	$2x$	$x+1/q_1-1/q_2$	

corrections in the right-hand column, to be used as indicated at the bottom of the table, are unimportant when s is small but are considerable when s is large.

EQUILIBRIUM BETWEEN SELECTION AND OPPOSING MUTATION

Mutations are very rare, and nearly all of them are harmful. Therefore they are to be considered as opposing selection, although perhaps at extremely rare intervals a favorable mutation does occur. The more abundant the desired genes are, the more of them are exposed to the risk of mutating to something less desirable. Hence the higher the frequency of the desired gene, the more strongly mutation tends to lower that frequency. That is shown in Figure 13 by the height of the straight line which shows "mutation pressure."

Selection will be a far more powerful force than mutation except when the undesired gene is very rare. This gives rise to an equilibrium value for gene frequency at a point where the undesired gene is already so rare that the few undesired genes eliminated each generation by selection are balanced by an equal number newly produced by mutation. A numerical example may make this clearer. If there is perfect selection against recessive gene *a* ($s = 1.0$) in a random breeding population of a million individuals, and if the mutation rate from *A* to *a* is such that in each generation one out of every million *A* genes mutates to *a*, then the equilibrium point for q_A will be about .999. At that point the proportion of *aa* individuals born will be 1 in 1,000,000, while about 1 in 500 will be *Aa*. Culling the *aa* individuals will in each generation eliminate two *a* genes from every 2,000,000 genes in that allelic series in that population: At the same time there will be 1,998,000 *A* genes exposed to mutation. A mutation rate from *A* to *a* of about 1 in 1,000,000 will provide two new *a* genes each generation to replace the two culled by selection, and the frequency of *A* will not change, even though selection for it continues.

A general formula for the value of *q* at the equilibrium point may be had by letting *s* equal the selection coefficient as before and *u* equal the net rate of opposing mutation. Then an undesired complete recessive is at equilibrium when its frequency is approximately $\sqrt{u/s}$. The corresponding equilibrium point for an undesired dominant is u/s and for an undesired gene where the heterozygote is exactly intermediate in undesirability is $2u/s$. All these frequencies at equilibrium will be low if *s* is large. Complete dominance shields the undesirable recessive from selection to such an extent that at equilibrium its frequency is $\sqrt{s/u}$ times as large as that of an equally undesired dominant. Since *s* to be detectable would usually need to be larger than .01, and *u* seems usually to be something of the order of .000,01 to .000,000,1, it is not at all surprising that undesired recessive genes should be anywhere from thirty to a thousand times as frequent as undesired dominant genes in a population which has long been under selection. This may be the major explanation for the widely observed fact that recessive genes, uncovered by inbreeding or otherwise, are nearly always less desirable than their dominant alleles.

ABUNDANCE OF RECESSIVE UNDESIRABLE GENES

While $\sqrt{u/s}$ is a very low equilibrium frequency for any one gene which is seriously undesirable, yet if the number of loci which can mutate to undesired alleles is several hundreds or a few thousands, as the evidence indicates, then the total number of lethal genes which can exist in the population is large. It might happen that nearly every individual would carry at least one lethal, although only rarely would the male and female which mate together both carry the same lethal gene. If so, the proportion of defective individuals born would be small, as long as the population is large and mating practically at random, but would increase sharply whenever inbreeding is begun. As a numerical example, suppose u is .000,001, s is 1.0 as it would be for a lethal gene, and h is zero as it would be for a completely recessive lethal which had no desirable effects at all. Then the frequency of a lethal gene would be maintained at about .001 in a very large and freely interbreeding population. Only about one in a million among those born would be homozygous and die. Yet about one individual in every 500 would carry the gene and would be capable of transmitting it. If there are 1,000 such loci capable of supporting lethals at equilibrium frequencies of .001, only about 14 per cent ($= .999^{2000}$) of all individuals will be entirely free from all lethals. The rest will carry one or more which they could transmit. Only about one individual in a thousand among those actually born will show any one of these 1,000 individually rare defects.

These figures will illustrate why lethal or otherwise undesirable genes may be so abundant in a population that inbreeding will be almost sure to uncover them in large numbers and yet, if the population is large and breeding at random, any one of those defects may be seen only rarely. They will also explain why genes against which selection has always been directed since time began still recur in appreciable numbers. Lethals are examples of such genes, although there is always the possibility that a gene now lethal in combination with the present genes of the species once may have been neutral or even advantageous at an earlier stage when the species had other genes.

The actual evidence on the abundance of lethals in farm animals is still scanty. It consists mostly of the considerable number of lethals for which the Mendelian basis has been discovered and reported already and of general observations concerning the effects of inbreeding. There is more evidence concerning natural populations of such organisms as Drosophila, although the question of whether the situation is the same among farm animals remains an open one. For example in one study (*Genetics* 26:25) of *Drosphila pseudoobscura* from the Death Valley region in California, over 15 per cent of all third chromosomes in wild flies carried lethals. In another population from Mexico and Guatemala the corresponding figure was 30 per cent. In another California population (*Genetics* 27:373) of 1,292 chromosomes the figure was 14 per cent. Another study (*Biological Symposia* VI:18) of New England, Ohio, and Florida wild populations of *Drosophila melanogaster* showed that 41 to 67 per cent of the second chromosomes carried lethals or semilethals. While such evidence is still meager, yet it seems to indicate that few individuals are absolutely free from all undesired genes. A small amount of the breeder's freedom to select will be used in combating the generally destructive tendencies of mutation.

SELECTION IN FAVOR OF THE HETEROZYGOTE

Selection can never fix the heterozygote. Examples of such heterozygotes are the Blue Andalusian fowl, the Erminette fowl, the cream color of guinea pigs, the yellow color of mice and the roan color of such breeds of cattle as the Shorthorn, the Blue Albion, the "race bleue" (in France), and the blue color inside the ears of Wensleydale sheep, and the Palomino color in horses. Selection of nothing but roans in Shorthorns would lead toward a ratio of 1 red: 2 roans: 1 white. Aside from a few exceptions possibly caused by other modifying genes, it would be possible to produce calves which all were roans by mating whites to reds, but that would be temporary if it were practiced all over the breed, because only roans would then be available for parents of the next generation.

Preference for the heterozygote over both homozygotes means that h in the general formula for Δq is negative. One of the two homozygotes may be preferred over the other, but if the heterozygote produces more offspring than either homozygote, such selection carries the frequency of the gene toward a stable equilibrium value, $\frac{1-h}{1-2h}$. As a numerical example we may take roan color, which in the Shorthorn breed is generally preferred over both white and red, although red is preferred to white. These preferences vary from region to region, white being in more disfavor in the southern and western parts of the United States than it is in the eastern Cornbelt or in Canada, and in more disfavor in the Argentine and South Africa than it is in Britain. If we assume that Wright's count of 8.6 per cent white, 43.8 per cent roan and 47.6 per cent red among 6,000 animals, equally distributed in the herdbooks of the United States, Canada and Great Britain, represents the equilibrium condition of the breed with respect to color, and if we further accept the monofactorial explanation of the inheritance of these colors, then, by setting Δq equal to zero a numerical expression can be had for the degree to which roan is preferred over red and to which red is preferred over white. That is shown graphically in Figure 14. The example illustrates how a population can cease to change while yet selection for a heterozygote continues and q has an intermediate value. The example is particularly instructive in showing how it can happen that the most highly preferred color (the roan) is not necessarily the most abundant when equilibrium is reached. This happens because red is preferred to white even more than roan is preferred to red.

Whether the heterozygote is often preferred over both the homozygotes is not yet clear. The cases for which the definite Mendelian basis is known are few but the phenomenon of heterosis, which is widespread and important, is believed by some to rest almost wholly on this. This is likely to be true if (1) each gene has several effects and if (2) the more favorable effect tends to be dominant over the less favorable effect, regardless of the other effects of the same gene. If this is generally true, ideal breeding systems for producing maximum vigor, health, and growth rates, as in animals destined directly for the market, should be based even more than at present on maintaining purebred but unrelated seedstocks and crossing these for the production of market and work stock.

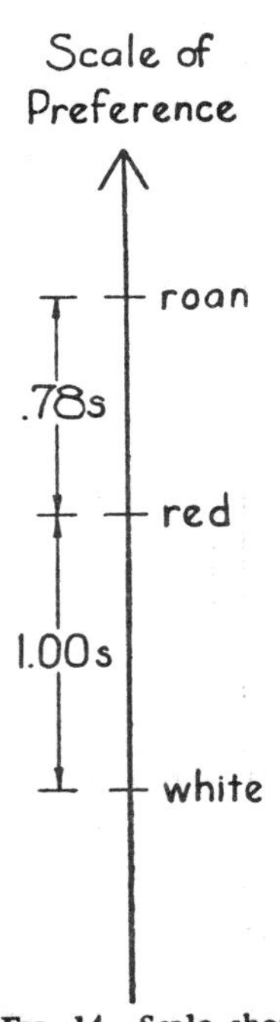

Fig. 14. Scale showing the average degree of preference for roan over red and for red over white among Shorthorn breeders if this preference for the heterozygote is the only thing holding these colors in constant proportions in the breed.

A preference for the heterozygote may be partly responsible for some lethal genes being as abundant as they are. The yellow mouse, the "creeper" fowl, extreme short leggedness in Dexter cattle, and probably the abnormally thick muscles of "doppellender" cattle, are examples of genes which are lethal when homozygous but have highly prized dominant effects when heterozygous. If a lethal gene has even one dominant effect, favorable enough to give the heterozygote a 1 per cent advantage over the more desirable homozygote, then its equilibrium frequency will be more nearly .01 than .001; nearly 2 per cent of all individuals would carry it, and one such lethal individual would appear among each 10,000 born. If the advantage of the heterozygote over the normal is 5 per cent, the lethal gene will have an equilibrium frequency of 1/21, nearly 9 per cent of all individuals would carry it, and about one individual in 484 would be lethal.

INTENSITY AND DIRECTION OF SELECTION MAY VARY WITH GENE FREQUENCY

The conditions under which and the purposes for which the breed is kept may be complex enough that there is need for at least a few of each genotype, just as in human societies there is an economic need and reward (an *ecological niche,* the biologist would say) for a few each of tailors, bakers, lawyers, doctors, etc., but if any one of these professions becomes overcrowded, its members are at a competitive disadvantage. If there is no pedigree barrier to the free interbreeding of types in similarly complex animal populations, the result is the same as if the heterozygote were preferred; namely, gene frequency is rather quickly carried to near the value which will furnish the optimum ratio between each of the two homozygotes and the heterozygotes under those conditions. In terms of the general formula this means that the size and even the sign of s and of h depend in part on q. Little is known definitely about whether this situation is rare or widespread and important, either in animal breeding or in nature. Presumably it will be frequent wherever ecological or economic conditions are highly varied.

SELECTION AND HOMOZYGOSIS

Selection changes homozygosis but little in any one generation. Such change as it does produce may be either to increase or to decrease homozygosis. If mating is random among those selected to be parents, $2q(1 - q)$ of the whole generation out of which the parents are selected will be heterozygous and $2(q + \Delta q)(1 - q - \Delta q)$ of the next generation will be heterozygous. The change in heterozygosis will be $2\Delta q(1 - 2q - \Delta q)$ which will depend on both Δq and q for its size and will be negative when q is larger than .5, provided that selection does have some effect and hence that Δq is positive. As numerical examples, consider first a case where $q = .2$ and selection is so effective that $\Delta q = .03$, and then a case where $q = .7$ and selection again is effective enough that $\Delta q = .03$. In the first case heterozygosis was .32 in the parental generation and rose to .3542 in their offspring. Here the successful selection decreased homozygosis by .0342. In the second case heterozygosis was .42 in the parental generation and fell to .3942 in the offspring. Here the successful selection increased homozygosis by .0258.

Referring back to Figure 3 it will be noted that $2q(1 - q)$ changes only a little with changes in q while q has values near the middle of its range. It does change rather rapidly with changes in q when q is near zero or 1.0, but those are the values of q at which selection cannot change q rapidly. Hence, under any but laboratory conditions, where selection might perhaps be extremely intense and directed entirely at the effects of one gene, selection has only tiny effects in any one generation on the homozygosity of the population. This is in marked contrast to the rather powerful effects it can have on the mean of the population when q is near .5.

Among the nonrandom mating systems, only inbreeding will increase homozygosity much. The amount of help or hindrance which selection will be to inbreeding in that respect will be so small in any one generation that for practical purposes it can be disregarded, although the accumulated effects may become important if selection is continued over many generations and if the inbreeding is mild.

SELECTION AND SEX-LINKAGE

Selection between sex-linked genes is more effective in the heterogametic sex than in the homogametic sex, both because the deceiving effects of dominance are absent and because the heterogametic sex shows the gametic ratio of sex-linked genes directly instead of the square of that ratio as the homogametic sex does. If s against a— individuals is the same as s against aa individuals, Δq pertaining to sex-linked genes in the heterogametic sex is exactly twice as large as it is for an autosomal gene which shows no dominance at all. In the homogametic sex, selection for sex-linked genes proceeds at exactly the same rate as selection for equally desirable autosomal genes.

MANY PAIRS OF GENES—SIMPLEST CONDITIONS

If there are n pairs of genes equal in effects, without dominance or epistasis, and if the characteristic is not affected by the variations in environment within that population, then the range between the most extreme individuals possible is $2n$ and the standard deviation is $\sqrt{2nq(1-q)}$ times the effect of one gene. Thus the range is $\sqrt{\frac{2n}{q(1-q)}}$ times the standard deviation. Individuals varying from the mean of a normal population by as much as two times the standard deviation in either direction are unusual (1 in 22), while those varying as much as three times are quite rare (1 in 370), and those varying more than four times scarcely occur at all except among truly enormous populations. Consequently, if the ideal is an extreme type and if more than three or four pairs of genes are involved, individuals homozygous for the desired genes may be so rare that they do not exist at all in the population from which the initial selections must be made. Perfect animals cannot be selected for parents simply because they have not yet been born! Instead the best of those available will be selected and, as this increases the frequency of the genes with the desired effects, each generation will average nearer to the desired goal than the preceding one did, but several generations of selection may be necessary before any ideal individuals appear. The rate of improvement from one generation to the next rises or falls with the standard deviation and therefore is generally maximum when gene frequencies are near .5. Improvement continues until the goal is eventually reached, or selection comes into equilibrium with opposing mutation rates. The distribution of the population becomes increasingly asymmetrical as the goal is approached. This is the situation usually pictured in generalized discussions of the results of selection. It is represented in Figure 15, where selection begins when the frequencies of all gene pairs are .5, as, for example, in an F_2 generation. Environmental effects and dominance change this situation chiefly in making progress per generation slower, and the changes in variability and symmetry less.

As n increases, Δq for each gene decreases, other things being equal. But, since the effects of more genes are involved, the rate of change in the population mean, which is proportional to $2n\Delta q$, is unaffected. Changes in things like the variability and homozygosity of the population, which depend on the rate of change in q, are made slower as n increases. But for the practical breeder the main difference resulting from whether a fixed amount of variability is caused by many genes each with small effects or by a few genes each with large effects is that in the former case the ultimate limit to which the population mean can be carried is much farther off, and he can expect progress per generation to be steadier and not to increase or decrease so much or so soon as in the latter case.

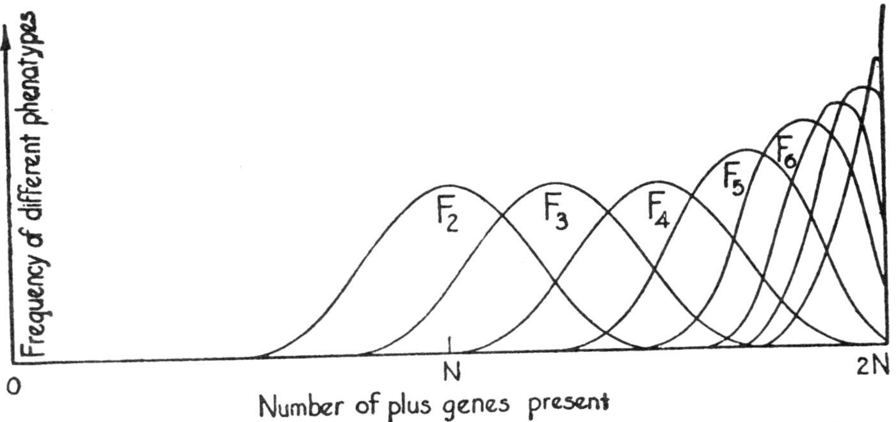

FIG. 15. Distribution of successive generations under intense selection toward an extreme, with few mistakes from dominance or from environmental causes and with no epistasis.

Doubtless some genes have large and some have small effects. Because of that, the variability of the population behaves as if n were smaller than the actual number of genes but larger than the number of those which have major effects. Δq will be larger for the genes with the larger effects than for genes with minor effects. When selection succeeds in making q for the more important genes approach such high values that they no longer contribute much to the genetic variability of the population, the situation becomes more nearly as if many genes each have minor effects.

SELECTION FOR EPISTATIC DIFFERENCES

Some genes have one effect in some combinations and another effect in other combinations. In combination with *Aa* or *AA*, *bb* may have an undesirable effect, and *aa* may have an undesirable effect when in combination with *BB* or *Bb;* but *a* and *b* may supplement each other's effects so well that *aabb* is as desirable as *AABB*. In such a case it would be meaningless to speak of *A or B* as desirable genes. They are desirable when together but undesirable when separate. Selection for net merit is against *A* when *B* is absent but for *A* when *B* is present. A partial analogy may be had by considering whether shoes help or hinder the speed of a man running a foot race. If he takes off one shoe, his speed will almost certainly be lowered; but, if he takes off the other one also, his speed will be raised again, perhaps even to a higher level than when he wore both shoes. Whether taking off a shoe makes him faster or slower depends in part on whether the other shoe is on or off!

There are practically an infinite number of ways in which genes could interact with each other so that the average effect of a particular gene substitution in a certain population might be nearly zero, even though it increased the desirability of some individuals and decreased that of others very much.

The general principle involved can be illustrated by the case of selecting for an intermediate, although in many cases it may not be apparent that the outward ideal is really produced by an anatomical or physiological intermediate. As an example we may consider the relation between speed and length of leg in a race horse. If a horse's legs are too short, his speed is lessened because his legs do not give his muscles leverage enough to manifest their full power in his stride. If his legs are too long, his speed may be lessened because his muscles and other parts are not well enough balanced to keep his legs under perfect control when racing. The maximum of speed may come neither with extremely long nor extremely short legs but with legs perfectly balanced with other parts so that the animal is a harmonious whole with all parts co-operating perfectly with each other. The genes which affect the length of leg might be entirely additive in their effects on length of leg, but they will not be additive in their effects on speed.

This simple situation is illustrated in Figure 16. The change from *a* to *A* may tend to make a horse's legs longer, almost regardless of how long they already are or of what other genes are present, but it will not consistently make the horse speedier or slower. We may speak of *A* as a gene which lengthens legs. We cannot consistently speak of it as a gene which makes a horse speedier. If substituted for *a* in a short-legged horse, it makes him speedier; but, if the same gene substitution is made in a long-legged horse, it makes him slower. Selection for speed is selection *for A* in short-legged horses and selection *against A* in long-legged horses.

The simplest scheme which will describe in Mendelian terms the consequences of selecting for an epistatic effect is to suppose that a characteristic is affected by two pairs of genes lacking dominance, equal in effect, and combining their effects by addition, but that the intermediate phenotype—the one with two plus genes—is considered the most desirable. If in each pair the gene with the plus effect has a frequency of .5, the nine possible genotypes will be grouped into five phenotypes as follows:

Genotypes:	1*aabb*	2*aaBb* 2*Aabb*	1*aaBB* 1*AAbb* 4*AaBb*	2*AaBB* 2*AABb*	1*AABB*
No. of plus genes in each animal	0	1	2	3	4
Ratio of numbers in each phenotype	1	4	6	4	1

Saving for breeding purposes only individuals from the middle phenotype would increase the proportion of that phenotype in the next gen-

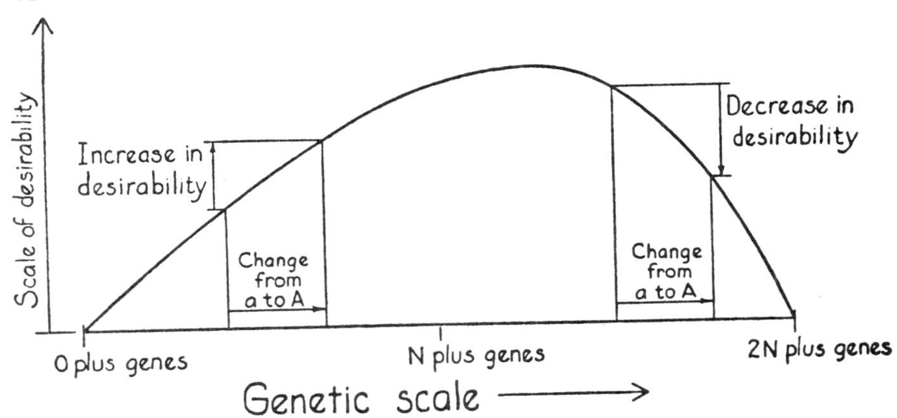

Fig. 16. Illustration of a simple case where the most desirable individuals are intermediates on the genetic scale. Whether the substitution of *A* for *a* increases or decreases the merit of the individual depends on the other genes which are present.

eration from 37½ to 50 per cent and would reduce the variability of the population. The breeder would appear to be making rapid progress. In the second generation of selection the percentage in the most desirable phenotype would increase from 50 to 56 per cent, and the variance which was reduced to 67 per cent of its original value by one generation of selection would be further reduced to 56 per cent of the original. Progress in the second generation is less than in the first. In the third generation of such perfect selection for the intermediate phenotype, the percentage of individuals in that phenotype would rise only from 56 to 57 per cent, and the additional decrease in variance would be only 2 per cent of the original amount. Progress would come nearly to an end with the second or third generation of such selection.

Not only is selection helpless to make much change beyond this point, but continued selection is necessary to hold the gains already made. If selection ceased at the end of the third generation, the percentage of individuals in the desired middle phenotype would fall in the next generation from 57 to 45 per cent and in another generation or two would be practically where it was (37½ per cent) before any selecting began.

In this example selection was neither for nor against *A* or *B;* it was for animals which had *any combination* of two of those genes. The result was a change in the gametic ratio because selection eliminated more of the individuals which would produce extreme gametes (*AB* and *ab*) than of those which would produce gametes (*Ab* or *aB*) containing only one of the plus genes. The unselected original population would have produced gametes in the ratio: 1 *AB* : 1 *Ab* : 1 *aB* : 1 *ab*. Saving for parents only those from the middle phenotype would cause the gametic ratio to change at once to: 1*AB* : 2 *Ab* : 2 *aB* : 1 *ab*. Continuing this selection for a second generation would cause the gametic ratio to become: 5 *AB* : 13 *Ab* : 13 *aB* : 5 *ab*. *A* is still just as abundant as *a* and no more so; the gene frequency was not changed, but only the ratio in which nonallelic genes are combined with each other. This is characteristic of selection for gene combinations rather than for genes—as long as the selection continues it distorts the gametic ratio from what it would be in an unselected population. But recombination of the genes when the selected parents produce their gametes is continually tending to carry the gametic ratio back toward what it would be if the genes combined at random. If two genes are not linked, the ratio of the gametes coming out of the selected parents differs from the ratio when the genes are combined at random only about half as far as did the ratio of the gametes which united to form those selected parents. With each additional generation, after selection ceases, the remaining difference between the actual gametic ratio and what that ratio would be under random combination tends to be halved. If the two genes are linked the rate of approach toward the random distribution is c (instead of one-half) of the remaining difference each generation, c being the percentage of recombination.

Selection for epistatic effects is somewhat like building a sand pile on the seashore exposed to each incoming wave. It is easy to build a little pile between waves, but each wave which rolls over it tends to flatten out the pile. When building is stopped, some traces remain after the first wave and perhaps even a few after the second and third, but soon practically all traces of the pile are leveled away. If building continues between waves, the pile can be built a little higher before the second and third waves than it was built before the first wave but soon a size is approached which can just be maintained, the building between waves being just enough to repair the leveling action of the preceding wave.

It should be emphasized that selection for an intermediate is not necessarily selection for heretozygosis. In the example just given, selection was for the *AAbb* genotype which is entirely homozygous, just as much as it was for the *AaBb* genotype which is entirely heterozygous. Intermediacy and heterozygosis are almost unrelated to each other, provided the characteristic is affected by more than two or three pairs of genes.

The existence of environmental effects and dominance to confuse in the selections, and the usual necessity of saving more than one phenotypic class in order to have parents enough, weaken the intensity of selection for epistatic gene combinations. Probably the Mendelian example just used showed more extreme effects than would often be met in practice, although the multiplicity of kinds of epistatic effects possible throws some doubt on the validity of that conclusion.

Also this example was somewhat artificial in its supposition that the frequencies of *A* and of *B* were both exactly .5 and would remain at that level. If one had been larger and the other smaller, selection would ultimately have made the whole population homozygous for the gene with the larger initial frequency and for the allel of the gene with the smaller initial frequency. The frequencies are in equilibrium when both are .5, but this equilibrium is essentially unstable. When disturbed by sampling variations, it would tend to depart from equilibrium at an increasing rate. Hence, the population would ultimately become either *AAbb* or *aaBB*. In more complicated epistatic situations the equilibrium might well be a stable one toward which each gene would tend to return when disturbed.

The general principle which the example illustrates is that, where a genetic intermediate is the goal, selection will carry a population rather quickly to the point where the number of plus genes will *average* nearly what is desired, but some individuals will have more of them and some will have less. Further selections can do little more than hold down the variation, unless the epistatic equilibrium is an unstable or moving one. If selection ceases, the average number of plus genes will not change but variability will at once increase, and the average merit of the population will decline sharply, most of that decline occurring in the first generation.

For *all* of the differences caused by the *Aa* pair of genes to be epistatic requires that the average effect of changing *A* to *a* shall be zero; i.e., that the cases in which the possessors of *A* have higher reproductive rates than the possessors of *a* shall be exactly balanced by the cases in which the possessors of *A* have the lower rates. Then the net selection pressure for or against *A* (the average *s*) would be zero, and selection would not tend to change the frequency of *A* in either direction. However, if selection changes the frequency of other genes which alter the difference between *AA* and *aa*, the proportion of genotypes in which *A* is at an advantage or disadvantage may change. This would then give *A* some average effect, and selection for or against it would begin. In short, *s* for *A* would partly depend on the frequency of genes in other allelic series, as well as on the physiological differences between *A* and of *a* themselves and on the kinds and frequencies of the different environments in which the population lives.

The general effect of epistatic interactions is to decrease the rate at which selection changes the frequency of a gene, but they may help gene frequency to drift about irregularly to an extent which may sometimes be important, especially in small populations being inbred.

AUTOSOMAL LINKAGE AND SELECTION

Autosomal linkage makes new combinations rarer. It is therefore a factor for stability, making it harder to get desired new combinations but easier to hold existing combinations of desirable genes while trying to add other genes to them. Although linkage is a drag on progress, it does not actively tear down any of the breeder's past accomplishments. It can be compared with friction in a machine which requires effort to overcome but helps keep the machine in position wherever it stops and can be useful in such parts of the machine as brakes and governors.

How linkage works can be seen by supposing an extreme case in which there is no crossing over. Then each chromosome would behave as one large gene with many effects, some favorable and some unfavorable. These effects would be distributed more or less randomly along the chromosome, but it would be a remarkable coincidence if two chromosomes of a pair were exactly equal in selective value. In the whole population the chromosomes of each pair would constitute an indefinitely extended series of multiple alleles. With any general tendency for dominance of the favorable effects, selection would favor the heterozygotes and tend toward an equilibrium at which the population would continue to keep all chromosomes which had any dominant favorable effects at all. But those chromosomes which had only a few favorable effects would be kept at a low frequency. As mutations occurred, the selective values of the chromosomes containing them would alter, and the equilibrium frequencies would shift.

Now if some crossing over takes place, the selective value of each chromosome will alter at a still more rapid rate. Any chromosome which loses more of the desirable genes than it gains by crossing over tends to be reduced to a lower frequency. One which gains more than it loses tends to be increased to a higher frequency.

Crossing over is continually tending to bring each pair of genes into random distribution with every other pair, so that the coupling phases of the double heterozygotes tend to become just as numerous as the repulsion phases. But whenever the genes are not in this equilibrium, the approach toward equilibrium is slower if there is linkage than if the genes were independent. Selection disturbs this randomness of the combinations of the genes with each other, as will be discussed in more detail in the section on selection and variability. The gametes coming from the selected individuals tend to include more of the intermediate combinations and fewer of the extreme combinations than if the same genes were combined entirely at random. If linkage exists this will persist longer and will build up to a wider discrepancy from the random distribution than if the genes are all independent.

If the population is large enough that the inbreeding effect can be ignored, the net result is that linkage will make the offspring of selected parents less variable, and this in turn will prevent the selected and the culled individuals in the next generation from averaging as far apart as they might otherwise. Linkage may constitute some reason for allowing two or three generations of interbreeding following a cross before selecting intensely to combine the desirable characteristics of two different races into one new one. That would give more time for the various genes to cross over so that their coupling and repulsion phases would have more chance to become equally abundant.

In selection for such epistatic effects as when the intermediate is more desirable than either extreme, linkage may play a still more active part in keeping the percentage of desired offspring higher than it would be otherwise.[3]

[3] For details about this see Mather's article in *Jour. of Genetics* 41: 159–93. 1941.

SELECTION AND THE VARIABILITY OF A POPULATION

Mass selection of the parents has little effect on the variation among the offspring, although of course the variation remaining among the selected parents themselves will be distinctly less than was in the population from which they were chosen.

Eliminating 10 per cent of the very poorest from a normal distribution will decrease the standard deviation of the remainder by 16 per cent, eliminating 20 per cent will decrease it 24 per cent, and eliminating 50 per cent will decrease it 40 per cent. Thus even a small amount of culling can make rather striking effects on the uniformity of the group of survivors. Probably this is the main cause of the rather widely held opinion that selection is an effective way of increasing uniformity. In most herds some selection is practiced, and the visitors see only the selected survivors of at least a little culling. When they do see a herd where, through the owner's carelessness or financial difficulties or other reasons, no selection is being practiced they see for the first time that rather rare sight—an entirely unselected population. They are likely to compare such a herd with the other herds they know, which are selected populations, rather than with the unselected offspring of selected parents. The latter is what should be done to find how selection of the parents really affects the uniformity of their offspring.

Sometimes in thinking about selection and variability we are contrasting two herds which are the products of selection in different directions. Since selection can shift the mean of a population considerably, even when it does not change the variability within that population, two herds started from the same foundation stock but selected toward different goals for two or three generations may differ rather sharply in their means. If so, it may appear by contrast that selection has made each herd uniform, whereas really it only increased the differences between herds without making much change in the variation within herds.

Sometimes when we think of selection and uniformity we are comparing the offspring of one selected sire with a whole breed or other large population. The offspring of one sire have some extra uniformity because they are half brothers and hence get half of their inheritance as samples from the very same genotype. By contrast individuals whose parents were not the same but merely were selected because they were much alike phenotypically may get widely different kinds of inheritance, since those parents will generally be less alike in breeding value than they are phenotypically.

Often when we think of selection and variability we are comparing the variation within one herd with variation within the whole breed.

Usually each herd has some environmental conditions which are different from those of other herds but tend to affect alike all the members of the same herd. These effects of common environment may often be enough to make each herd distinctly less variable phenotypically than a population composed of a fair sample from all herds of the breed.

All these things may be mistaken for the effects of selection. One or more of them often are. It is not surprising that many persons, without having seen experiments or herds where the actual contrast is only that the parents were a selected group in the one case and a random sample of their population in the other, should have inferred that selection would increase uniformity distinctly. This opinion is still widespread, notwithstanding the fact that actual experiments have contradicted it and that these have been published. As long ago as 1907 E. D. Davenport wrote[4] "We often speak of 'fixing' the type by selection, meaning by that the reduction of variability. All recent studies, however, go to show that we do not greatly reduce variability by selection, however much we alter the type." and "The principal function of selection, therefore, is to *alter the type, not to reduce variability, . . .*"

Selection of the parents alters the variability of the next generation in two ways from what it would have been if the parents had been unselected. First, it changes gene frequency, and that automatically has some effect on variation. Second, the gametes from selected parents contain a somewhat larger proportion with intermediate combinations of desirable and undesirable genes than would be the case if the very same genes with the same actual frequencies were combined entirely at random.

The first effect is almost always very slight and may either increase or decrease variability. It has already been discussed in connection with selection and homozygosis (page 131). Variance goes up and down in proportion to a term which always has $2q\ (1 - q)$ as a factor, the other factors depending on whether there is dominance and upon the nature of the mating system, if that is not random. Successful selection will increase variability if q is small at the start but will decrease variability after q is much larger than .5. But the change is very small when q has values in its middle range, and Δq will be small when q is near zero or 1.0. Therefore, this effect of selection in changing variability may be plus instead of minus but is exceedingly small in any one generation.

The other effect of selection in producing an excess of intermediate gametes, as compared with what there would be if the same genes were combined with each other entirely at random, is the same process already discussed (pages 136 and 137) in connection with selection for a

[4] Pages 534 and 537 in *Prin. of Breeding*. Ginn & Company.

genetic intermediate except that here it is usually less extreme, since individuals are discarded from only one end of the curve and not from both. As an extreme Mendelian example, consider again the case on page 135 and suppose that only the two phenotypes on the extreme right were saved for breeding. The frequency of *A* and of *B* among the gametes they produce would both be .8. The actual array of gametes from those parents as contrasted with what would occur if the same genes were combined strictly at random would be as follows:

Gametes	*ab*	*Ab*	*aB*	*AB*
Actual frequency	.00	.20	.20	.60
Frequency if random	.04	.16	.16	.64

Although the example is extreme and selection is assumed to be without mistake, the departures from the random distribution are small.

This excess of intermediate gametes tends to correct itself in subsequent segregations and recombinations of the genes. Whatever difference there is between the actual gametic array and what it would be if the genes were combined at random tends to be halved with each succeeding generation, so far as concerns any two unlinked genes, and to lose *c* of its amount in each generation if the genes are linked, *c* being the percentage of recombination.

This effect of selection through narrowing the gametic array is generally slight. It depends for its size on the intensity of selection as well as on the square of the heritability of the characteristic being selected. As a numerical example, if only the best half of a hitherto unselected and random breeding population is saved for breeding, the standard deviation of the offspring will be 17 per cent less than the standard deviation of the population from which their parents were taken if the characteristic being selected is not affected at all by environment, dominance, or epistasis. But if heritability is 50 per cent, the reduction in standard deviation will be only 4 per cent. If heritability is as low as 30 per cent the reduction in standard deviation will be less than 2 per cent. If the culling could be so extraordinarily intense that only the best 10 per cent were saved for parents, the corresponding reductions in standard deviations for those three levels of heritability would be 24 per cent, 5 per cent, and 2 per cent, respectively. For characteristics with heritabilities much under 50 per cent, the reduction in the variability of the offspring caused by selection of the parents is thus only a tiny amount.

The reduction in variability proceeds only a little farther in following generations if selection is continued. As in selection for epistatic effects, most of what can be done is done in the first generation or two of selection. Further selection only does enough to cancel the tendency for the genes to recombine at each segregation to produce a gametic array which would be more nearly random. When selection ceases, this slight reduction in variability caused by selection having made the gametic array nonrandom disappears quickly, most of it going in the first generation produced from unselected parents.

SUMMARY

The primary effect of selection is to change gene frequency. Its outward results are consequences of that. Conditions which modify the *rate* at which selection changes populations, but do not change the ultimate goal, are:

1. The proportion needed for replacements may be so large that not all of the undesired homozygotes can be discarded at first.

2. Selection must be between individuals, which are pairs of genes, rather than gene by gene.

3. Environment may duplicate or hide the effects of genes, and dominance may cause two or more genotypes to be indistinguishable, so that the breeder makes mistakes which cause his selection to be less intense than it would be if he knew the genotypes perfectly.

4. The amount of selection that can be practiced depends in part on the amount of genetic variability which is present, that is, upon the gene frequency, and on the mating system if that was not random.

Conditions which modify the ultimate goal which selection can attain, as well as the rate at which that is approached are:

5. Selection becomes progressively feebler at eliminating the undesired genes as those become rare. Hence, as an undesirable gene becomes nearly extinct from a population, the power of selection to make it still rarer comes into equilibrium with opposing mutation rates, even when the latter are very low. Generally this equilibrium frequency of the undesired gene is so low that it is not of much importance in practical breeding, but a tiny fraction of the breeder's efforts is required for combating mutation.

6. When the heterozygote is more desirable than either homozygote, selection ceases to change gene frequency while yet the gene frequency has an intermediate value which may be rather far from either 1.0 or zero.

7. If economic or ecological conditions provide a useful place in the population for at least a few individuals of each of the homozygous types, and if the population is freely interbreeding, this has the same effect as if the heterozygote were preferred. Progress in changing the population by selection comes to a halt when gene frequency reaches whatever value will most nearly provide that proportion.

8. Epistatic effects tend to lower the rate at which selection changes gene frequency because selection for a gene in some combinations tends to be balanced by selection against the same gene in other combinations. If all the variation caused by a certain gene is epistatic, the net selection pressure for or against that gene is zero. Selection then merely tends to keep the frequency of the gene at this equilibrium point.

In any one generation selection has very little effect on homozygosity or variability of the population.

The number of genes responsible for a given amount of genetic variability does not affect the amount of progress which selection can make in the next generation, but if the gene number is large the rate of progress will not change so much or so soon, and the ultimate limits to which selection can change the population will not be so near as if the genes which cause this same amount of genetic variation are few.

Autosomal linkage lessens the effectiveness of selection slightly by making the array of gametes from selected parents less widely diverse than it would be if the genes were independent.

Selection for sex-linked genes is roughly twice as effective in the heterogametic sex as is selection for autosomal genes. In the homogametic sex, selection is equally effective for sex-linked and for autosomal genes.

REFERENCES

The classic work on selection is Darwin's *Origin of Species,* which, however, was written before the mechanism of inheritance was discovered. It seems to have been wrong chiefly in assuming that inheritance was blending in nature and, therefore, that hereditary variations must be seized by selection almost at once after they occur, else they would be "swamped" in the subsequent matings and lost. Hence, also, it assigned too much importance to mutation. Also, the qualitative distinction between hereditary and nonhereditary variations was not entirely clear. Sexual selection probably was overemphasized. R. A. Fisher's *The Genetical Theory of Natural Selection* shows how Darwin's conclusions are modified or extended by modern genetics. Sewall Wright's "Evolution in Mendelian Populations" (*Genetics,* 16:97–159, March 1931) and also his "The Roles of Mutation, Inbreeding, Crossbreeding, and Selection in Evolution" (*Proc. Sixth International Congress of Genetics,* 1:356–366) treat extensively of the interplay of selection, mutation rates, and inbreeding systems. It is his notation which is mostly followed here. Wright's articles in the *Journal of Genetics,* 30:243–266, are at this writing the most comprehensive study yet published of the genetic consequences of selection for epistatic effects.

CHAPTER 12

How Selection Changes a Population—The Outward Results

Although the genes are the units of inheritance, the animal is the smallest unit which can be selected or rejected at any one time. Selection may be between still larger units such as families, inbred lines, breeds, races, etc., but that is optional. The breeder may study the different characteristics of each animal as separately as he will, and may like some of its characteristics very much and dislike others of its characteristics at the same time, but what he *does* with the animal applies to all its characteristics, the admired ones as well as the disliked ones.

The animal is selected or rejected for breeding according to the breeder's opinion of how much its meritorious characteristics outweigh its weaknesses, and in comparison with the other animals which are available for him to use in case this one were rejected. It is thus convenient, when considering the general consequences of selection as the breeder sees them, to consider selection as being made for net merit as if that were a single characteristic. Of course net merit is a compound characteristic affected by many genes, but so too are most measurable characteristics, such as weight, wither height, egg production, litter size, etc. Net merit is also likely to change in definition as economic conditions change, or when one characteristic in the breed improves so much that variations in it become less important than they once were. Also breeders will not entirely agree on the ideal toward which they are striving and on the actual importance of different variations. The yardsticks for measuring net merit are thus somewhat elastic, changing a bit from time to time and from place to place and according to the varied purposes for which the animals are to be used. These are important practical difficulties in measuring net merit for each animal in an objective way so that all would agree on the merit of each animal. Yet the general idea of net merit is as easily understood as the idea of obtaining an individual's net income by subtracting his losses in some enterprises from his gains in the others, or obtaining an individual's net worth in a financial statement by adding his various assets and subtracting his liabilities from them.

Figure 17 shows two diagrams of the way selection might take place. The kind of selection pictured in *A* corresponds to that actually practiced for important traits in stock breeding where many different traits must be considered. Some animals which are mediocre or even

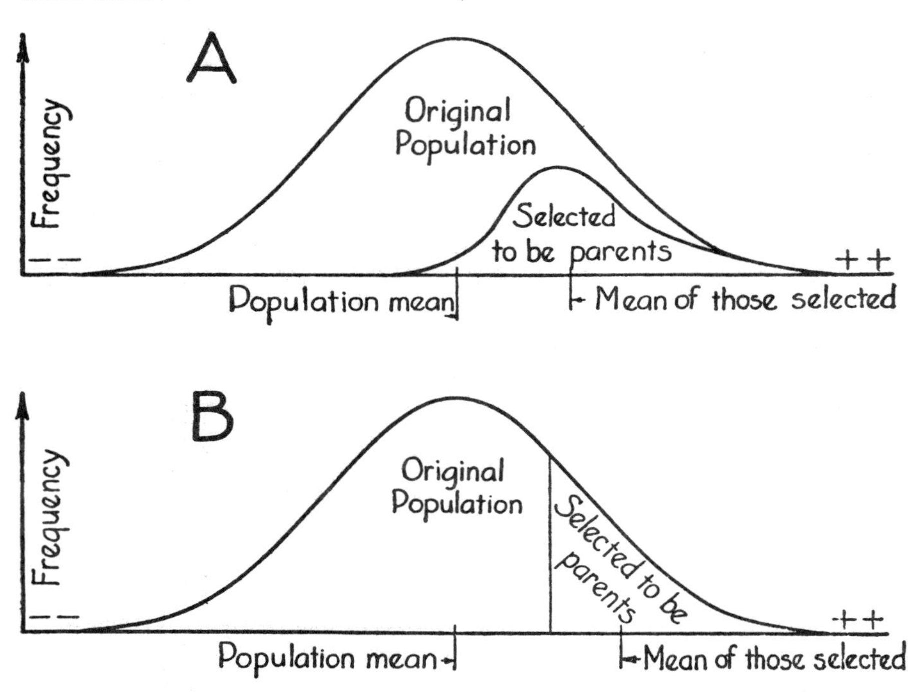

FIG. 17. Two ways in which the merits of those chosen to be parents by rather intense selection might be distributed with respect to the merit of the original population from which they were taken. The better individuals are to the right, the poorer to the left. *A* indicates the usual kind of selection where at least a few mistakes are made and where some attention must be paid to characteristics other than the one for which merit is indicated here. *B* is the most extreme form of selection conceivable. No mistakes are made and selection is entirely for the characteristic for which degrees of merit are indicated along the horizontal scale.

inferior in the characteristic pictured are saved because they are unusually desirable in several other characteristics or because the breeder is careless or confused. That pictured in *B* is the extreme kind of selection which might be practiced in a laboratory experiment on selection for one trait alone, disregarding all others. The selection practiced in livestock breeding can be like that pictured in *B,* if the net merit of the animal as a whole is the characteristic which is measured along the horizontal axis in *B*. The kind of selection pictured in *B* is, of course, more effective if the percentage saved is the same as in *A*.

THE SELECTION DIFFERENTIAL

The most useful measure of the intensity of the selection actually practiced is the difference between the average of those selected to be parents and the average of the whole population in which they were born. It is convenient to speak of this superiority of the selected parents as the "selection differential." As a numerical example, if a herd of gilts in their first litters farrow an average of 8.6 pigs and we select for large litter size intensely enough that those which are kept for further breeding averaged 9.5 pigs in the first litters, then the selection differential for litter size at this particular culling was 9.5 minus 8.6 or .9 pig. The selection differential which can be attained is sharply limited by the fact that enough offspring must be saved to replace the parents in any breed stationary in numbers. More than that must be saved in a breed which is increasing in numbers. Reproductive rates and percentages of deaths and other losses from controllable and uncontrollable causes differ among species of farm animals. Table 11 shows what are believed to be

TABLE 11
ESTIMATED REPLACEMENT RATES WHICH LIMIT BREED IMPROVEMENT

Kind of Animal	Average Interval Between Generations (i.e., Average Age of Parents When Their Offspring Are Born)	Percentage of Progeny Reared Which Are Needed for Replacements in a Population Static in Numbers	
		Females	Males
Horses	9 to 13 years	35 to 45	2 to 4
Beef cattle	4½ to 5 years	40 to 50	3 to 5
Dairy cattle	4 to 4½ years	50 to 65	4 to 6
Sheep	4 to 4½ years	45 to 55	2 to 4
Swine	About 2½ years	10 to 15	1 to 2
Chickens	About 1½ years	10 to 15	½ to 2

reasonable figures for the usual percentage of offspring which must be saved for replacement purposes. The vital statistics of farm animals are not well enough known to make these estimates as accurate as they should be. Of course, there is some variation in the replacement rates from year to year and from farm to farm as conditions of health and management vary.

Table 12 shows for selection such as that pictured in *B* of Figure 17 how much the parents can average above the whole population from which they were selected.[1] It thus gives *the maximum* selection differen-

[1] Table 12 is for a *normally distributed* population. Most animal breeding populations are nearly enough normal that these figures are sufficiently accurate to be useful. Where a few are culled from the long "tail" of a distinctly skew curve, the gain will be more than is indicated in the upper lines of Table 12, but gains from the heavier culling shown in the bottom lines will be less.

tial which could be attained in a whole breed if the percentage of the population which must be saved for replacement were the only limiting factor. The selection differential in Table 12 is expressed in terms of standard deviations, so as to be applicable to all kinds of characteristics.

An example will show how Tables 11 and 12 are used. The standard deviation of the weights of fleeces shorn in the same year from a group of Rambouillet sheep described in Technical Bulletin 85 of the United

TABLE 12

SELECTION DIFFERENTIAL (IN TERMS OF STANDARD DEVIATIONS) ATTAINABLE BY VARIOUS INTENSITIES OF SELECTION

Percentage of Population Saved	Selection Differential
.90	.20
.80	.35
.70	.50
.60	.64
.50	.80
.40	.97
.30	1.16
.20	1.40
.10	1.75
.05	2.06
.04	2.15
.03	2.27
.02	2.42
.01	2.67

States Department of Agriculture is about 1.86 lbs. (*See* Table 2 in that bulletin.) If a hitherto unculled group of such ewes and rams were to be culled solely on fleece weight at a single shearing, the 50 per cent of the ewes which it would be necessary to save would excel the flock average by .80 times the standard deviation, or 1.49 pounds of wool. The 3 per cent of the rams with the heaviest fleeces would excel the average for all rams by 2.27 times the standard deviation, or 4.22 pounds of wool. Since inheritance is practically equal from sire and dam, this would make a selection differential of 2.86 pounds of wool at that one shearing, so far as the next generation is concerned.

INCREASE TO BE EXPECTED IN THE POPULATION MEAN

The next generation would be expected to be about as variable as the preceding one, but to average outwardly whatever their parents averaged genetically. In case the genes all combined their effects additively and the existing environmental variations did not affect the characteristic at all, the genetic average of the parents would be the same as their phenotypic average, the offspring would average whatever their selected parents did, and the increase in the population mean per generation would be equal to the selection differential (*see* Figure 18). Actually the permanent improvement in the population average each generation will be only a fraction of the selection differential. That fraction has for its numerator the additively genetic variance (Chapter 7) and for its denominator the actual variance; i.e., the fraction is $\frac{\sigma_G^2}{\sigma_G^2+\sigma_D^2+\sigma_I^2+\sigma_E^2}$ which for brevity we may call the "heritability" of the differences which existed in the parental generation before selection began. In addition, epistatic differences will cause temporary gains which in amount are less than half of $\frac{\sigma_I^2}{\sigma_G^2+\sigma_D^2+\sigma_I^2+\sigma_E^2}$ of the selection differential. These gains from selecting for epistatic differences tend to disappear in future generations as the genes recombine. They can be maintained only by continued selection.

In the example of selection concentrated on fleece weight, the increase in average fleece weight per generation would be 2.86 pounds per generation in the impossibly extreme case in which all differences in the parental generation were additively genetic, 1.43 pounds per generation in case heritability of differences is 50 per cent, and only .95 pounds per generation in the more probable case that heritability of differences in shearing weights is about one-third. The *annual* increase in the flock average would be about one-fourth or fifth of these increases *per generation,* since the interval between generations, the average age of the parents when the lambs are born, is about four or five years.

This example is somewhat artificial for farm animals in that it assumes that all selection would be practiced at one stage in each generation. Actually, only a few of the rams born would be kept as rams even until the first shearing. Many of the ewes would be culled after the second shearing, others would be culled after the third shearing, others after the fourth, etc., some culling taking place all through the lifetime of that band of sheep and much of the culling being based on things other than fleece weight. All these things operate to lessen the intensity of the culling which could be done at any one time and to render difficult the measurement of the intensity of the selection actually practiced. The example illustrates the general principles that the intensity of selection possible is sharply limited by the necessity for replacements and that the intensity of selection actually practiced is to be measured in terms of the difference between the average of those saved for parents and the average of the generation in which they were born. The method of computing the selection differential in this example is fairly well suited in actual practice to the case of an animal, or some of the annual plants, where the generations do not overlap and where nearly all the selection is practiced at one stage in the life cycle.

CONSEQUENCES OF INCOMPLETE HERITABILITY

Some individuals are mistakenly saved or rejected for parents because the effects of environmental variations make them appear phenotypically better or worse than their genetic values. These environmental effects are not transmitted to their offspring. Selection of the phenotypically superior tends automatically to keep among those saved more than a fair share of those which appeared phenotypically better and less than a fair share of those which appeared phenotypically worse than they were genetically. The environmental effects are left behind when they reproduce. Their genes segregate from these combinations to recombine in the unselected offspring to give a nearly fair picture of the genetic worth of the parents.

Similarly where the favorable genes tend to be dominant the heterozygotes will have been made to appear phenotypically better and the homozygotes relatively worse than corresponds to their average breeding value. The dominance deviations of a parent are not transmitted as such to its offspring, since they are caused by the interaction of a pair of allelic genes and only one gene out of each allelic pair can be transmitted in any one gamete.

Epistatic effects, being dependent on combinations of nonallelic genes, are transmitted to a portion of the offspring, that portion being progressively smaller the more complex the combination. If *A* and *B* are not linked but together have an effect which neither of them has separately, that effect would be transmitted in about one-fourth of the gametes from an *AaBb* individual, whereas the additive effects of *A* would be transmitted in about half of the gametes. An epistatic effect requiring the joint presence of three nonlinked genes, *A, B,* and *C,* would be transmitted in only about one-eighth of the gametes from an *AaBbCc* individual, etc. Even when such epistatic effects are transmitted, the gene combinations responsible for them tend to segregate in later generations, this process tending to continue until the genes are combined at random. Hence the partial but transitory gains from selecting for epistatic differences.

Figure 18 shows what would be expected to happen in an experiment on selecting for a perfectly hereditary character, with the selection intensity such that in the high line only the upper half in each generation were saved for parents, while in the low line only the lower half in each generation were saved for parents. Obviously, even with such an impossibly extreme case as perfect heritability (no effects at all by environment, dominance, or epistasis), it will require several generations of selection before all overlapping between the two lines ceases. Among the offspring of selected parents there will always be some poorer than the poorest of the parents although, if heritability were perfect, the average of the offspring would equal the average of their selected par-

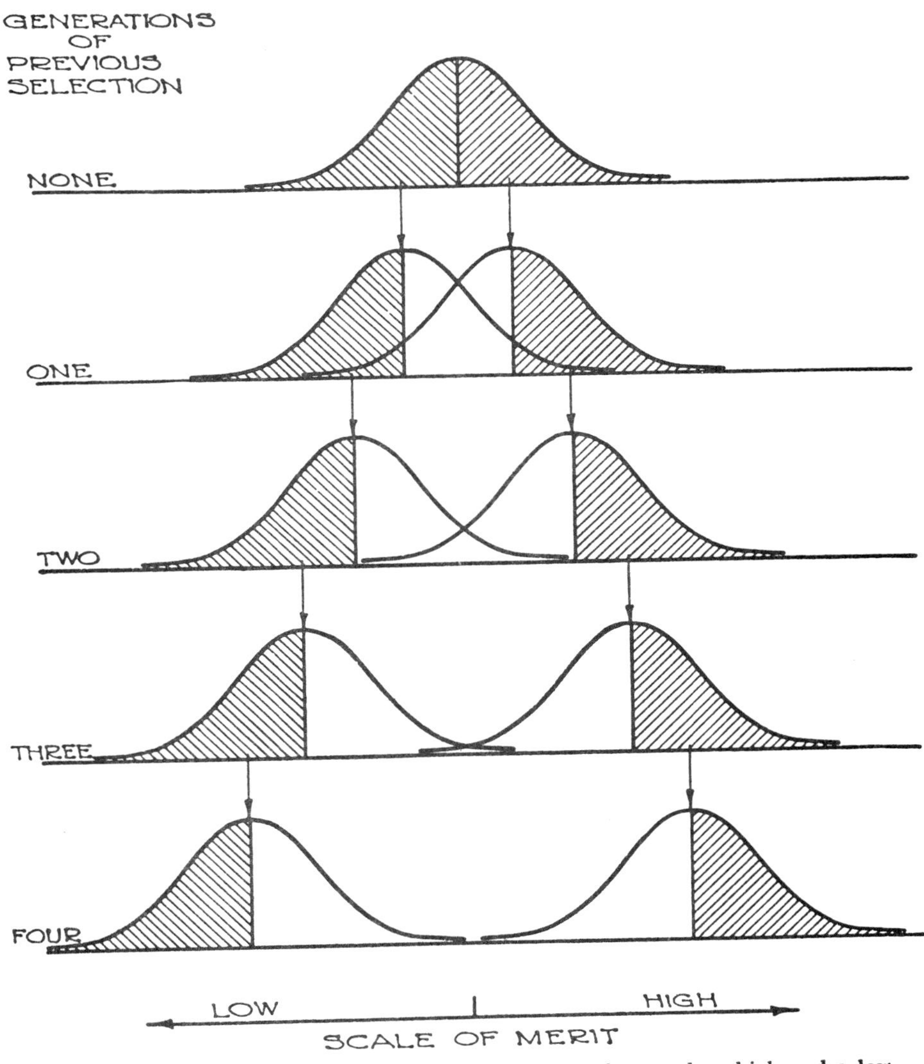

FIG. 18. The results expected when selecting simultaneously a high and a low line for a perfectly hereditary characteristic. In the high line the high half and in the low line the low half in each generation are saved as parents.

ents. The selected parents will not be much more or less homozygous than the average of the population from which they came. They will produce some gametes worse than are typical of them as well as some which are better. When two inferior gametes unite, the result is an offspring poorer than the poorest of those individuals which were saved to be parents.

Figure 19 is the more realistic diagram of what would happen under the same conditions except that the characteristic is only 20 per cent hereditary. The offspring average only 20 per cent as far as their parents above the mean of the generation from which those parents were selected. In other words the offspring "regress" 80 per cent of the way from the average of their selected parents back toward the average of the generation from which their parents were taken. It will require many generations to pull the high line and the low line so far apart that they will not overlap at all, although steady progress is made in every generation if the high line selections are always from high line parents and the low line selections are always from low line parents. If no such a pedigree barrier were put between the high and low line, it might not be possible to separate the two lines completely. With a perfectly heritable characteristic, the animal's phenotype would show its net merit accurately, and paying any attention at all to parents would introduce mistakes into the selections. Somewhere between the extremes of perfect heritability and very low heritability there must be a point of unstable equilibrium where it would be doubtful whether selection in opposite directions with no attention to parentage could ever split the population into two entirely separate lines.

CHANGES IN THE RATE OF IMPROVEMENT BY SELECTION

Whether the rate of progress increases in succeeding generations, or remains steady, or decreases, depends on changes in the amount of genetic variance, provided the percentage of culling remains about the same. The genetic variance changes about as $q\,(1 - q)$ does. It therefore increases as the frequency of the favorable genes goes from low values toward .5 and declines as it goes from .5 toward 1.0. The decline in the rate of progress even then will be very slow, especially if heritability is not high. It is not often, except when the amount of epistatic variance is large, that the rate of progress by selection will decline sharply after only a generation or two. The rate of progress may even increase sharply after a few generations if the epistatic relations are of a threshold kind such that at the start most of the population was below some threshold above which it must rise before the genetic differences could express themselves freely. Something of this kind seems to have happened in Payne's selection experiments for bristle number in Drosophila (Indiana University Studies V, No. 36. 1918) and perhaps in Goodale's selection for body weight in mice (*Journal of Heredity* 29:101–12, 1938.)

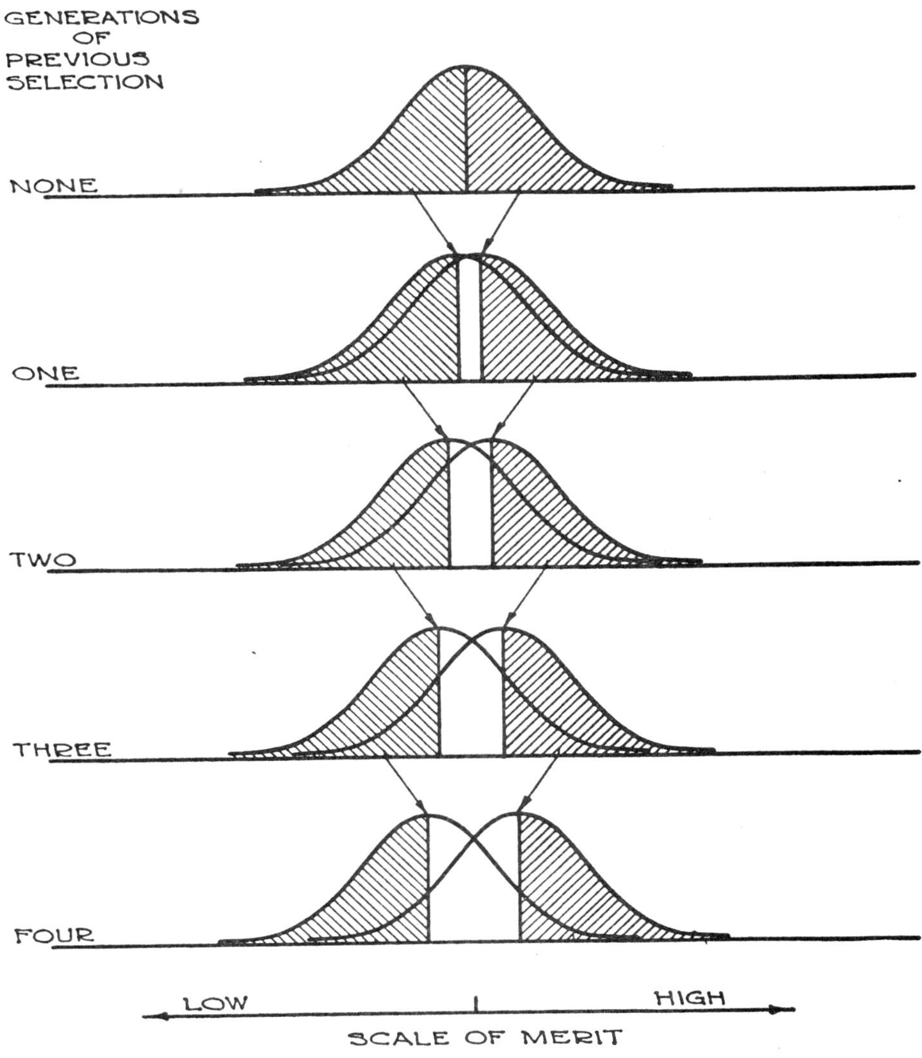

FIG. 19. The results expected when selecting simultaneously a low line and a high line for a characteristic only 20 per cent hereditary. Selection is entirelv on the individual's own characteristics except that its parents must have belonged to the same line it did; i.e., there is a pedigree barrier against exchanging animals from one line to the other, no matter what their individual characteristics are. The intensity of selection is such that in each line half are saved to be parents.

SELECTION FOR EPISTATIC EFFECTS

The kinds of gene interaction possible are so numerous that they defy cataloguing, but the outward results of selecting for them seem to be typified by what happens when selection favors an intermediate degree of a characteristic affected by many pairs of genes. It may not be apparent to the breeder that selection really is being directed toward a genetic intermediate. In that case he will see that selection improved his stock distinctly in the first generation, increasing the percentage of desirable animals and sharply decreasing the percentage of extremely undesirable ones. But he will see that selections after the first generation or two have little additional effect and that he has to keep selecting to hold what he has gained. It will appear to him that there is an inherent tendency for improved stock to deteriorate or "run out" and that much of his efforts are spent merely in combating that tendency to degeneration.

So many of the experiences of stock breeders fit this description that it is reasonable, even on this ground alone, to infer that in many characteristics the ideals of stock breeders are directed toward what are really genetic intermediates, whether they appear so outwardly or not. In considering breeders' opinions on the necessity of continued selection, even to hold the gains already made, it is necessary, of course, to allow for some psychological bias. The breeder is usually making his comparisons between the *selected* parents and their *unselected* offspring; he thereby includes the regression due to dominance and to environmental effects in the evidence for his belief that there is an inherent tendency to degenerate. Often he has built unjustified hopes of the degree of success which would result from his selections and is unduly impressed, therefore, by the extent of his disappointments. Practical breeders do not often get to see the results of deliberate selection in the undesired direction. Experiments with selection in both directions sometimes show that, after the first generation, progress in one direction is about as slow as in the other (Cornell Station Bulletin 533).

From what is known about the physiology of form and function, it is reasonable to suppose that many of the genes which affect comparatively simple anatomical traits like the length of bone, stature, or even weight, may have effects which are simply additive or nearly so. It is just as reasonable to suppose that physiologically complicated characteristics like milk production, health, vigor, fertility, speed, economy of gain, etc., are dependent for their maximum expression on a harmonious balancing of the magnitudes and functions of many different organs. If that is so, then it must often happen that in selecting for maximum production in these economically important characteristics the breeder really is selecting for balanced or intermediate sizes of lungs, of heart, of digestive tract, etc. As a purely mechanical illustration, consider how with an automobile the maximum mileage per gallon of gasoline is not obtained at the very slowest speeds and certainly not at the very highest speeds. Moreover the value of the driver's time or the urgency of the errand may make the most desirable speed something other than that which is most economical of gasoline. In most stock judging there is much emphasis on symmetry and "balance" in the animal as a whole. Perhaps this is more justified than would be the case if each gene were consistently desirable or consistently undesirable, as is inferred in many discussions of applied genetics.

A good example of a characteristic which is optimum at an intermediate value is the thickness of the back fat in hog carcasses to be sold in the bacon trade. In Sweden since 1938 the optimum thickness of fat over the middle of the back has been considered to be 29 to 31 mm.[2] When the thickness is already near this optimum, an increase or decrease of one millimeter changes the carcass value only a little. But when the fat is already extremely thin, or much too thick, then one more or one less millimeter in thickness makes a large change in the value of the carcass. For example, an increase of one millimeter would change the carcass score as follows when the initial thickness is as shown:

Initial thickness	Change in score
13 mm.	3.5 points increase
19 mm.	2.2 " "
27 mm.	.5 " "
33 mm.	.6 " decrease
36 mm.	1.3 " "
44 mm.	3.1 " "

Thus a gene which will increase backfat thickness one millimeter would be highly desirable in a population where the carcasses range between 14 and 22 millimeters in thickness. Most of its effects in that population would be additive and selection for the relatives of those which have the best carcasses would increase the frequency of that gene. In a population which averages 30 mm. in thickness such a gene would lower desirability of its possessor in about as many cases as it would increase desirability. Its average effect would be zero, all of the individual effects it actually makes would be epistatic, and selection would

[2] For details see: Activities of the research stations for testing swine breeding stock during 1937 (Translated title), Bul. 487 from "Centralanstalten för försöksväsendet på Jordbruksområdet" Stockholm, 1938.

not tend consistently either to increase or decrease its frequency. In a population in which the thickness already ranges from 36 to 44 mm., such a gene would be undesirable, nearly all of its effects would be additive, and selection would tend to lower its frequency.

The emphasis laid on symmetry, balance and proportion in most animal husbandry judging, the physiological and mechanical relations between the functioning of an animal and the dimensions of its parts, the fact that so many chemical reactions in metabolism are of a threshold nature, and similar considerations, indicate that situations in which the intermediate is favored over either extreme are rather common, although there may be some strictly linear relations, too. Probably there are many situations in which the regression of desirability on genotype is curvilinear but the curvature is slight enough within the limits of that population that a straight line comes fairly close to describing the facts and selection would change gene frequency a long way before reaching an optimum or some threshold beyond which further changes in gene frequency would have no effect on average outward desirability.

The idea of a desirable intermediate may be extended, and indeed must be extended, to cover cases where two or more intermediates widely separated on the genetic scale may each be more desirable than the genotypes immediately adjacent to them and yet need not be exactly equal in their own desirability. Desirability for the purposes of the animal breeder (or "fitness," if the problem is being considered from the evolutionary point of view) is such a complex thing that there must be many cases where a certain magnitude of a characteristic fits its possessor better for a certain purpose than magnitudes just a little larger or a little smaller would, and yet a magnitude very distinctly larger or smaller would fit it better for some other purpose or ecological niche. A crude illustration of that is milk production in cattle. There are regions, especially in the corn belt, where both specialized beef production and specialized dairy production can be profitable systems of farming. The most desirable milk production for a cow used in the specialized beef farming is just enough to feed her calf well. More than that would lead to some trouble with spoiled udders, etc. But the peak of desirability in specialized dairy production is far different. There may, of course, be other farms where the physical resources and the aptitudes of the owner make an intermediate milk production most desirable. In that case there might exist, even in the same region, several different peaks of desirability in milk production. Another example is size in horses. In most of the United States there is not much demand for a horse which is too big to be a children's pony but too small to be a good saddle horse for a grown person. If it were small enough (like the Shetlands) or large enough (like the American Saddle Horse), it might have a high cash value; but, if it is one of the "in-between" kinds, there may be few who will want it. Again, if it were between the ideal for saddle horses used for pleasure and the ideal for cavalry horses, or if it

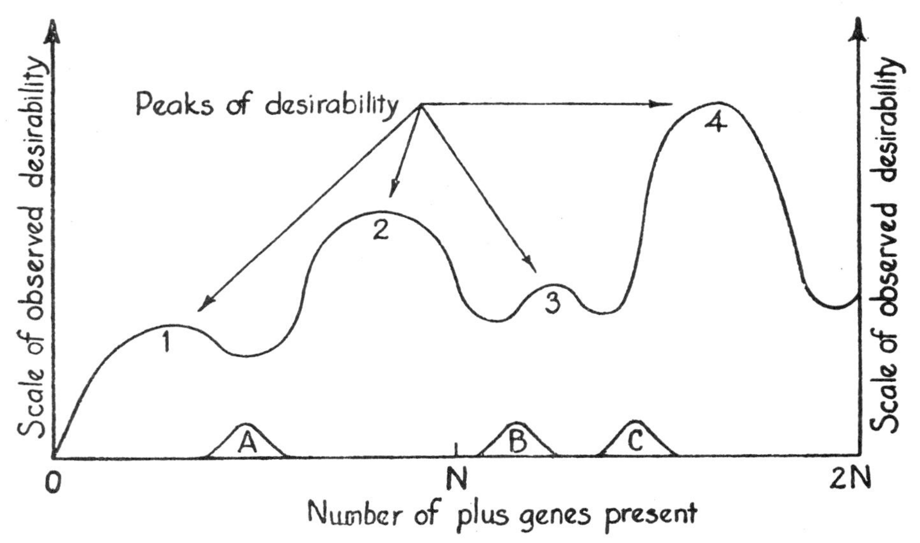

FIG. 20. Illustrating the case of several different genetic intermediates (1, 2, 3, and 4), each of which is more desirable than the genotypes which are most nearly like it. *A*, *B*, and *C* are populations whose averages are at different places along the genetic scale.

were too big for a cavalry horse, it would be at a disadvantage compared with those which were just the right size. Figure 20 shows such a situation diagrammatically, so far as it can be pictured for variation along one dimension.

If the animal's position on the horizontal scale can be seen or measured, the situation offers no new complications over the general case already described for selection directed toward an intermediate. But if only its position on the vertical scale of outward desirability can be observed[3] and the horizontal scale is long enough for the peaks of desirability to be distinctly separated from each other by deep valleys,

[3] Thus, in the example about length of leg and speed in race horses, one might have abundant records on the actual racing speed of many horses but no information at all about the lengths of their legs. Then one would know in detail whether they were fast or slow (their outward desirability) but would know nothing about whether their legs were long or short (their position on the horizontal scale). Two horses might be equally slow, one because its legs were too short and the other because its legs were too long, but a man knowing only the record of speed would not know whether they were alike or far apart on the genetic scale.

then a new kind of complexity occurs. That is shown by *A, B,* and *C,* which represent different positions along the genetic scale in which the genotypes of a population might happen to be distributed when selection began. If a population were in position *A,* immediately in the valley of undesirability between peaks 1 and 2, chance would play a considerable part in determining whether it would be pulled by selection into peak 1 or into peak 2. The probabilities would favor the latter if the population were a freely interbreeding one, since the slope toward peak 2 is the steeper. If some of the population started each way, crosses between the two groups would probably permit the more intense selection toward the higher peak to pull the whole population that way. If the population were in position *B* at the start, partly under the shallow peak 3, the population would probably move on the genetic scale to a position under the center of peak 3. If selection were very intense, it might remain in peak 3 indefinitely because it would have to go against the direction of selection to move in either direction. But if selection were weak or if peak 3 were narrow compared with the variability of the population along the genetic scale, a few individuals might get scattered far enough to the left or right by Mendelian segregation that they would be past the adjacent valleys of undesirability and on the slopes toward peaks 2 or 4. If those individuals were isolated enough to breed largely among themselves, they and their descendants might fairly soon rise to the heights of peaks 2 or 4 and by occasional crosses back with the rest of the population might pull the whole population over to their peak. If, however, the occasional individuals which are different from their population in enough genes to be past the valleys interbred freely with the whole population, their offspring would probably be pulled back into the general population because the mates would usually be near the population average.

Whether the population would remain in peak 3 would therefore depend on the balance between selection, the degree to which the population tends to separate into rarely interbreeding groups, and the height and width of peak 3 and the depths of the valleys surrounding it. The more intense the selection, the more the population would tend to be held in that peak. The only force tending actively to get it out of that peak to where it might perhaps find the road to a higher peak is chance at segregation causing gene frequency to vary in a random direction. That is a very weak force in large freely interbreeding populations but may become powerful in a population highly subdivided into small groups which rarely interbreed. This latter condition leads to some mild inbreeding which, under some circumstances, may be necessary to get a population out of a peak where selection has carried it. If the population were at position *C* when selection began, selection would be almost certain to carry it to peak 4. There would be no need of inbreeding to help in that.

This may explain some of the surprising effects sometimes observed when crossing distinct strains or races. Crossing race *A* and *B* would give a population with gene frequencies putting it nearly in the center of peak 2 and would be considered a "lucky nick." On the other hand, crossing a race already in peak 2 with one already in peak 4 would give a race with gene frequencies which would put it near the much lower peak 3. The unfavorable effect might not show in the first generation of the cross, since each race contributes a full set of its genes, and there is often enough dominance of the favorable genes to furnish a margin of safety. But if the first crosses were interbred, the decline in merit from F_1 *to* F_2, when the combinations of genes which worked well together were scattered, might be extreme. How often this actually happens is open to question but it points toward a general principle, valid when epistatic effects are important, that in attempting to perpetuate the good qualities of an individual it should be mated to members of the same race or local strain rather than to equally good individuals from unrelated races where the gene frequencies and combinations may be widely different. This is the principle of "linebreeding" (Chapter 23). Crosses with unrelated or distantly related races may sometimes be advantageous, too, in originating new lines with desirable combinations but should always be considered experimental and venturesome, rather than a dependable general practice.

Desirability for the purposes of the breeder, or fitness in terms of evolution, depends on many different characteristics, each of which may have intermediate peaks of desirability. The interdependence of these on each other's magnitudes increases tremendously the possibilities for such peaks of desirability where all adjacent genotypes which might be reached easily by Mendelian segregation in a freely interbreeding population are less desirable. Figure 20 is a rough sketch of that, showing variability in only one characteristic. Figure 21 shows the interplay of variation in two dimensions on desirability, which is pictured as the height of an irregular surface. By observing the increased complexity of Figure 21 over that of Figure 20, one can imagine how complex the situation may be in reality where adaptation or desirability depends on variation in *n* dimensions and *n* is a large number!

While it appears impossible ever to know enough about the genes and their physiology and their interplay with environmental circumstances to know a population's position exactly in all *n* dimensions of its adaptability, or to explore the surrounding terrain clearly enough to know whether there is a higher peak of desirability nearby which might be reached, yet the general consequences of selection in such a situation are fairly clear. Selection when first applied will quickly (in terms of generations) carry the population up the steepest slope of increasing desirability to the nearest peak in that direction. Selection cannot carry

FIG. 21. Contour diagram showing how the level of desirability may depend on genetic variability in two different characteristics. Plus signs indicate peaks, and minus signs indicate hollows. Selection can carry a population up a slope of continuously increasing desirability but cannot carry it down a slope and across an intervening valley to reach a still higher peak (After Wright in *Proc. Sixth Internat. Cong. Genetics*).

the population across a deep intervening valley of lessened desirability to reach a peak of higher desirability on the other side. Only chance wandering of gene frequency against selection can do that, and this chance wandering is a very feeble force except when there is considerable inbreeding. The terrain is apt to be extremely rugged and irregular in places where two or more genes which individually have very minor effects may in certain combinations have extremely important effects on desirability. This gives rise to surprising "nicking" effects which can hardly be seized by selection alone if they are dependent on more than three or four genes which separately have undesirable effects.

As a result of the power of selection to carry populations into these peaks and its inability to get the population out again, most characteristics which have been under selection for many generations may be expected already to be in some of those peaks. It is only when selection in a new direction is just beginning that the position of the population is as apt to be on a slope ready for rapid progress as it is to be in a peak. As ideals change with economic or other conditions, peaks will sometimes change to valleys or the reverse, thus releasing the population from its peak and permitting rapid progress in some new direction for a time. To some extent this surface of desirability is always changing, like that of the sea, so that the peaks do not remain peaks forever. Yet the rate of such change may perhaps be so slow that, except for changes in ideals which are caused by changes in economic conditions, it should be likened more fairly to geological changes in the heights of mountains and plains.

SELECTION FOR MANY CHARACTERISTICS AT ONCE

The practical animal breeder must consider many different characteristics in his selections. Some of these are independent of each other or nearly so; others are positively correlated so that selection for one of them brings with it a little improvement in the other, although, even if x and y are correlated rather closely, selection for x indirectly by selecting for y is less effective than selecting for x directly if that is possible. Others are negatively correlated with each other. This makes it a little harder to select for both of them at once than it would be if they were independent. Some characteristics are much more important than others. This needs to be taken into account in balancing excellencies in one respect against deficiencies in other respects when deciding whether to keep or cull the animal. The fact that several things must be considered lowers the intensity of selection possible for each of them, but there is no escape from that so long as all those things have something to do with the net desirability of the animal to the breeder or to his customers.

Culling may be done in at least three general ways. The first or tandem method is to select for one characteristic at a time until that is improved, then for a second characteristic, later for a third, etc., until finally each has been improved to the desired level. The second method is to cull simultaneously but independently for each of the characteristics. This amounts to establishing for each characteristic culling levels, below which all individuals are culled, no matter how good they are in other characteristics. The third method is to establish some kind of a total score or selection index to measure net merit. This would be done by adding the animal's score for its merit in x to its scores for merit in y, in z, etc. Then those with the poorest total scores would be culled. Figure 22 illustrates the total score method where two characteristics are involved.

The tandem method is by far the least efficient of the three, even when the characteristics are not affected by any of the same genes and it can be assumed that the improvement made in the first one will not be lost later while selecting for improvement in the others. Selecting for one thing at a time will improve that one thing faster than can be done by any other method of selection, but while that is being done the other things must wait. Where other things must be improved also, the improvement made in the first characteristic while it was under selection must be divided by the whole number of generations necessary to improve them all in order to get the average rate of improvement in that one thing. In the simple case in which n characteristics are independent and equally important, the average improvement per generation in each will be only one nth of the improvement which is made in it in the generations when it is the sole object of selection. In this case the selection index method is $\sqrt{n}$ times as efficient as the tandem method.

The selection index method is more effective than the method of independent culling levels because it permits unusually high merit in one characteristic to make up for slight deficiencies in the other. When culling by the total score method under the simple conditions of n independent and equally important characteristics, selection for each characteristic will be $\frac{1}{\sqrt{n}}$ as intense as if all the efforts of selection could have been concentrated on that characteristic alone.

Under the method of independent culling levels, if w is the fraction of the population which must be saved for breeding, then the intensity of selection for each characteristic is the same as if selection were directed at that alone but the fraction which must be saved were $\sqrt[n]{w}$. For example, if length of body and soundness of feet and legs are uncorrelated in swine, and a breeder must save 10 per cent of his gilts for breeding purposes, he can save the 10 per cent with the very longest bodies if he selects for that alone, or the 10 per cent with the soundest feet and legs if he selects for that alone, but only 1 per cent of his gilts will be in the best 10 per cent in both respects. If he pays equal attention to both things, he must save all gilts which are in the 32 per cent which are longest and are also in the 32 per cent with best feet and legs if he is to save 10 per cent of his gilts altogether. If he takes quality also into account, and if quality is not correlated with body length or with excellence of feet and legs, he will have to save from among the best 46 per cent (instead of 32 per cent) in each trait in order to have 10 per cent of the best in all three respects. Four traits would increase this to 56 per cent instead of 46 per cent. If the traits are positively correlated with each other, the intensity of selection for single traits will not fall off

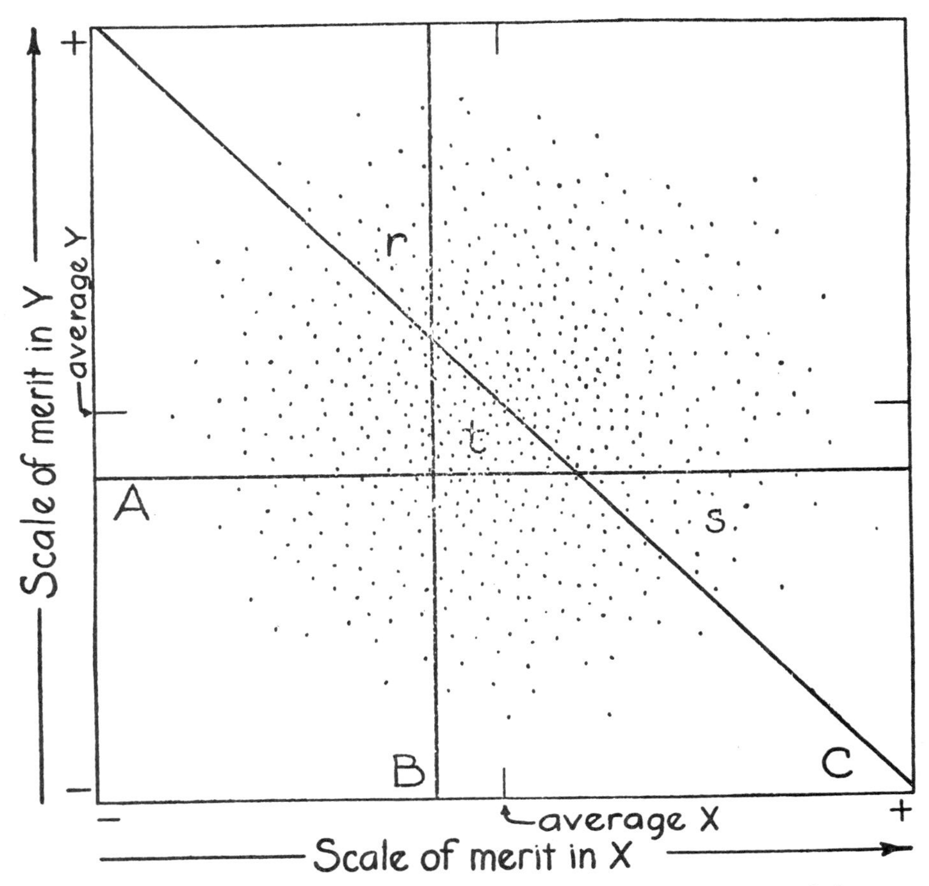

Fig. 22. Superiority of culling on total score as compared with culling independently first for one characteristic and then for another. Each dot represents an individual. Merit in *x* is not correlated with merit in *y*. *A* represents the level of merit in characteristic *y* below which all individuals are culled when *y* is considered independently. Similarly, *B* represents the independent culling level for merit in *x*, all individuals to the left of *B* being culled. *C* represents a level of culling on total score, with *x* and *y* being regarded as equally important, which would result in keeping an equal fraction of the population. Animals in areas *r* and *s* would be kept when culling on total score but would be discarded when culling independently on *x* and *y*, while the reverse would be true of the animals in area *t*. The fate of the animals in the other four areas would not be altered by changing the method of culling. If *x* and *y* were not equally important the example is still valid, but the slope of line *C* would be different (Brier *et al.*, pages 153–60, in *Proc. Amer. Soc. An. Prod.* for 1940).

quite so fast with increasing n as these formulas indicate; but, if the traits are negatively correlated, the intensity of selection for each will fall off a little faster.

Here lies the real damage done by paying attention to "fancy points" in selection; namely, that the more attention given to them, the weaker selection must be for the utility purposes, which are more important after all. The arguments about selection on type or on production have their justification not in any antagonism between the two but in the fact that the more things to which the breeder pays attention, the weaker his selection will be for the average one of them. The breeder *must* pay attention to many things; but it is to his interest to keep the less important in the background by every means possible, lest he allow these to weaken the effectiveness of his selection for truly important things.

The method of independent culling levels cannot be as efficient as the selection index method, when the culling levels and the proportions to be given each characteristic in constructing the index are both such as would give maximum efficiency. The method of independent culling levels has the practical advantage, which may be important under some circumstances, that culling on each characteristic may be done whenever that characteristic develops and without waiting to measure or score the later characteristics. Where a selection index is used, some of this advantage of culling on independent levels can be had by culling the very worst in each characteristic as that develops, but leaving the doubtful cases to be decided later by the selection index.

It would, of course, be a remarkable coincidence if all of the characteristics to which attention is to be paid were of equal importance. In all three methods of selecting for several characteristics, one runs the risk of paying too much attention to something which really is of minor importance. It cannot be emphasized too strongly that increasing the emphasis on one thing automatically reduces the opportunity for culling on something else.

CONSTRUCTING SELECTION INDEXES

The principles of constructing selection indexes designed to make maximum improvement, are those of multiple regression where it is desired to predict as accurately as possible an unknown or "dependent" variable from two or more known (independent) variables. In this case the dependent variable is the animal's net genetic merit or breeding value, while its various characteristics or even the characteristics of its relatives are the independent variables.

Four bits of information are needed for each characteristic. 1. The average amount which a given variation in that characteristic actually raises or lowers the net phenotypic merit of the animal. This we may call the *importance* of the characteristic. 2. The *heritability* of each characteristic is important because it is the average fraction of phenotypic improvement we get in the offspring for each unit of phenotypic merit in the selected parent. 3. *Genetic correlations* between that characteristic and the others may arise if some of the same genes affect two or more of them. These will mean that selection for characteristic x will help or hinder improvement in characteristic y, as compared with what would happen if they were independent. 4. *Phenotypic correlations* between that characteristic and the others will exist if some of the same environmental incidents have affected them. This, in combination with differences in heritability, may even lead to some characteristics being useful mainly as indicators of the kind of environment under which more important characteristics developed.

A quantitative example of the idea of relative importance is the finding by Winters (*The Empire Jour. of Exp. Agr.* 8:259–68, 1940) that one pound of wool is worth 3.4 pounds of lamb. The relative importance of each characteristic may need to be established separately for each kind of animal, each region, each type of farming, and almost for each breeder. Naturally this job is never done permanently but needs to be reviewed whenever the market demands and premiums make any large and presumably permanent change. Discussions about what is "the right type" to breed mainly concern the relative importance of different characteristics, although they have in them something of the other three bits of information also.

Heritability can be approximated by doubling the intrasire regression of offspring on their dams. This requires data on several hundred pairs and is not likely to be within the reach of the individual breeder, but from his general observations he can get a rough idea of the relative heritability of two characteristics by observing whether the offspring generally tend to resemble their parents rather closely or slightly in each of these.

The genetic correlation between two characteristics on the same animal can be measured by observing a large number of pairs of closely related animals and correlating characteristic x in one member of the pair with characteristic y in the other. This requires large numbers, and the estimates have high sampling errors. It is the bit of information most likely to be lacking when an index is to be constructed.

The environmental correlation between two characteristics can be had by correlating the two characteristics on the same animal and subtracting their genetic correlation obtained as above.

Occasionally it is profitable to pay more attention to a highly hereditary characteristic which is of limited economic importance than to a slightly hereditary one, the variations in which affect the value of the individual animal more strongly. This is because heritability is the fraction which one gets of what he reaches for when selecting the parents, and importance measures the value of that for which he is reaching. One will have more net profit by getting 50 per cent of something which is worth 20 cents than by getting 30 per cent of something which is worth 25 cents! It is a question of deploying the available efforts and resources so as to secure maximum returns.

Because of this relation between heritability and importance, the different characteristics in a selection index should be given emphasis in proportion to their heritability times their importance, rather than in proportion to either one alone. This will be completely true if the characteristics are not correlated. If the characteristics have strong environmental and genetic correlations among themselves, this might alter the size and could even reverse the signs of the attention to be paid to each characteristic when constructing a selection index. This is just a special case of the well-known fact that in a multiple regression equation, strong correlations among the independent variables can alter greatly the net regression coefficients from what they would be if all the independent variables were uncorrelated with each other. Emphasizing each characteristic in proportion to the product of its heritability and its importance is as good an approximation to the proper weights as is possible in most cases.

The few studies yet made seem to indicate that if heritability and importance are known with rough accuracy, the *efficiency* of the index based on them will not be changed much by the intercorrelations between the variables, although the relative importance and even the sign of some of the weights may be changed much. As an example in a swine breeding experiment at the Iowa Agricultural Experiment Station, a selection index was first constructed using only the importance of the items and the observed phenotypic correlation between litter mates. It was: $I = 1/3\ W + S + P + .303\ \overline{W} + 1.667\ \overline{S}$ where W is weight at 180 days, S is score for market desirability, P is productivity of dam, $\overline{W}$ is average weight and $\overline{S}$ is an average score of the litter in which the pig was born. L. N. Hazel's analysis[4] of the data which were then collected in the next few years of the experiment indicated that the most efficient index would have been:

$$I' = .3W - .5S + .5P + .270\overline{W} + .605\overline{S}$$

[4] *Genetics* 28:476–490. 1943.

The coefficients in the two indexes are markedly different, especially in the reversed sign for the coefficient of S, yet the efficiency of I was .364 and that of I' was .404, which is larger, but not greatly so. By efficiency is meant that progress by selecting exactly according to I would make progress .364 as fast as could be made by the same percentage of culling if the genotype of every pig were known exactly.

SUMMARY OF RESULTS EXPECTED FROM MASS SELECTION FOR NET MERIT

Mass selection is expected to cause the average of each generation to exceed the average of the preceding generation by an amount (M) which is equal to the heritability fraction $\frac{\sigma_G^2}{\sigma_O^2}$ of the selection differential (S), the latter being the average merit of those selected to be parents minus the average of the whole generation from which they were taken.

There is also a little deterioration from mutation, but this is too small to be considered further in problems of animal or plant breeding, although it may be important in evolutionary considerations.

When selection is first begun there will also be some temporary gain from epistatic effects. This will be something less than half of $S \times \frac{\sigma_I^2}{\sigma_G^2 + \sigma_D^2 + \sigma_I^2 + \sigma_E^2}$. It will tend to disappear with each generation of segregation and recombination and thus, unless constantly renewed by fresh selection, tends to disappear soon after selection is relaxed.

The obstacles to rapid progress naturally fall into two groups: (1) Circumstances or practices which make S small and (2) circumstances or practices which make σ_G^2 small or σ_D^2, σ_I^2, or σ_E^2 large, thereby lowering heritability. Although M may perhaps be small, it will not be zero, provided both S and σ_G^2 have positive values.

Among things which may make S small are:

1. Perhaps only a small fraction can be culled. Remedies: Anything which will improve the health of the herd and reduce deaths from disease or accident; earlier breeding and quicker rebreeding of females, and anything else which will increase the number of offspring raised annually per 100 females.

2. The population may be so uniform that the difference between those selected and those rejected cannot be large. Remedy: Sometimes one can change the environment so that genetic differences may express themselves more fully and be magnified. Outcrossing may help.

3. The breeder may be careless about his cullings or changeable in his ideals. This makes the selection more like discarding at random or at least more like the top than the bottom of Figure 17. Remedy: More

care in deciding on the ideals, more attention to detail, planning the cullings and selections well in advance of the time when they must actually be made.

4. The measures or yardsticks of individual merit may not be definite or simple. This has the same effect as 3. Remedy: Clearer and more quantitative definition of goals, more systematic scoring, grading or classifying at regular ages or dates, simple but systematic records of production where possible.

5. The breeder may be trying to pay attention to too many things in his selections, thus weakening the intensity of selection for the more important things. Remedy: Resolutely keeping minor things in the background, perhaps using a selection index.

Among things which make heritability low are:

6. σ_G^2 may be small. Remedies: An outcross to a relatively unrelated stock having some desirable characteristics which are absent or rare in the breeder's own herd may restore genetic variability. Probably most breeds of farm animals still have enough genetic variability within them that crossing with other breeds is not necessary for making a large amount of further improvement. Introducing blood from outside the breed cannot be done in most breeds under the prevailing standards of purebreeding, although outcrossing within the breed is always possible. Inbreeding a population without discarding any of the lines tends toward doubling σ_G^2 but puts most of it between lines and tends to extinguish genetic variance within lines. When the better lines are selected and the poorer lines are discarded, the genetic variance then remaining is likely to be smaller than was in the population when the inbreeding began. Sometimes it may be possible to alter the environment enough to magnify the outward differences between genetically different individuals, particularly where thresholds are involved, but it may be difficult to find the environmental changes which will do that.

7. σ_E^2 is usually large except for a few characteristics such as colors and things which are fairly simple anatomically, such as the dimensions of the bones, shape of head, set of ear, etc. Remedy: Keep the environment alike for all individuals as far as is economical, correct for the effects of the more important differences in environment which did occur, use lifetime averages where those are practicable, and give some attention to the merits of relatives and progeny.

8. σ_D^2 may be large. This can be an important obstacle only where most of the undesired genes are recessives already rare, and in pairs of genes where the heterozygote is preferred over both homozygotes. Remedy: Consider the relatives. The collateral ones usually give more help in this respect than the ancestors do. The progeny are still more

informative, especially if they are inbred.

9. σ_I^2 may be large. This seems likely to be important only where the population has already been under selection for many generations. Remedy: Consider the relatives and progeny. Inbreed enough to form distinct families, only rarely making crosses between them. Breed within the family as long as its average merit is good.

These ways of overcoming partially the obstacles to progress by mass selection will be considered separately in the following chapters. The purpose of this chapter has been to describe and explain the results of unaided mass selection. It alters the population mean almost in proportion to n times the amount it changes gene frequency, n being the number of genes affecting the characteristic.

The rate of improvement per generation may increase or decrease in later generations but is not likely to change rapidly unless there is a considerable amount of epistatic interaction.

Only rarely is mass selection *completely* ineffective, as when selection is for a heterozygote, when selection has already carried the population into a stable epistatic peak, or when selection is within an entirely homozygous line. Often, however, the rate of progress by mass selection is slow and could be made more rapid by a judicious use of relatives and progeny or by more careful control or consideration of the environment.

REFERENCES

The classical cases of selection experiments with higher animals are those of Castle on the hooded pattern in rats, mostly published from about 1914 to 1920, and Pearl's experiments on selection for high egg production in the poultry flock at the Maine Station a little earlier. In plants the Illinois Experiment Station selections for high and low oil content and for high and low protein content in corn kernels and Johannsen's selections in Denmark for seed size in beans are famous among experiments on selection. For brief accounts of these and other actual experimental studies of selection, see:

Babcock, E. B., and Clausen, R. E. 1927. Genetics in relation to agriculture. pp. 221–31 and 538–50.

Castle, W. E. 1930. Genetics and eugenics. pp. 236–45.

Goodale, H. D. 1938. A study of the inheritance of body weight in the albino mouse by selection. Jour. Heredity 29:101–12.

Sinnott, E. W., and Dunn, L. C. 1925. Prin. of genetics. pp. 339–46.

Winter, F. L. 1929. Mean and variability as affected by continuous selection for composition in corn. Jour. Agr. Res. 39:451–76.

CHAPTER 13

Aids to Selection—The Use of Lifetime Averages

Many of an animal's important characteristics vary in their expression from one time to another. Familiar examples are: the amount of milk and fat which a cow produces in different lactations, the number of pigs a sow farrows in different litters, the weight of the fleeces which a sheep produces from year to year, the number of eggs a hen lays in different years, the speed with which a horse runs a race, the amount of pull which a draft horse exerts, the degree of fatness of most animals, and many things about an animal's action, temperament, and health. Examples of things which change but little from one time to another are coat color in most farm animals and the dimensions and shapes of bones after maturity is reached.

Most of the variations from one time to another are due to variations in the environment which prevails at the time the observation is made or which did prevail just previously. Internal conditions, such as the temporary state of health, are part of the environment as meant here. So far as the peculiarities of the environment are known and their effects can be estimated, the proper procedure is to correct the animal's production (or score or other figure which is used to represent its appearance or its performance) for the effects of those peculiarities of environment. In this way production records or scores may be "standardized" to what they would probably have been if environmental conditions had been the same as those chosen for standard. It is common practice to correct dairy records for age and for times milked per day. Sometimes they are corrected also for the date at which the next conception occurred, length of preceding dry period, for season of freshening, for weight of the cow, or for the fat percentage of the milk. Likewise, in considering the speed of race horses, allowance is often made for age, for the condition of the track, and for the weight carried. There is almost no limit to the number of such corrections which might be made in cases where many details about the environment or management are recorded. But it is impossible to know all about the environment. Moreover, the correction factors used will not be exactly correct for every individual, even when they are correct on the average for the whole population. Since it is therefore impossible by the use of correction factors to make all standardized records exactly what they would actually have been under the standard conditions, and since some effort or time is required to use each correction factor, the law of diminishing returns usually makes it scarcely worth while to correct for more than two or three of the most important environmental conditions.

Each standardized record can be considered as equal to the real ability of the animal under the standard conditions plus or minus some error for incomplete or inaccurate correction for conditions which were not standard. If the corrections have been made by a method which on the average is fair for that particular population of records, then the error remaining in any record chosen at random is just as likely to be positive as it is to be negative. The corrected record of the same animal in the next lactation or next year or at the next inspection will be the same real ability of the animal plus or minus another error for incomplete or inaccurate correction to standard environmental conditions. So far as temporary environmental conditions are concerned, these errors remaining in the corrected records will be independent of each other. Hence, if all the records of the animal are averaged together, some of these will have positive errors, and others will have negative errors which will tend to cancel the positive ones. This makes the amount of error in an average less than it is in single records although, of course, it would be too much to expect that the errors would cancel each other *exactly* so that the average would be entirely free of them. The effect of the averaging can be pictured as follows, where $\pm$ indicates that the error is as apt to be positive as it is to be negative and Σ means "the sum of the":

First observation = animal $\pm$ first error
Second observation = animal $\pm$ second error
Third observation = animal $\pm$ third error
. . . .
. . . .
*N*th observation = animal $\pm$ *n*th error

Σn observations = $n \times$ animal $\pm$ Σ errors

Dividing by *n*, we get:

$$\text{The average observation} = \text{the animal} \pm \frac{\Sigma \text{ errors}}{n}.$$

The average of the *n* observations differs from the real ability of the animal only by one *n*th of the sum of those errors which did not happen to cancel each other. As *n* becomes larger, there is more chance for positive and negative errors to cancel. Thus the proportion of error in the average becomes smaller if the errors were really random.

Allowance for the reduced variability of averages must be made when comparing animals which do not each have the same number of records in their averages. For example, let us suppose three cows have the following corrected averages in a herd which averages 400 pounds of fat:

A's only record is 600 pounds.

B has an average of 565 pounds for two lactations.

C has an average of 560 pounds for four lactations.

Which cow probably has the highest and which the lowest real producing ability? The 600 pounds is the highest figure, but this is for a single lactation in which conditions might possibly have been much better than we thought. In other words, it indicates that the cow was a good producer; but we are not sure how much faith it merits. The fact that it is so far above the herd average makes us suspect that its excellence was not due to the cow alone. This cow is somewhat in the position of a prospective employee who bears a letter of very high recommendation, but that letter is written by a man about whose veracity the prospective employer is in doubt! If the records are taken at face value, cow C is the poorest producer of the three; but her record is an average of four different lactations, and it is less likely that she would have had much better environment than we thought in all four of her lactations. Such good luck might have happened to A or perhaps even to B. All three cows in this example are probably high producers, but we need some rule or formula for estimating the real productivity of each if we are to make the least error in estimating which of them is the highest producer, as we might want to do if we were trying to buy one of them or to choose between their sons.

In making such an estimate we need to know something about how "repeatable" these records are. If a cow tends to produce almost exactly the same amount each lactation, just as she is practically the same color every year, the first lactation would tell almost the whole story and would be almost as reliable as the average of four. On the other hand, if dairy records were only slightly repeatable, the first record would be only an indication, not very dependable, and the process of averaging four records would remove much of the error but would also reduce the variability. The measure of repeatability needed is the "coefficient of correlation" between records made by the same cow in this herd of cows. With that coefficient (r in the following equation) and the herd average and the records of each cow, we can estimate the real productive ability of each cow under the conditions standard in that herd. The equation for this prediction, where the cow's average is based on n records, each corrected for the known environmental circumstances, is as follows:

Most probable producing ability of the cow $= \frac{nr}{1 - r + nr} \times$ (her average record) $+ \frac{1 - r}{1 - r + nr} \times$ (the herd average). Another way to state the same formula is that the most probable producing ability of the cow = the herd average $+ \frac{nr}{1 + (n - 1) r}$ times (her own average minus the herd average). The fraction, $\frac{nr}{1 + (n - 1) r}$, shows how much we trust the cow's own average as an indication of her real producing ability. When we know nothing about the cow we can make no better estimate of her producing ability than that she is an average cow of that herd. When she has one record, that gives us an indication of what she will produce in future lactations but, if r is small, this one indication is not very reliable. So we trust it a little but not very far. When she has two records we trust what they indicate about the cow a little more. As n increases still more we come nearer and nearer to trusting the cow's average completely. Consequently, we have less and less use for the herd average.

The use of lifetime averages makes selection more efficient simply because it reduces the amount of variation caused by temporary circumstances, and therefore lessens the number of mistakes made. That is shown graphically in Figure 23. Because the heritability fraction increases with n, the breeder actually gets a larger fraction of what he reaches for in his selections. This advantage is partly offset by the fact that the lessened variation among averages prevents him from reaching so far. The very highest averages are not as high and the very lowest averages are not as low as the highest and the lowest single records, respectively. The net result of the large increase in heritability and the small decrease in the selection differential which can be attained with the same percentage of culling is that progress per generation when selecting on an average of n records is $\sqrt{\frac{n}{1 + (n - 1) r}}$ times as much as if selections were made on only one record per animal. Table 13 shows the values of this fraction for a few selected values of n and r. Obviously, the method of averaging many records or observations is most useful and most needed for characteristics for which r is low. Each additional record contributes less additional information than the preceding one did; therefore, much of the entire usefulness of the method of averaging can be had while n is still as low or 2 or 3, although each additional record adds something more to the accuracy, especially when r is very small.

In records of yearly milk and fat production, considering only cows which are in the same herd, r is usually somewhere between 1/3 and

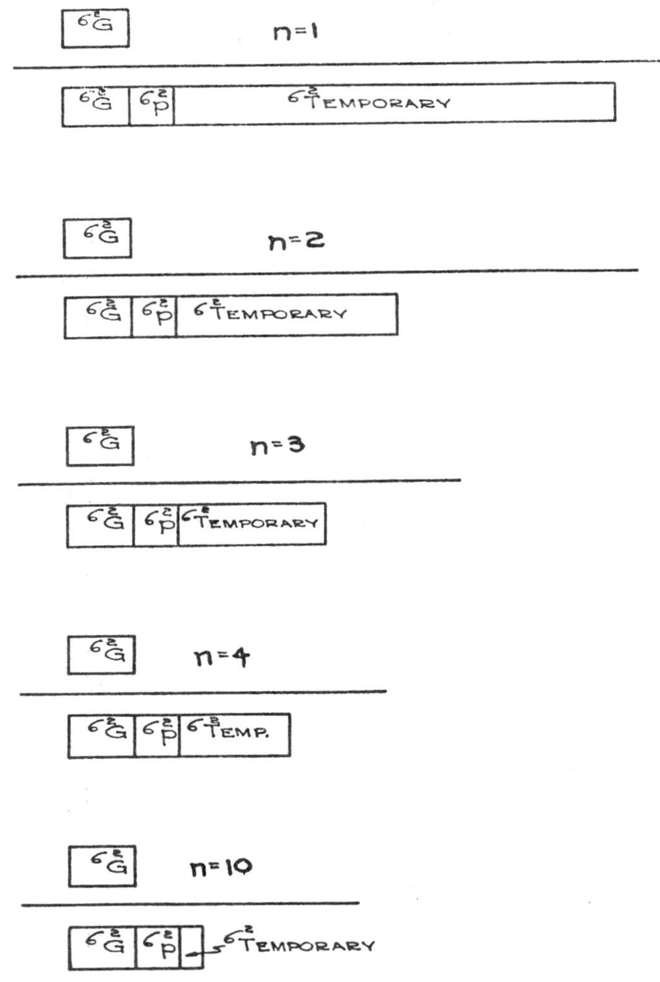

FIG. 23. Diagram showing how the heritability of differences between averages increases as the number (n) of records in each average increases. Drawn to scale for the case in which heritability of differences is .12 when n is 1 and repeatability of single records is .20. That means the case in which 80 per cent of the variance between animals with one record each is caused by temporary environmental circumstances. σ_G^2 is the additively genetic variance between individuals. σ_P^2 is the variance due to permanent but nontransmissible differences between individuals. These include differences due to dominance deviations, epistatic deviations, and to such effects of environment as are permanent for each animal but differ from one animal to another. As n increases, the variance due to temporary things falls to one nth of its value in single records.

1/2. We shall not be far wrong if we take 2/5 as the general figure to be used in the preceding equation, although a higher figure would be justified in herds where management has been unusually standardized and corrections for the known environmental circumstances have been unusually complete. In fact, r is the fraction of the total variance among the corrected records which is due to permanent differences between cows; and $1 - r$ is the fraction of the variance caused by temporary circumstances which vary from one record to another of the same cow.

TABLE 13

PROGRESS WHEN SELECTING BETWEEN ANIMALS WITH n RECORDS EACH, AS A MULTIPLE OF THE PROGRESS WHICH COULD BE MADE BY SELECTING BETWEEN THEM WHEN THEY HAD ONLY ONE RECORD EACH

n	r								
	.1	.2	.3	.4	.5	.6	.7	.8	.9
2	1.35	1.29	1.24	1.20	1.15	1.12	1.08	1.05	1.03
3	1.58	1.46	1.37	1.29	1.22	1.17	1.12	1.07	1.04
4	1.75	1.58	1.45	1.35	1.26	1.20	1.14	1.08	1.04
6	2.00	1.73	1.55	1.41	1.31	1.22	1.15	1.10	1.04
10	2.29	1.89	1.64	1.47	1.35	1.25	1.17	1.10	1.05

More rigid control of the environment will naturally make r higher. That is, r is a description of conditions in a particular population and is not a fundamental biological constant. If we use the fraction 2/5 in the preceding example, the equation simplifies to: The cow's ability $=$ $\frac{2n}{2n+3} \times$ (her own average) $+ \frac{3}{2n+3} \times$ (herd average). That gives the following for estimates of the real producing ability of the three cows: A = 480 pounds; B = 494 pounds; C = 516 pounds. C is probably the best and A the poorest of these three, so far as the evidence goes; but all three are good cows, and the differences between them are small enough that we should not be greatly surprised if another lactation or two would change their order.

Figure 24 shows graphically the results of such computations for butterfat production in an actual herd. The numbers along the vertical scale are the barn numbers of the cows. Such a graphic scale of estimated productive ability is a convenient help in making decisions about culling the cows or saving bull calves from various cows. It must be kept reasonably up to date, of course, if it is to be useful. At any particular time some of the cows will have incomplete records which indicate their producing ability but do not merit as much confidence as completed records. In making Figure 24, the records estimated from lacta-

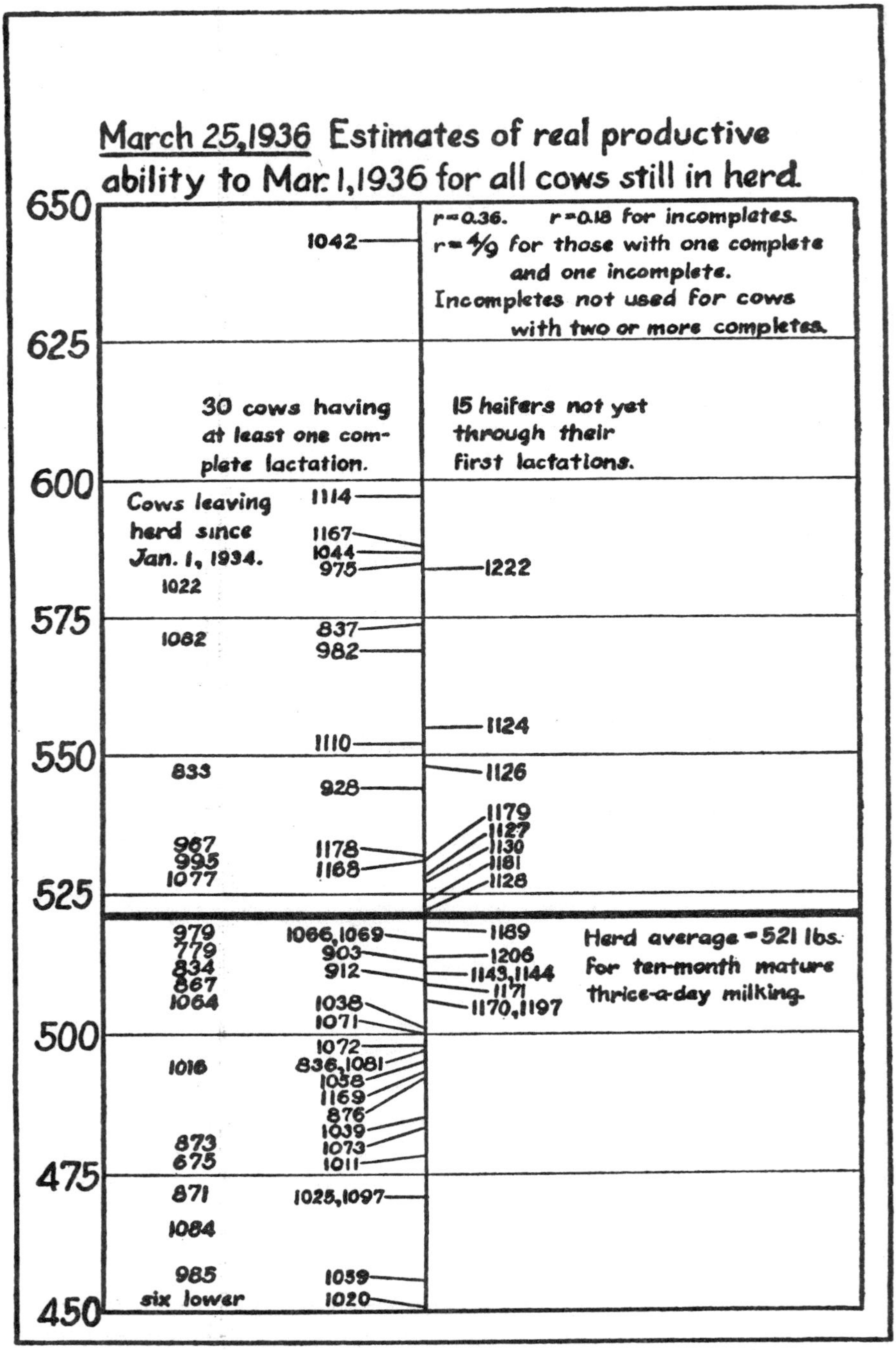

March 25, 1936 Estimates of real productive ability to Mar. 1, 1936 for all cows still in herd.

r = 0.36. r = 0.18 for incompletes.
r = 4/9 for those with one complete and one incomplete.
Incompletes not used for cows with two or more completes.

30 cows having at least one complete lactation.

15 heifers not yet through their first lactations.

Cows leaving herd since Jan. 1, 1934.

650
1042
625
600
1114
1167
1044
975
1222
1022
575
837
1082
982
1124
1110
550
833
1126
928
1179
1127
967
1178
1130
995
1168
1181
1077
1128
525
979
1066, 1069
1189
779
903
1206
834
912
1143, 1144
867
1171
1064
1038
1170, 1197
1071
500
1072
1016
836, 1081
1058
1169
876
1039
873
1073
675
1011
475
871
1025, 1097
1084
985
1059
six lower
1020
450

Herd average = 521 lbs. for ten-month mature thrice-a-day milking.

Fig. 24. Estimated producing abilities of many cows with unequal numbers of lactations, based on all information available at the date when the scale was made.

tions still incompleted have arbitrarily been given about half as much confidence as completed records in the case of cows which have not yet completed their second record. The incomplete records were not used at all for other cows. The heifers with only an incomplete lactation were placed by themselves on the right to emphasize further the uncertainty about their ability. The cows at the extreme left were included to show whether those which left the herd either through death, disability, or voluntary culling, had really averaged less in productive ability than those which remained.

Many other examples of the use of averages might be given. In each of them it is necessary to know something about the repeatability of the characteristic. In fertility of swine the repeatability of the number of pigs born to the same sow in different litters is about one-sixth. In the weight of fleeces shorn from sheep, the repeatability is about .5 to .6. The corresponding figure for the fleece weights of Angora goats is about .4. Some of the earliest studies on repeatability of production records in farm animals were made on the records of first and second year egg production in egg-laying contests. Most of those figures were of the order of .45 to .60. All of these are computed on the basis of records made within one herd or flock.

The method of averages can be extended to include scores or any other ratings of type which can be expressed numerically. The repeatability of such type ratings is usually low enough that there would be a material gain in the accuracy of selections by obtaining and averaging type ratings of the animal at different times in its life as compared with relying upon the opinion of one judge, no matter how expert, who saw the animal only once. The repeatability of type ratings of dairy cows at intervals of one year was .34 to .55 in the only study[1] yet reported. The opinions which the same judge would hold of the animal if he inspected it at different times in its life usually vary more than do the opinions of several judges who might inspect the animal at nearly the same time. Probably this varies considerably with the class of animal and with the ages at which the inspections are made.

The method of averaging repeated observations, of course, is limited to characteristics which can be observed more than once. It cannot be applied to such things as growth rates, age to sexual maturity, or to carcass qualities which can be observed only upon slaughter of the animals. It is also limited in usefulness for egg production, where such a large part of the total economical lifetime production of the bird is made during its first year.

Of course the costs of waiting to cull until more records are avail-

---

[1] *Jour. of Dy. Sci.* 25:45–56, 1942.

able need to be considered, too. Besides the actual loss incurred by keeping any animal which is not actually paying its way, such waiting tends to make the interval between generations longer and thereby to lower the progress *per year*. This will partly offset the gain it makes by increasing the progress *per generation*. The costs of waiting to cull would vary with the animal and with economic circumstances, being higher with hens, for example, where the second year's production is lower than the first, than with cows where the second record is generally some 12 to 15 per cent higher than the first.

In the case of cows the first lactation will be several months along and the heifer may be with calf again before it is certain that the production in her first lactation will be very low. Often it will not cost much then to keep her enough longer for the first three or four months of the second lactation to confirm or disprove what the first lactation indicated. A practical rule for many herds is to cull in their first lactations only those with extremely low production, keeping the moderately low and doubtful ones through the flush of production in their second lactations. They will usually pay at least their feed costs for that, since so much of the production comes in the first half of the lactation.

In many cases the decision to keep or cull must be made while the animals are young if economic loss is to be avoided. For example, with range cattle or sheep the heifer calves and ewe lambs can usually be sold to better advantage and with lower feed costs at or soon after weaning age than if they are kept much longer. The ranchman could cull them more accurately if he could wait until they grow up and until he can have rated their type several times, but the gain from doing this may not be enough to pay for the loss he will take in lower sale prices for those which are finally culled after they are too old to be sold as heifer calves or lambs.

One practical problem involved in using lifetime averages is what to do with records thought to have been made under abnormal conditions for which no satisfactory correction is known. In principle such records should be omitted. The practical difficulty is how to decide fairly when conditions really were abnormal. Some circumstances are definite enough that they offer no difficulty. For example, in Denmark records are omitted for years in which the cow aborted or had foot-and-mouth disease. No other omissions are permitted. Other circumstances are not definite enough to permit a clear decision. For example, a cow may have had a bad attack of milk fever at the beginning of her lactation. The owner may believe that she did not recover soon enough to produce normally during that lactation. But how is one to be certain? To base the decision on the size of the record opens the door to all kinds of biases. The guiding principle should be to omit no record except when the circumstances are so definite that no doubt can exist. Those circumstances must be something other than the size of the record of performance itself.

Basing selections on lifetime averages will automatically foster some selection for longevity and real "constitution," since breeders will tend to save for sires only the sons of females which have proved themselves by several records of production. When selections are based on lifetime averages, it will hardly be possible for a heifer or gilt with only one record of production to get her son saved to head a purebred herd unless that one record was truly phenomenal.

SUMMARY

The use of an average of many repeated observations as a basis of selection is one of the most effective ways of overcoming mistakes and confusion which would otherwise result from the effects of temporary environmental conditions. The method is inexpensive, requiring only the existence of the records, the time needed for averaging them, and whatever it costs to postpone culling until two or more observations have been made. It can be made to foster some selection for longevity and constitution. In using such lifetime averages, allowance must be made for the lessened variability of averages which are based on many records. If this is not done, it will appear that most of the extreme producers, both high and low, are individuals with only one or at most two records. The method is needed most for things which are least constant from one time to another in the animal's life. Much of the gain from using it comes with the second record, but if r is small the gain from waiting for a third or even a fourth record may be considerable. The method does not help at all to keep the breeder from being deceived by permanent effects of environment, such as permanent stunting when young, nor by the consequences of dominance and complex gene interactions. Placing much reliance on selecting animals by their lifetime averages will naturally lead men generally to buy breeding stock from herds which they know well and in which they have had several opportunities to study the animals.

CHAPTER 14

Aids to Selection—Pedigree Estimates

"The bull gives no milk, of course, yet will not a bull descended from several generations of high-producing dams produce, when mated with a highly productive cow, calves which possess this characteristic to a still higher degree?"—Bergen, in 1780.

The decision of whether to reject or keep an animal for breeding may be modified by what its relatives are or do. Wherever that is done, the intensity of individual selection is reduced. The average individual merit of those selected must be lowered every time an animal which would be rejected on account of its own individuality is kept for breeding because it has unusually excellent relatives, or every time an individually excellent animal is rejected because its relatives are of low merit.

By paying a reasonable amount of attention to the relatives it may be possible to increase the genetic *accuracy* of the selections more than enough to offset the decrease in the *intensity* of selection for individual merit. There is real danger of doing damage by paying too much attention to the relatives, since they are not a perfect guide to the individual's breeding value either. The proper balance between paying too much and paying too little attention to the merits of the relatives shifts from case to case according to how well the merits of the relatives are known, how closely they are related to this animal, how well the individual merit of this animal is known, and how highly hereditary the characteristics are. An understanding of the principles and practical difficulties involved will help in using the approximate rules which in actual practice must guide us in estimating an individual's breeding value from what we know about the merit of its relatives.

From the genetic principles involved, relatives of an individual may all be grouped in two classes: those which are related to it through its parents and those which are its descendants. The former group is considered here collectively under the general term of "pedigree," which is the subject of this chapter, while the latter group will be considered in the next chapter under the term of "progeny test."

Attention to pedigree can make selection more effective only because individual selections are not perfectly accurate. We never know exactly what genes an individual has. Our knowledge of that is especially scanty when the animal is still immature, as many are when first selections must be made. If we estimate its inheritance from its own appearance or performance, we will make some mistakes on account of the effects of environment, dominance, and complex interactions of genes. If we estimate its inheritance by paying attention to the individuality and performance of its relatives, we may avoid some of those mistakes; but we run the risk of introducing other mistakes, because those relatives will not have exactly the same genes as this animal does. Moreover, our estimate of the genes in those relatives is itself subject to error from our being confused by the effects of dominance, environment, and complex gene interactions on those relatives.

GENERAL PRINCIPLES WHICH LIMIT THE USEFULNESS OF PEDIGREES

The biological basis for the usefulness of pedigrees lies in the fact that an individual gets half of its inheritance from its sire and half from its dam. If we knew what inheritance these parents had, we could estimate what inheritance this individual probably received from them. Because the parents were heterozygous for some of their genes, such estimates cannot be perfectly accurate. Chance at Mendelian segregation plays a part in determining what the parents transmit to any one offspring. An additional limitation on the use of pedigrees is that we will rarely know exactly what genes the parents did have, although we will often know more about their inheritance than about the inheritance of each of the offspring, because the parents have had a longer time in which to demonstrate their characteristics or performance. Also, because some of the ancestors will have had other offspring, they will be to some extent progeny-tested, whereas most individual selections must be made before any progeny are available.

Even if we knew exactly what genes each parent had—and no amount of pedigree study could tell us that much—the sampling nature of inheritance limits the average likeness between the inheritance of an individual and the inheritance of either parent in a random breeding population to a correlation of + .50. On the same basis the correlation between an individual and the average of its two parents is + .71. If much inbreeding, or mating of like to like where ideals are diverse, is being practiced, these correlations will be larger; but in any actual population they will be far from perfect. If we are trying to predict the average merit of many offspring, the effects of chance at segregation will tend to cancel each other. The correlation between the inheritance of one parent and the average kind of inheritance of n of its offspring approaches $+ 1.00$ according to the formula $\sqrt{\dfrac{n}{n+3}}$ if the other parents of those offspring are a random sample of the breed. The correlation between the average inheritance of both parents and the average inheritance of n of their offspring approaches $+ 1.00$ according to the formula $\sqrt{\dfrac{n}{n+1}}$.[1] These correlations are between genotypes and not between the directly observable characteristics of the animals, but the formulas explain what seems at first to be a contradiction between the facts that pedigrees cannot be highly accurate for estimating what an individual animal will be and yet can be very accurate in predicting the average qualities of a large number of offspring from one pair of parents.

If we knew exactly what genes the sire and dam each had, nothing would be gained by considering more distant ancestors or collateral relatives. Figure 25 shows a Mendelian example of that with respect to one pair of genes.

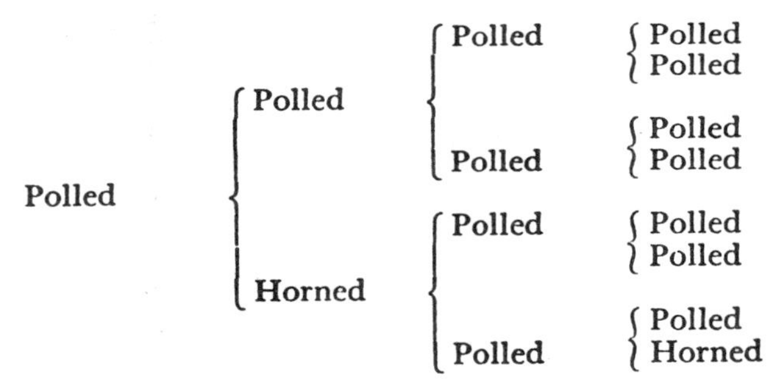

FIG. 25. A pedigree of a Polled Hereford, illustrating the general principle that study of remote ancestors tells nothing more about an animal's inheritance when the genotype of an intervening ancestor is known with certainty.

Insofar as we are correct in thinking that horned cattle are *pp* and polled cattle are either *PP* or *Pp*, it is certain that the dam in Figure 25 was *pp* and that this offspring of hers is *Pp*. No amount of study of her ancestors would add anything to our knowledge of that. That brings us to the question of how much attention to pay to each ancestor in the general case where we know something about the individual merit of most ancestors but are not entirely sure of the genotype of any of them.

[1] *See also* page 260.

The answer to that question depends on five different circumstances: First, how closely the ancestor is related to the subject of the pedigree; second, how completely the merit of each ancestor is known; third, how well the merit of intervening ancestors is known; fourth, how highly hereditary the characteristic is; fifth, how much environmental correlation there is between ancestor and subject and between different ancestors.

The general fact that an animal gets half of its inheritance from each parent would naturally lead to one form of what is generally called "Galton's law"—of which more is said in chapter 20—namely: In estimating an individual from knowledge about one of its parents, it should be estimated as equal to one-half of that parent plus one-half of the breed average; if it is being estimated from a grandparent, it should be estimated at one-fourth of that grandparent plus three-fourths of the breed average; with a great grandparent, it should be one-eighth of that great grandparent plus seven-eighths of the breed average, etc. The importance of the ancestor in such a prediction equation is halved with each additional generation which intervenes between the individual and its ancestor. This "law" in general is sound in a random breeding population, provided only one ancestor is being considered and provided some conservatism is practiced by basing a smaller share of the estimate on the ancestor and a correspondingly larger share on the breed average wherever the characteristic is not highly hereditary or there is uncertainty that the information about the ancestor really is a true picture of its inheritance. Two ancestors cannot both be used in a single prediction of this kind if one of them is an ancestor of the other. We could combine information about the sire and the maternal grandsire, estimating the animal at one-half of the sire plus one-fourth of the maternal grandsire plus one-fourth of the breed average, with still more emphasis on the breed average as we are less sure that what we know about the sire and grandsire is really typical of their breeding value. But we could not combine information about the sire and the paternal grandsire in the same way because, to a considerable extent, the things which could be estimated from knowledge of the grandsire are the same things which could be estimated better from knowledge of the sire himself. The use of both in a single prediction would be using some of the information a second time.

Czekanowski[2] has explored the question of the comparative importance of sire, grandsire, and great grandsire in the special case when (1) all three are in the same line of descent, (2) no other ancestors are con-

[2] Zeit. f. Ind. Abst. u. Vererbungslehre 64:154–68. 1933.

sidered, (3) exposure of relatives to the same kind of environment has contributed nothing to the correlation between them, and (4) the merits of all three ancestors are equally well known. His figures for the amount of attention (the net regression coefficient) to be given each of these three ancestors for several values of the correlation between parent and offspring are as follows:

	Correlation Between Parent and Offspring						
	.10	.20	.25	.33	.40	.44	.50
Sire	.095	.186	.232	.312	.381	.431	.500
Grandsire	.039	.059	.062	.059	.046	.030	.000
Great grandsire	.016	.020	.018	.012	.006	.002	.000

The figures show how little is to be gained by considering the remote ancestors when the merit of the intervening ones is equally well known. For highly hereditary characteristics the sire's own appearance or performance is almost a perfect guide to the net worth of his inheritance. For slightly hereditary characteristics, relatively more attention should be paid to the remote ancestors, but in that case the predictive value of pedigree is low anyhow, no matter how used. It is doubtful whether Czekanowski's figures have any practical usefulness beyond demonstrating these general principles. In actual practice some ancestors will be much better known than others. For example, if the sire in Czekanowski's problem were still too young for his mature characteristics to be unmistakable or if the grandsire were thoroughly progeny-tested, less attention should be given to the sire and more to the grandsire than these figures indicate.

Collateral relatives in the pedigree are a progeny test which furnishes evidence about the genotypes of their parents or grandparents. Although a grandparent is generally as good an indicator of an individual's inheritance as a half brother or sister, an individual can have only four grandparents but may have a much larger number of half brothers and sisters. In case it does, an estimate of its inheritance based on the appearance and performance of its half brothers and sisters may be much more accurate than an estimate based on its grandparents. The evidence furnished by collateral relatives should be used in the pedigree according to what it shows about the kind of inheritance which the mutual ancestor had and according to the completeness of the evidence. For example, in a dairy pedigree where the production of a large number of paternal half sisters is known, the sire may be considered reasonably well proved, and there is little to be gained by studying his ancestors. If the dam in the same pedigree is still a young cow in her first lactation, considerable information could be gained by considering the maternal grandparents, although consideration of the paternal grandparents would be of little use.

The chief danger in pedigree selection is that it will do more harm by lowering the intensity of individual selection than it does good by making the selection more accurate. Pedigree is not often as accurate a basis for estimating breeding value as individuality is,[3] although it will occasionally be so in a hitherto unselected population, especially if the individuals being selected are still too young to have had much chance to prove their characteristics at the time the selection must be made. Pedigree selection is rarely as useful in animal breeding as it can be at times in plant breeding, because there is almost nothing in animal breeding to correspond to the distinct and uniform lines or families which exist in those plant species, such as wheat and cotton, where self-fertilization or some other intensive form of inbreeding is the rule. In plants in which self-fertilization is possible, an individual may have only one parent, only one grandparent, one great grandparent, etc., and each of these may be more thoroughly progeny-tested than is possible where reproduction must be bi-parental.

PRACTICAL LIMITATIONS ON THE USEFULNESS OF COMMERCIAL PEDIGREES

Often pedigrees contain little information except the names and numbers of the ancestors. To use them at all, you must find from other sources how excellent or poor those ancestors were. That is less true of dairy and poultry pedigrees than of others and is being improved in the pedigrees of most kinds of livestock; but, as printed today, many pedigrees are only a meaningless genealogical jumble of names and numbers to one who does not know from other sources how meritorious those ancestors were. This limitation does not matter much to the man who is thoroughly familiar with his breed. Perhaps all purebred breeders should aim at that goal, but an enormous number of potential customers and even many breeders simply have not time nor opportunity to keep familiar with the merits and deficiencies of the prominent

[3] In a random breeding population in the limiting case most favorable to confidence in the pedigree—the case where the genotypes of sire and dam are perfectly known—the correlation between an animal's breeding value and its own individual appearance or performance is $\sqrt{2a}$ times as large as the correlation between its breeding value and the average genotype of its parents, where a is the additive genetic fraction of the variance in individual appearance or performance. This relation indicates the basis for the general principle that pedigree can become more dependable than individuality for characteristics which are slightly hereditary (i.e., where a is less than 1/2), provided the parents are so thoroughly known that there is little doubt about their genotypes.

breeding animals of their breed. Pedigrees would be more useful to these men if more information were included about the merits of the ancestors, even if that meant only printing the pedigree as far the parents and grandparents. In justice to those who print the pedigrees, it should be emphasized that in most cases there is no information to print because no systematic plan of measuring and recording the merit of individual animals has ever been in operation.

This is no new idea. As long ago as 1832 the following was written in *The Thoroughbred Horse in Prussia:* "If they do not contain production tests, such herdbooks will be useless and without interest, since they would contain only names of which no one knows anything and which mean nothing."

Such information as is given in the pedigrees is usually selected to show the animal in the most favorable light possible. Actual falsehoods in printed pedigrees are rare because of the heavy penalties in loss of business which are exacted of a breeder who is suspected of dishonesty; but plenty of "filler" is still used, although general practice in this respect is improving. An example of "filler" is shown in Figure 26. The information printed under an ancestor's name applies in many cases to half sisters of its parents or grandparents or even to more remote relatives. The following statement was found listed under the paternal grandsire in a pedigree of a bull used in a large dairy herd in Iowa in 1937. "His dam is a granddaughter of that noted transmitter, Sir . . . ., sire of . . . ." The actual records mentioned here were made by cows separated by six Mendelian segregations from the bull in whose pedigree they appeared, and related to him less than 2 per cent. By contrast Figure 27 shows a pedigree in which each bit of information is listed under the ancestor which it most directly concerns.

Generally the ancestors were selected individuals from among their contemporaries. That is especially likely to have been true of the males and hence of their ancestors. Some regression to the breed average is to be expected for that reason, even if the pedigree information is supplied by an utterly impartial and accurate agency. The intensity of the selection practiced in choosing those ancestors is not usually known.

In most pedigrees little information is given about collateral relatives. That is beginning to be remedied in dairy cattle and poultry pedigrees by the inclusion of progeny tests wherever those are available. Occasionally in the pedigrees of meat animals one finds comments on the winnings or performance of brothers or other collateral relatives. Information about collateral relatives may be much more highly selected than information concerning ancestors, since there can be no choosing of ancestors to be mentioned but only selection of data con-

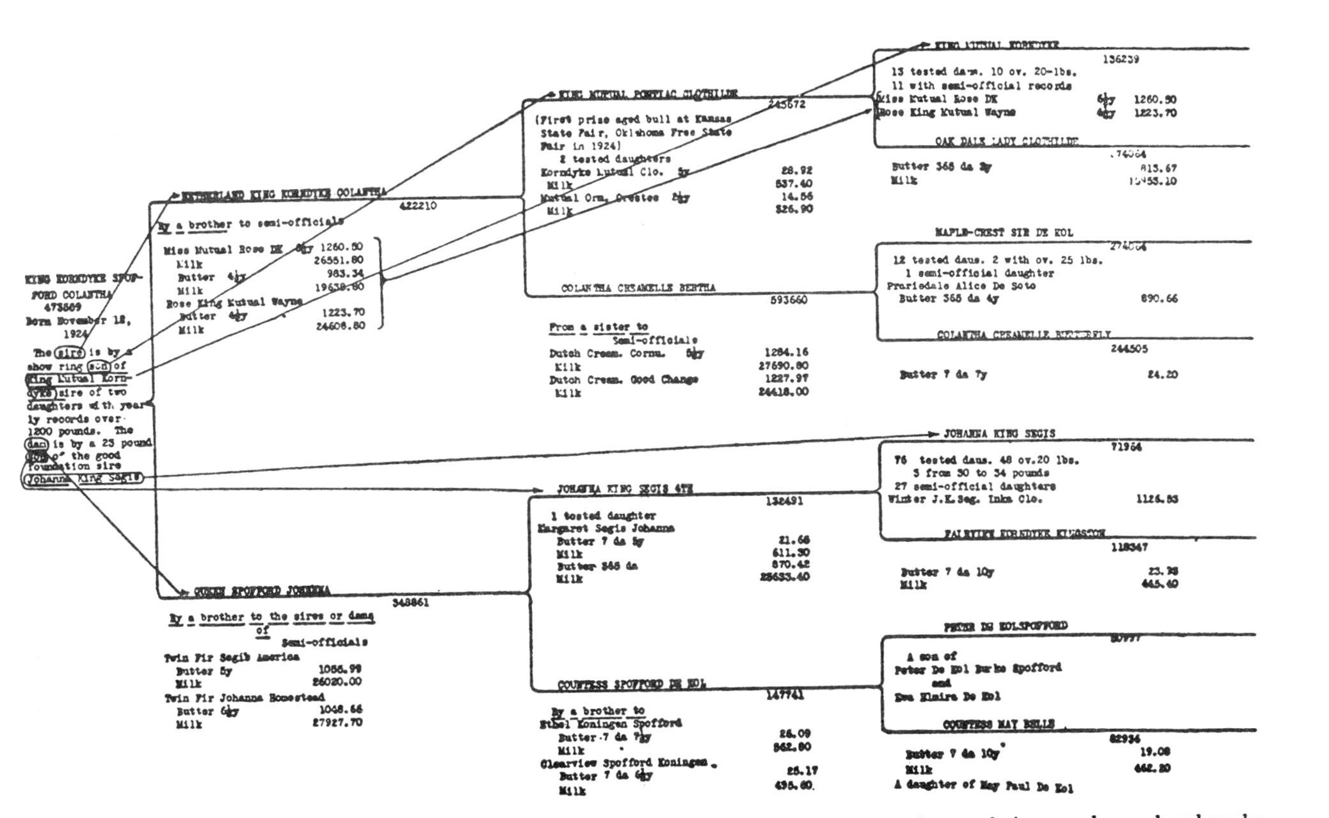

FIG. 26. A pedigree showing a moderate amount of "filler." Favorable items are repeated several times and are placed under animals rather distantly related to those on whose inheritance the items throw most light. (From Iowa Agr. Ext. Service, Bul. 162.)

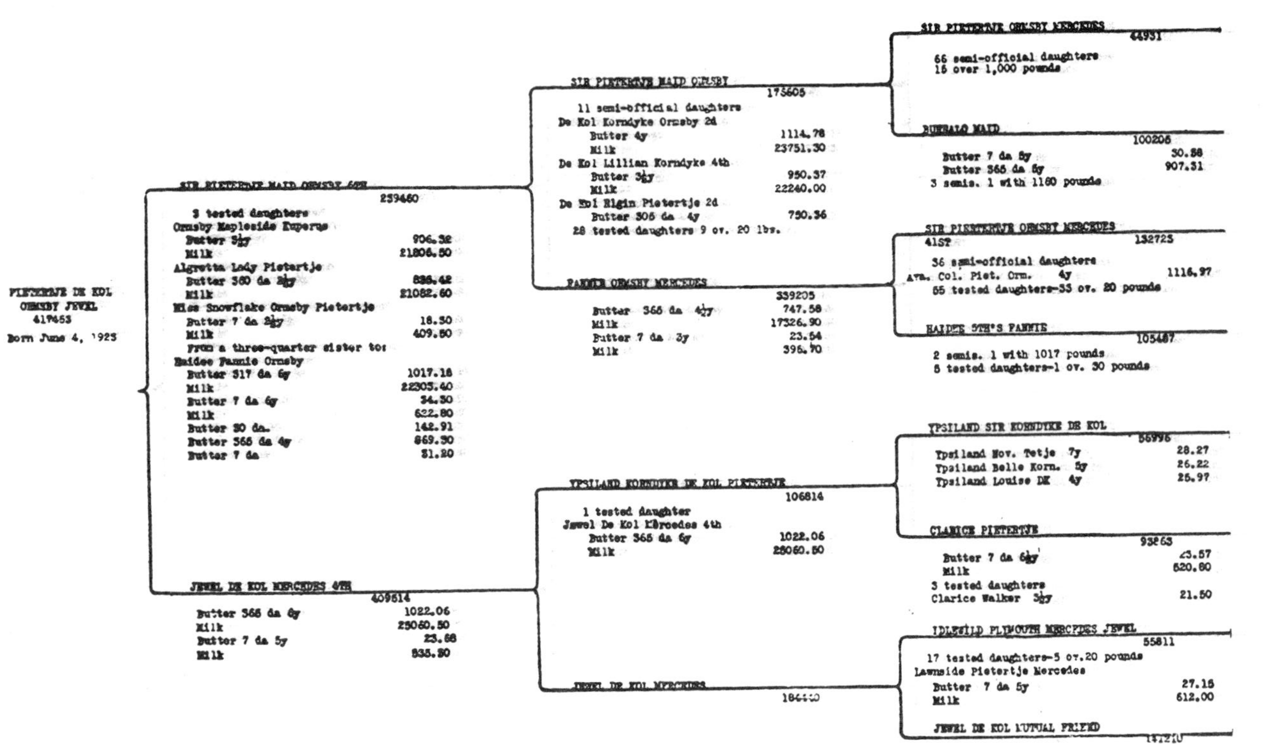

FIG. 27. A pedigree where the information is located under the animal to which it applies and where the information about most ancestors is reasonably complete. The only duplication is the dam's record, which is reported under her and also under her sire. (From Iowa Agr. Ext. Service, Bul. 162.)

cerning them; whereas among the collateral relatives there may be choice of which individuals shall be mentioned and also selection of the best records of those individuals. One often finds in dairy pedigrees such comments as "20 A.R. daughters, two with over 1,000 pounds of fat." Since the records of 18 of the 20 are omitted and the number of daughters not tested is not stated, such a statement tells little more than that this sire was used extensively and was regarded highly enough that 20 of his daughters were tested officially.

Usually no average pedigree of the breed is available for comparison with the one being considered. That lack would not bother the man who has studied enough pedigrees of his chosen breed to get a fairly definite idea of what a typical pedigree is, but many people simply cannot spare the time for such study or do not have access to enough pedigrees to learn what is typical for the breed.

In a sense individual selection is pedigree selection for the next generation. When the parents have been selected for individuality, the offspring which would have had the worst pedigrees are not permitted to be born! Much of the use which might be made of pedigree selection in a hitherto unselected population is simply not available to the breeder who is pursuing the same goal which was pursued by those who bred and selected the parents of his animals. This sets further limits to the usefulness of pedigree selection under ordinary practical circumstances. There are some exceptions to this general situation. One is the fact that fewer males than females are needed for breeding. After the males are born there can be much pedigree selection among them, although there cannot be much among females. This is what a cattle breeder does when he decides he will save a calf from a certain cow if it is a heifer but not if it is a bull. When a breeder's ideals differ distinctly from those of many of his fellows and he lays much importance on things which they consider unimportant and vice versa, the pedigrees of their animals offer him considerable opportunity for selection if he purchases from them. Also, when the ideal of a breed changes, there is momentarily a reappraisal of the merits of the ancestors. While that is happening there is opportunity for pedigree selection on the new basis for a generation or two. Selection among the parents does not reduce the variability among their offspring nearly as much as it does the variability among the parents themselves. Hence in a population which has already been under selection for a generation or two there is much more opportunity for selecting the offspring on their individuality than on their pedigrees.

ERRORS WHICH MAY BE CORRECTED BY PEDIGREE SELECTION

Pedigree selection helps to overcome deception by environment because it does not often happen that the ancestors were all kept under the same environment.

Pedigree study would be of some help in overcoming deception by dominance if full information about the collateral relatives were presented, but that is rarely done. If the recessive is rare and only the direct ancestors are given, pedigree is almost no help in this respect. For example, pedigrees do not help in eliminating red in the black breeds of cattle, since all ancestors are black and the existence of red sibs or aunts or uncles or other collateral relatives is not mentioned. The parents of red calves and something more than half of the sibs of red calves are heterozygotes. Culling them because of their relationship to the red animal would make the genes for red scarcer than can be done by culling the red alone, but present pedigrees do not give the information which would permit that. Nor is it likely that breeders would report such information if it were to be used to reflect unfavorably upon animals they have for sale. Pedigrees help a little in overcoming deceptions by dominance in such cases as the polled Herefords where the distinction between horned and polled is shown in the pedigree.[4]

Pedigree studies may help considerably in overcoming the deceiving effects of complex gene interaction. An unusually good animal from poor or mediocre stock on both sides always suggests that this animal is the result of a lucky combination of several genes all of which are necessary in order to permit each to manifest its desirable effects. The occurrence of such an animal is often called "nicking." In that case the animal would almost certainly be heterozygous for many of the genes whose cooperation is required. There would be small chance of its transmitting enough of those genes in each gamete to cause this excellence to reappear in many of its offspring. Likewise, a poor individual with excellent ancestors and relatives on both sides will be suspected of lacking just a few genes which are necessary for the successful operation of the other genes for the excellent qualities of his family. It will be suspected that he has many good genes which do not manifest their presence in him. If he were used for breeding, many of his mates would transmit to their offspring the necessary genes he lacks. He may be preferred to a better individual from a poorer family, although, of course, he would be discarded if there were enough good individuals in his own family.

RULES FOR THE USE OF PEDIGREES IN SELECTION

In evaluating a pedigree one should estimate what kind of inheritance the sire and the dam had and then average those two estimates.

[4] The probability that a polled calf from polled parents is homozygous for the gene for polledness is equal to $\frac{1 + s + d + sd}{3 + s + d - sd}$ where s is the probability that the sire is homozygous and d is the probability that the dam is homozygous. Probability is expressed on a scale ranging from zero for absolute impossibility to 1.0 for absolute certainty.

Wherever there is uncertainty about an individual's inheritance, conservatism requires that it should be estimated closer to the average of its breed or herd than the evidence about it, if accepted at face value, indicates it to be. Thus, if there is much evidence about the sire but only a little about the dam, the estimate should give much weight to the sire but most of the weight which would have gone to the dam should be placed on the breed or herd average. The estimate of the sire's inheritance, or of the dam's inheritance, is made similarly by averaging the inheritance of its sire or dam and then modifying that average by what its own individual merit seems to be, using the general principle that the individual's own merit should receive more weight than its pedigree under most circumstances. For characteristics which are but slightly hereditary or in cases where the ancestors' merits are much better known than the individual's merit—as, for example, where the ancestors are well progeny-tested or have long lifetime records while the individual is still immature—it may be correct to attach more weight to pedigree than to individual merit.

The more certainly an ancestor's merit is known, the less weight should be placed on its ancestors. This is difficult to place on a quantitative basis, but Czekanowski's table will give a rough idea of about what that relation should be when two ancestors in the same line are equally well known. If the breeding value of the nearer ancestor is known with much more certainty than that of the more remote ancestor, there is scarcely any use to pay attention to the more remote ancestor.

If there is reason to think that much sex-linkage is involved in a particular case, the proper procedure is to give the sire more weight in estimating the inheritance of female offspring and to give the dam more weight in estimating the inheritance of male offspring. (See page 255 for reasons.) The principle of conservatism in cases where the merit of the two parents is not equally known can also give rise to unequal weighting. Thus, in dairy pedigrees where the sire is too young to be progeny-tested but the dam's production is known, much weight can be placed on the dam, but only a little on the sire on account of the uncertainty about the sire's side of the pedigree. In another dairy pedigree where the sire is well progeny-tested but the dam is a young heifer in her first lactation, the situation may be reversed, with more weight being put on the sire's side of the pedigree, only a little on the dam, some on her pedigree, and much on the breed average for conservatism in her case.

Because the ancestors are in most cases selected animals and because the information about them is selected to show them in a favorable light, allowance should be made for regression toward the breed average. There seems to be no quantitative rule for doing that except the general one that the more extreme the selection is believed to have been among the parents or in the information presented concerning them, the more allowance should be made for regression.

SUMMARY

The accuracy of pedigree selection is limited because of the sampling nature of inheritance wherever genes are heterozygous. This makes it impossible to be exactly sure of what an individual offspring will be, even if one were in the extreme position of knowing exactly what inheritance its sire and dam had.

The accuracy of pedigree selection is further limited in practice because one is never entirely sure of the inheritance the parents had, since that, too, must be estimated from their own appearance or performance or from that of various of their relatives.

A limited amount of attention to pedigree will make the selections more accurate, but it necessarily cuts down on the intensity of the individual selection which can be practiced. Too much attention to pedigree may do more harm by decreasing the intensity of individual selection than it will do good by increasing accuracy, since pedigree selection is not perfectly accurate either.

The use of pedigrees is based on the principles: that inheritance is approximately equal from sire and from dam, and that, wherever one's knowledge of the inheritance of the relatives is uncertain, one should be conservative and estimate that the individual is somewhat nearer to the breed average than the performance of its relatives.

Rarely should the pedigree receive as much weight as the animal's own appearance or performance, although that can happen for slightly hereditary characteristics if the merit of parents and grandparents is much better known than the individual's own merit, either because those ancestors are well progeny-tested or because the individual is still too young to show its merits as unmistakably as its parents and grandparents do.

Collateral relatives constitute progeny tests of some ancestor. They therefore make the estimate of its inheritance more certain. If an animal is well progeny-tested, little will be gained by studying the ancestors back of it in the pedigree.

The proper amount of attention to pay to different relatives depends on the closeness of their relationship to the subject of the pedigree, upon how many of them there are, upon how completely the merit of each other relative is known, and upon how highly hereditary the characteristic is.

Because the ancestors are usually selected individuals, some regression toward the breed average is to be expected. Moreover, the information about the ancestors is usually selected to make them appear in the most favorable light. For that reason more regression is to be expected.

Serious defects in the practical use of pedigrees include the fact that the merit of collateral relatives such as sibs, uncles, cousins, etc., is rarely mentioned. If it is mentioned, it usually concerns only certain individuals selected to make the pedigree more attractive and, moreover, is selected information about those.

In most commercial pedigrees, other than those for dairy cattle and poultry, little information of any kind is included except the names and identifying numbers of the animal. Such a pedigree is useful only to the extent that one knows or can find from some other source how meritorious or mediocre those ancestors were.

In a breed which has been selected steadily toward the same goal for two or more generations, there often is only a little room to practice pedigree selection because the worst individuals among those which might have become ancestors were eliminated by individual selection. Therefore, the individuals which might have had the worst pedigrees, if there had been no selection, simply do not get born.

The general conclusion regarding pedigree selection is that it should usually be a minor accessory to individual selection, being permitted to sway the balance in making decisions which are fairly close on individual merit. It is most needed for characteristics which are not very hereditary, for characteristics which only one sex can manifest, and in selections which must be made while the animals are yet too young to show clearly in their own appearance or performance what their individual merit is.

The kind of errors in individual selection which are most likely to be remedied by pedigree selection are those arising from the immaturity of the individual and from mistaking differences caused by environment and epistatic interactions for differences in breeding value. It helps remedy errors caused by dominance when fairly full information about collateral relatives is included, but is not of much help in this respect when only the ancestors are described.

REFERENCES

Marshall, F. R. 1911. Breeding farm animals. pp. 93-104.

Wriedt, Christian. 1930. Heredity in livestock. (Especially the chapter on "Valuation of Pedigree," pp. 168–72. Also pp. 6–10.) New York: The Macmillan Company.

CHAPTER 15

Aids to Selection—Progeny Tests

"The quality of a ram can usually be determined from his conformation and from his get." "You may judge them by their get if their lambs are of good quality."—Varro. The Husbandry of Livestock, the first century B.C.

"One may consider the ancestors as thoroughbred only when all the progeny are thoroughbred."—Thaer, in 1806.

By progeny test is meant estimating the individual's heredity by studying its offspring. The general idea is an old one, as is indicated by Varro's comments some 2,000 years ago. As long ago as 1826 Andre recommended progeny testing as one of the main purposes for keeping herdbooks for sheep. On page 299 in the USDA Yearbook for 1894, the proving of bulls and the continued use of sires of proved excellence are urged.

The principles of the progeny test come from the sampling nature of inheritance. Each offspring receives from the parent a sample half of that parent's inheritance. Each additional offspring receives another independent sample from the same source. If one can find out what was in several such samples, he will be fairly sure of what was in the parent. A crude analogy may make the case clearer. If a barrel contained 100 apples, of which a certain number were ripe while the rest were green, and if we could not get in the barrel to count the apples but could count a sample half of them taken at random, we could tell within certain limits what proportion of the apples in the barrel were ripe. There would be a high degree of uncertainty about our estimate because this one sample might by chance have contained considerably more or less than a fair share of the ripe apples. If the sample half could be placed back in the barrel and thoroughly mixed with the others and then a fresh sample half could be counted and if that process were repeated often enough, we could finally come to the point of knowing how many apples were ripe and how many were green in the whole barrel, although it would take several such samples for us to be sure that we were within three or four of the right number.

The first practical difficulty encountered in using the progeny test is that we do not know exactly what genes the offspring have. We can be deceived by the effects of environment, dominance, and complex interactions of genes in the offspring, just as we could in estimating the parent's genes from its own appearance or performance. There is this important difference, however: There are several of the offspring and the deceiving effects of environment, dominance, and epistasis have an opportunity to cancel each other. Thus, there is a chance for us to know the average inheritance of the offspring with more certainty than we know the inheritance of the parent. The parent is only one individual and in it there has been no chance for the plus and minus errors to have been canceled by averaging.

The second practical difficulty encountered in using the progeny test is that each offspring also has received half of its inheritance from its other parent. Since we usually do not know exactly what was in that parent and will be still farther from knowing just what it contributed to this particular offspring, we are often in doubt as to whether a certain good quality in one offspring came from its sire or from its dam. One way of overcoming this difficulty partially is to use all the available information to estimate the other parent's inheritance; then allow for that according to the general rule that the phenotype of an unselected offspring will tend to average half way between the genetic values of its two parents. Let X represent the unknown genetic value of the parent being progeny-tested and Y be our estimate of the genetic value of the other parent of an offspring which has a phenotype, P. Then the most probable value of P is $\frac{X+Y}{2}$ and the equation, $X = 2P - Y$ furnishes an estimate of X, although not a very dependable one when there is only one offspring, since Mendelian sampling variations, environmental effects, etc., can have made P in a particular case deviate widely from the general rule. Also, of course, we may be rather wide of the mark in our estimate of Y. Another way of overcoming this difficulty consists of progeny-testing an animal by breeding it to a large number of different mates, in the hope that the merits and defects of those other parents would just about cancel each other. Any general difference, then, between the average of the progeny and the average of the breed could be credited to the common parent. This method might, of course, lead to large errors if the other parents were so selected that their average merit was distinctly different from the breed average.

Theoretically, a third way of not being confused by what the other parent transmitted is to use a special "tester strain" so constituted genetically that their contribution to the inheritance of the offspring will not hide the inheritance received from the parent being progeny-tested. Occasionally this can be done in actual practice. Examples are the use of red cows to test the homozygosis of bulls belonging to black breeds, or the use of horned cows to test the homozygosis of polled bulls.[1] This method has its principal use in overcoming the effects of dominance but can also be used to prevent deceit by epistatic effects if the genetics of the situation is known clearly enough that suitable tester animals can be found or produced. Some of the most brilliant research in genetics owes its success to the devising and use of such tester strains. In economic animal breeding, however, this method cannot often be used unless animals already produced for other purposes can be used for testers. The production and maintenance of a suitable tester strain would be expensive, if possible at all. The plan is not practical for females because of the limited number of young they produce and the economic impossibility of using them for several gestation periods to produce to the tester males offspring which would then have to be discarded because of the inheritance from their sire, no matter what they proved about the genotype of their dams. The simplest form of the plan could not be used to test for lethals since there are no tester animals homozygous for those, and it would be difficult to maintain a strain of farm animals heterozygous for a lethal gene. Generally the best that can be done in such a case is to breed the suspected sire to a large number of his own daughters although that requires more offspring than if he could be mated to known heterozygotes.[2] Even where tester animals are available, as in the case of red and horned cattle, it would not be often that the same animals could be used to test for more than a few genes. A sire being progeny-tested for many genes would have to be used on one set of tester animals to test him for some of his genes, on another set to test him for a few more of his genes and on still other sets to provide an adequate progeny test for many pairs of genes, simply because a single strain of animals suitable for testing all genes would not exist. From these and other considerations, it seems likely that the progeny test in actual practice will nearly always have to be made incidentally by studying the offspring produced by mating the sire to the females to which he would be mated anyhow for other reasons.

---

[1] If a black bull is mated to red cows and produces even one red calf, he is known to be heterozygous. If he produces only black calves, he is probably homozygous, the assurance of that increasing with the number of calves thus produced. The laws of chance governing the case are such that if a heterozygous black bull sires one calf from a red cow, that calf is just as likely as not to be black; but, if he sires two calves, there is only 1 chance in 4 that both will be black; there is 1 chance in 8 that three such calves will all be black, 1 chance in 16 with four calves, 1 chance in 32 with five calves, etc., the chance being $(1/2)^n$ where n is the number of calves sired. Hence, if a black bull has sired more than five or six calves from red cows and none of those calves are red, the chance that such a bull was really heterozygous is exceedingly small.

[2] Berge, S., 1934, *Nordisk Jordbrugsforskning*, pp. 97–114.

It is believed by some, especially in the corn breeding work, that the most accurate tests of the general combining power of different strains can be made by testing them on poor or weak strains rather than on medium or good ones. Theoretically, this seems likely to be true wherever nearly all the favorable genes are completely dominant and the best strains have in them nearly all of those desired genes; or wherever being "best" is largely a matter of having a considerable margin of unused merit, a factor of safety so to speak, which would not ordinarily be needed. Crosses of the various strains to tester material known to be weak in many ways might produce first crosses low enough in average merit to show the differences in reserves or factors of safety among the strains being tested. Such differences would not be detected if the strains being compared were crossed on varieties which were themselves good enough that nearly all the first crosses would be above the threshold below which these extra margins of safety are needed. The practical importance of using poorer-than-average stock for tester strains is uncertain. It is not likely to be as useful with animals as with plants because of the high cost of maintaining tester strains of animals. Attempts to collect tester stocks by selecting poor individuals wherever they are found would encounter practical difficulties in that the supposedly poorer individuals would not be as poor genetically as they are phenotypically. It would be difficult to discount for that correctly.

A third practical difficulty in using the progeny test is that the offspring of a given individual are apt to have been born at somewhere near the same date and to have been reared under much the same environmental conditions. If there was anything unusual about that environment and if proper allowance for that was not made, we will credit or blame the heredity of the parent for something which was really caused by the environment of the offspring. This is probably the most important general limitation on the accuracy of the progeny test, and there seems to be no automatic way of overcoming it. One can merely study as closely as possible the environment under which those offspring were tested and make such allowance as he thinks fairest for any conditions which were not standard.

The principles involved in the progeny test of a sire are illustrated in the equations below. In these equations, "first error of appraisal" includes all mistakes made in correcting the records of "first dam" and of "first offspring" to standard environmental conditions and all mistakes made in allowing for the effects of dominance and epistasis on their records. These equations show how it is that the average error from Mendelian sampling and the average error from mistakes of appraisal become smaller as the number of offspring in the progeny

$$\text{1st offspring} = \frac{\text{sire} + \text{first dam}}{2} \pm \text{1st Mendelian error} \pm \text{1st error of appraisal}$$

$$\text{2nd offspring} = \frac{\text{sire} + \text{2nd dam}}{2} \pm \text{2nd Mendelian error} \pm \text{2nd error of appraisal}$$

. . . .

. . . .

$$N\text{th offspring} = \frac{\text{sire} + n\text{th dam}}{2} \pm n\text{th Mendelian error} \pm n\text{th error of appraisal}$$

$$\text{Average offspring} = \frac{\text{sire}}{2} + \frac{\Sigma\ \text{dams}}{2n} \pm \frac{\Sigma\ \text{Mendelian errors}}{n} \pm \frac{\Sigma\ \text{errors of appraisal}}{n}$$

$$\text{Sire} = 2 \times (\text{average offspring}) - \text{average dam} \pm 2 \times (\text{average Mendelian error}) \pm 2 \times (\text{average error of appraisal})$$

test increases. They also show why it is important that these individual errors shall be as likely to be plus as they are to be minus. If that is the case, the plus and the minus errors will tend to cancel each other so that their average approaches zero as the number of offspring becomes large. The errors from Mendelian sampling are certain to be thus unbiased, but that may be far from true of the errors in appraising. Because the offspring are apt all to have been reared under similar environment, any peculiarity about that environment for which we have not corrected perfectly is likely to have made most of the uncorrected environmental errors plus or most of them minus. In such a case, the average environmental error does not tend toward zero but toward a figure determined by the kind and size of the systematic error made in standardizing the records. Since such an error is doubled in getting the estimate of the sire and does not tend toward zero as more offspring from the same herd are tested, this is probably the major source of error in most progeny tests—at least if there are enough offspring (more than four or five) to make the errors from Mendelian sampling small. Errors in appraising the effects of environment on the dams can also fail to be random, but this is not so likely to be extreme as in the case of the offspring, since the dams are less likely to have all been kept and tested under the same environment, especially if lifetime records are used for them. Errors in allowing properly for the effects of dominance and epistasis are apt to be more important in the case of the dams than they are in the case of the offspring because the offspring are usually unselected or nearly so, while the dams are often somewhat selected. If dams were selected, considerable regression toward the average of the breed should be allowed in estimating their real breeding value from their records.

It is important that the offspring be an unselected sample. Otherwise the progeny test is biased by the selection practiced in choosing which progeny to include. This bias is difficult to measure and discount. To omit the poor offspring is unfair and misleading.

The result of the progeny test contains a term for the average merit of the dams. If the dams were known to be typical of the breed, that term will be the breed average and, since it will be the same for all sires tested, will not affect the comparison of two sires. If the dams can be assumed to have been a random sample of the breed, only a little error is introduced by ignoring the dams in comparing two sires. If the dams were not a typical or random sample of the breed, then neglect of this term for the average merit of the dams can lead to serious error.

Increasing the number of offspring in the progeny test reduces the errors of Mendelian sampling and the random errors in accordance with the law of diminishing returns. That is, each additional offspring reduces these errors but makes less reduction than the preceding one did, so that further reductions in those errors are slight after the merits of the first few offspring are known. Increasing the number of offspring in the progeny test does *not* reduce the errors from *systematic* mistakes in correcting for environment, dominance, or epistasis. If such systematic errors are large in the population to be studied, little gain in net accuracy is to be had by increasing the number of offspring much past three or four.[3] Additional offspring do contribute some information, but this is so little, compared with the large systematic errors remaining, that efforts to increase still further the number of offspring in the progeny test without first correcting for the systematic errors may be described by the Biblical allusion about straining at a gnat and swallowing a camel.

Requiring too many offspring for proving a sire can actually lower the rate of progress by causing a smaller number of sires to be proved and therefore preventing intense selection among them after they are proved. As a numerical example, if 100 heifers born each year can be used for proving sires, we can prove five sires per year if we require 20 daughters each, ten sires with 10 daughters each or 20 sires with five daughters each. If we must keep for future use the five best among the sires proved each year, we would in the first case have to save them all, regardless of their proof. Nothing would have been accomplished by the proving. In the second case we would need to keep the half which had the best proof. In the third case only the best fourth need be saved. Progress would be fastest in the third case, although the "proof" would be less accurate. On the average the extra errors in the proof would be more than offset by the opportunity to cull more intensely. For maximum progress the optimum number to prove each sire would be about three daughters if selections were entirely on progeny test and if one could neglect the extra costs of maintaining the larger number of bulls

[3] Lush, Jay L., 1931, The number of daughters necessary to prove a sire, *Jour. of Dy. Sci.*, 14:209–20.

while waiting for their proof. The existence of such costs makes the practical optimum number higher, the amount of that shift depending on the costs as well as on how extensively each bull is to be used after he is proved. The essential point here is that it is no use to prove a sire unless something is done on account of that proof—something which would not be done otherwise. Making the proof highly accurate can actually retard progress if it is done at the expense of proving fewer sires, as *must* be the case if the number of daughters required per sire is increased in a cow population of fixed size.

Diallel crossing is a method of progeny-testing introduced into genetic literature by J. Schmidt[4] as a means to avoid setting any value on the dams and to avoid correcting for the environmental circumstances. The breeding values of two males are compared by breeding them at different times to the same females and then comparing the average merit of the two sets of progeny. By referring to the equations used on page 198 to illustrate the principles of the progeny test, it will be seen that if the equation for one sire is subtracted from the equation for another sire, the difference between the real breeding values of the sires will be *twice* the difference between their progeny averages plus or minus terms for differences in the dams, the environments, and the Mendelian sampling errors. Now if the dams are the very same individuals, as they are in diallel crossing, the term for the difference between dams will disappear. If the tests are carried out under the same environmental conditions as, for example, when the two sires are used contemporaneously, half of the females being bred to each sire the first time and then each female being bred to the other sire the next time, then the difference in the environmental conditions approaches zero also. The error from Mendelian sampling can be made small by having a large number of offspring. This leads to the very simple rule that, if this plan can be followed, the difference in the breeding value of two males is simply twice the difference in the average merit of their offspring. The plan has practical difficulties in such cases as dairy cattle, where one can measure production only in daughters and cannot be sure of getting from each cow a daughter by each sire, but it seems to have advantages enough to deserve wider use than it has yet received. It can be extended with more difficulty to the simultaneous comparison of more than two sires. That is being done experimentally with swine[5] but how it will work in practice is not yet certain. The case of four boars requires eight groups of sows which are indicated by letters A to H in the following mating plan:

---

[4] 1919. Compt. Rend. Lab. Carlsberg, 14, No. 6.

[5] Kudrjawzew, P. N., 1934, Polyallele Kreuzung als Prüfungsmethode für die Leistungsfähigkeit von Zuchtebern. Züchtungskunde, 9 (No. 12) :444–52.

MATING PLAN IN POLYALLEL CROSSING

	Boar Number			
	1	2	3	4
First season......	A and B	C and D	E and F	G and H
Second season....	H and C	B and E	D and G	F and A

The diallel comparison of boars 1 and 4 is provided by the progeny of sows in groups A and H. The comparison between boars 1 and 2 is provided by the progeny of the sows in groups B and C, etc. There is no direct comparison between boars 1 and 3, but each of these boars is compared directly with boar 2 and also with boar 4. Two indirect comparisons of 1 and 3 are possible from that. The "ring" arrangement provides that, if for any reason the comparison between two boars is not completed, there is still a "chain" arrangement by which any boar may be compared with any other unless there is a second failure in some other comparison.

In an unselected population where individual merit is equally well known for all animals, one offspring is as reliable as a parent in indicating what an animal's inheritance is. But an individual can have only two parents, while there is no such biological limit on the number of offspring it can produce! Because of that, the progeny test can surpass the pedigree as an accurate guide to an individual's inheritance, provided the offspring are half-sibs, and can equal it if they are all full sibs. In an unselected population with the outward merit of each animal equally well known and no environmental correlations between relatives, a progeny test based on three offspring where the other parents can be assumed to have been a random sample of their breed, is equal in accuracy to all that could be learned by studying the animal's pedigree. Two offspring are enough to make the progeny test attain the same level of accuracy if the merits of the other parents are known and discounted. These simple conditions never prevail exactly. Usually the individual merits of the offspring are not as certainly known as the individual merits of the ancestors because the latter are older and have had a longer time in which to manifest their qualities. The offspring are more apt to have been reared and tested under the same environmental conditions which, not being perfectly discounted, are apt to make the progeny test less reliable than under the simple conditions. A circumstance which operates in the opposite direction is that the ancestors were usually selected to some extent, whereas the offspring usually are almost or quite unselected. An extreme example of this effect of selection among the ancestors concerns the occurrence of red in black breeds of cattle. Since all red animals are culled from the pure breed, the pedigree would not be of any help in locating the heterozygous animals (unless perhaps it were one of those few pedigrees which contained information about some red collateral relatives), but the production of even one red offspring would be positive proof of heterozygosis.

These usual exceptions to the simplest theoretical conditions leave the question of when progeny test generally becomes more accurate than pedigree somewhat uncertain in actual practice. Perhaps it is just as well to think of pedigree and progeny test as about equal when there are two or three offspring, although that does not do justice to the usefulness of the progeny test where the ancestors have been highly selected. On the other hand, such a rule does not do justice to the usefulness of pedigree where the offspring have all been raised under some environmental peculiarities about which we do not know or for which our corrections have not been entirely unbiased.

The point at which the progeny test becomes more accurate than the individual merit of the animal itself is of special interest. A numerical solution has been given[6] for the case of a hitherto unselected population where the individual merit of the parent and of each of its offspring are equally well known. In that case at least five offspring are necessary for slightly hereditary characteristics where there is no correlation between the offspring for any other reason than that they are related through this one parent. If the characteristic is highly hereditary, more than five will be necessary. If there is a correlation between the offspring for other reasons, more offspring will be necessary; and if that correlation is as high as + .25, and if correction for it is not made, it will be impossible for the progeny test to become as accurate an indicator of the parent's heredity as the parent's own merit is. In actual practice the parent's individual merit will usually be known better than that of the offspring; but on the other hand, the parents will usually be to some extent the survivors of previous cullings for individual merit and for pedigree. If there has been much selection of that kind, the practical situation is that the possibilities for culling by pedigree or by individuality have been partially exhausted before the progeny test becomes available. The progeny test is thus a fresh opportunity to cull from an entirely new direction. Hence, in a population of selected parents, the progeny test is relatively more useful than the above figures indicate.

Progeny tests are most useful for traits which can be expressed

---

[6] Lush, Jay L., 1935, Progeny test and individual performance as indicators of an animal's breeding value, *Jour. of Dy. Sci.*, 18:1–19.

in only one sex (e.g., milk production, egg production, prolificacy, etc.). In such cases study of the individual animal of the sex which cannot express the trait does little if any good. Selection in that sex is limited to the basis of pedigree and progeny tests, although individual merit is also available as a basis for selection in the other sex.

The fundamental genetic effect of progeny tests is that they permit selections to be more accurate and hence more effective because they prevent the breeder from being deceived as much by the effects of environment, dominance, and epistasis as he might othewise be. They do not alter any genetic process.

All the initial selections must be made while the animals are still young and many of the final selections before they are old enough to have any progeny of known merit. Therefore, the progeny test is applied only to animals which have already met minimum standards of pedigree and individuality. Some loss is incurred by culling without a progeny test some which would have proven to be better genetically than their pedigree or individuality indicated, but it is utterly impossible to test the progeny of them all. The ideal is to select on pedigree and individuality to some extent but not so much that all one's freedom to select will have been exhausted before any of the offspring can be examined. Within limits dictated largely by costs and convenience but partly by the accuracy of selecting on different bases, the breeder should strive to sample enough of those with good pedigrees and individuality that he can still do considerable culling when the results become apparent.

Breeders have always made general but somewhat unsystematic attempts to use the progeny test. They have done this by seeking sires and dams or the sons of sires and dams which had produced unusually good offspring. "Get of sire" and "produce of dam" classes have been included in shows in nearly all countries for many years. In recent years a pronounced interest in progeny testing has developed, especially in dairy cattle and in poultry. In meat animals the widest systematic use of the progeny test has been the Danish system of testing litters of swine[7] which has also been adapted and used widely in Sweden, Canada, The Netherlands, and Germany.

In a certain sense all sires and many dams become progeny-tested eventually, but usually they are dead by that time[8] and the information can be used only in pedigree estimates.

[7] Lush, Jay L., 1936, Genetic aspects of the Danish system of progeny-testing swine, Iowa Agr. Exp. Sta., Res. Bul. 204.

[8] In a survey of the ages of 35,000 dairy bulls in Michigan, it was found that 94 per cent were slaughtered before they reached three years, although a bull cannot be proved before he is five years old. Michigan Quart. Bul. 15 (No. 3):143. See also Iowa Sta. Bul. 290, *The ages of breeding cattle and the possibilities of using proven sires.*

Where there is sex-linkage, offspring show more about the inheritance of the opposite-sexed parent than they do about the inheritance of the parent of their own sex. This is not often important, but some allowance of that kind can be made where sex-linkage is suspected.

SUMMARY

1. "Progeny test" is a general term for estimating the breeding value of an animal by studying the characteristics of its offspring.

2. The things which may keep the progeny test from being perfectly accurate are: (1) The sampling nature of inheritance makes it possible for the parent to transmit to its offspring inheritance which is better or is worse than is typical of it; (2) the offspring receives half of its inheritance from its other parent, and the breeding value of that other parent may be distinctly different from the average of the breed; (3) environmental effects, dominance deviations, and epistatic deviations may deceive us in our estimate of the real merit of the offspring or of the other parent.

3. Errors coming into the progeny test because of the sampling nature of inheritance may be made as small as we please by getting a sufficiently large number of offspring. Where there are five or more offspring the errors from this source are usually small in comparison with errors from the other sources.

4. Where the other parents can be considered a random sample of the breed or are known to be typical of the breed, errors in the progeny test from neglecting the merit of those other parents are zero or approach zero rapidly as the number of offspring increases. Where the breeding merit of the other parents is distinctly above or below the breed average, allowance for that can best be made by estimating the breeding value of the parent being tested as equal to twice the average merit of its offspring minus the average merit of the other parents of those offspring, with some allowance for regression toward the breed average or herd average.

5. Errors coming into the progeny test through not making fair allowance for environmental conditions which were not standard for the offspring are usually the most serious limitation on the accuracy of the progeny test. Random errors of this kind are reduced rapidly by increasing the number of offspring and thus allowing the plus and the minus errors to cancel each other. Systematic errors are not thus reduced and are likely to be important. The only remedy for this is the general one of studying closely the conditions under which the offspring were reared and tested and making the best allowance one can for those. When errors of this kind are important, not much information is

gained by increasing the number of offspring past three or four.

6. The progeny test can become more accurate than a pedigree estimate in a population as a whole, when there are more than three offspring, but this depends on whether the individual merits of the offspring are as certainly known as the individual merits of the ancestors, on how much environment the offspring have had in common and on how much the variation among the ancestors has already been reduced by selection among them.

7. In a hitherto unselected population, individual merit is usually more dependable than progeny test unless: (1) there are at least 5 offspring, (2) there is no environmental correlation between the offspring, (3) the characteristic is not very highly hereditary, and (4) the individual merits of the offspring are known at least as accurately as the merit of the parent being tested. In actual practice, (2) and (4) are not usually fulfilled, but they may be more than offset by the fact that selection on the basis of pedigree and individuality may already have exhausted much of the possibilities in such selection, while the progeny test comes as a fresh source of evidence from a new direction.

8. Progeny tests are most useful for characteristics which only one sex can express or which, like many carcass characteristics, cannot be measured until the animal is dead.

9. Progeny tests are next most needed for characteristics which are only slightly hereditary and for which individual selection is therefore not very accurate.

10. The chief practical limitation on progeny tests is that they come so late in the animal's life that most of the decisions about culling or using an animal for breeding must already have been made. Therefore, progeny tests have their widest use in making pedigree selections more accurate by pointing to the sires and dams whose offspring are most likely to be worth saving for breeding.

CHAPTER 16

Selective Registration

In nearly all breeds of livestock in the United States every animal which has a registered sire and a registered dam is eligible to registry unless it has one of the few disqualifications which exist in some breeds. Of course, many eligible animals do not get registered; but the decision to register or not to register is left entirely to the breeder. It is occasionally proposed that each animal ought to be inspected and approved by authorized judges or ought to comply with certain minimum standards of production, where those are appropriate, before it is finally accepted for registration. This is selective registration.

This would be a special application of mass selection. No new principles of inheritance are involved. The diagrams in Figure 28 show how various proportions of the population might be affected by such a plan. Diagram *A* shows the impossibly extreme case where the official inspection could be perfectly accurate in eliminating the animals with the lowest breeding values. Diagram *B* shows, with areas of the same size, how the real breeding values might very well be distributed in the case where individual merit is only moderately correlated with breeding value. Some of those which would be eliminated by selective registration would be superior to some of those which would be accepted. That is inevitable wherever outward individual merit is not perfectly correlated with breeding value. The proportions of the different areas in these diagrams are not particularly important. Those vary from breed to breed and from time to time and are different for males and females.[1]

Both diagrams indicate that the animals not now registered include many which are of higher breeding value than some of those which are registered. It is probable that the average breeding worth of those registered is higher than the average breeding worth of the eligible ones which do not get registered, although the difference may not be as extreme as is pictured here. Selective registration would increase the intensity of selection for those genes which would usually make their possessors appear more desirable to the inspectors. This would increase

[1] In the cattle breeds the number of males registered in recent years has generally been about one-half to one-third as large as the number of females registered.

slightly the rate of improvement in the breed, as far as mass selection can do that and as far as the ideals set up to guide the inspectors are in agreement with real merit. Whether the amount thus gained would be sufficient to make selective registration worth the money and effort it would cost is open to some doubt.

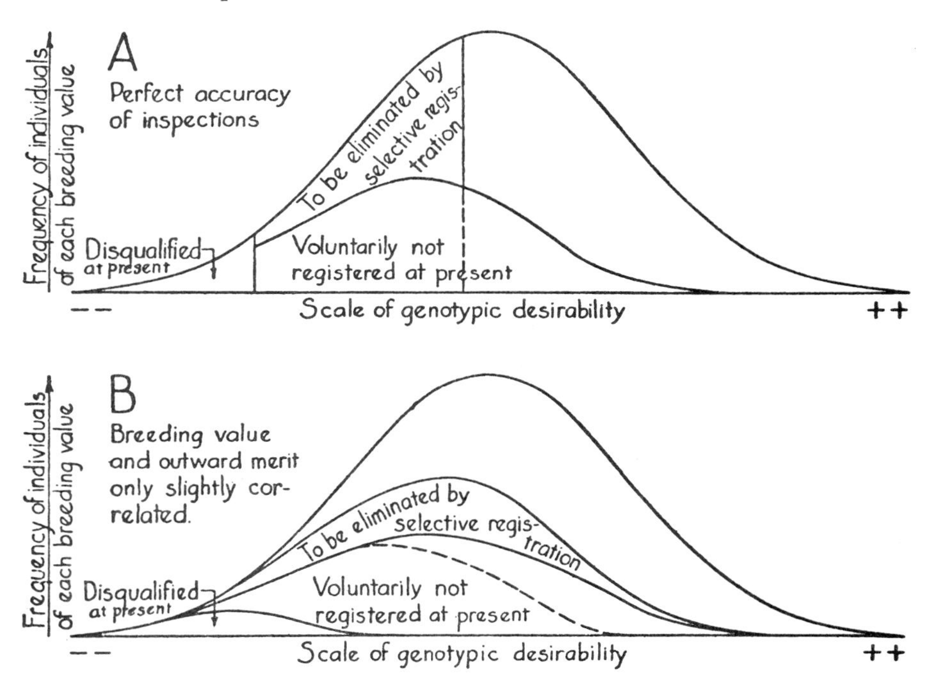

FIG. 28. How the distribution of a population might be affected by selective registration. Corresponding areas are the same size in the two diagrams. The areas in each diagram are exclusive of each other. *A* presents the impossibly extreme case where the inspections could be perfectly accurate. *B* presents a more probable situation where many mistakes would be made by the inspectors although they would be right oftener than they would be wrong. If many were eliminated by the inspectors, it might be necessary to register some of those voluntarily not registered at present. Those would come from the areas just to the right of the dotted lines.

Selective registration would have some tendency to unify the standards of selection within each breed, since each breeder would have to be guided in part by the ideals of the inspectors. Those might not always be superior to his own. This control over standards of selection might do actual harm by preventing individual breeders from trying out their own ideas as freely as they would like. To understand that, one has only to imagine what would have happened if there had been an official system of inspection in the Shorthorn breed when the Cruickshanks were founding the "Scotch" type or in the Poland-China breed when Peter Mouw was forming the "big type." The help which the inspectors would give many breeders in their selections might more than offset this. Moreover, the inspectors would probably be lenient enough to allow each breeder considerable leeway for his own standards. It remains an open question whether the average progress of the breed will be faster in the long run when each breeder is entirely free to work out his own ideas than when his standards of selection, even within rather broad limits, are partly forced on him by a majority of his fellow breeders working through selective registration administered by the breed association.

Selective registration requiring visual inspection of each animal would be costly in the United States because of the great distances over which most breeds are scattered. There would be some expense connected with maintaining the records of the inspections and administering the work, although the overhead costs of administration might be only a small addition to the regular costs of operating the secretary's office. There would probably need to be provision for appeal where the breeder was not satisfied with the inspection given his animals. From time to time ill feeling would arise when some breeder felt that the inspector was being deliberately unfair to him. It might be necessary to make provision for registering the offspring of rejected animals if those offspring were so meritorious as to prove the inspectors wrong.

Breeders of grade animals might be willing to pay increased prices for animals which were certified as to individual excellence as well as to purity of breeding. This might go far enough to cause some of the public to infer that all registered animals of the breed were of about the same merit. Anything which would cause the potential customer not to study the animals before he makes his selection, or which would tend to minimize attention to the breeder's reputation, has its possibilities of doing harm. It seems unlikely that selective registration would be carried far enough to make this danger important.

One commercial aspect of the situation is that some owners make more strenuous efforts than others to sell males to other breeders of purebreds. A certain amount of overhead in showing, advertising, and sales efforts is required to sell any large fraction of the males produced. In some cases those who make such efforts try hard to sell nearly all the males they produce, while others make little effort to sell males except to breeders of grades. The herds concerned usually do not differ that much in real breeding merit. The adoption of selective registration might restrict the efforts of those who are most active in selling males more than it would those who are not now active at that. Doubtless other unforeseen effects and complications will develop when selective registration is put into practice. Any association attempting it will need a trial period of several years to gain the experience necessary for making the plan operate successfully when applied to the whole breed.

From 1942 to 1944, the American Jersey Cattle Club admitted bulls to registry only if their paternal sisters or their dams had met certain production requirements or if the sire was a "star" bull. The sire becomes a star bull by his ancestors having met certain production requirements or if at least ten of his paternal sisters and his dam have been classified officially as to type and have scored high enough. There are four grades of starring, the four-star bulls having the most promising pedigrees, three-star bulls next, etc. In addition there are the unstarred bulls which have not enough production or type ratings among their near relatives to earn them even one star. Since 1944 the compulsory features involved in selective registration have been dropped but the starring plan is retained. Since 1944 the Ayrshire association lists progeny-tested sires in three classes of apparent merit but what the breeder does with this listing is optional with him.

SELECTIVE REGISTRATION IN OTHER COUNTRIES

Selective registration is practiced in many of the continental European countries. The rules for operating the plan vary but in general are somewhat as follows. The inspectors work under the direction of the farmers' co-operative societies or of the breed association. In either case there is usually some financial support from the government, accompanied by a small amount of governmental supervision to see that such money is spent in accordance with the laws appropriating it. The breed association maintains tentative registry books and permanent registry books. Soon after an animal is born it is entered in the tentative registry and at an appropriate time an inspector certifies to this animal's individual merit. It may then be placed in the permanent registry, or in some cases it may remain in the tentative registry as long as it lives and be placed in the permanent registry only after its death. Naturally the animals are not inspected until they are grown or at least well along toward maturity so that the inspector can be fairly certain of their mature individuality. Tentative registry is therefore necessary to keep the records straight. In the case of dairy cattle it is usually required that females shall produce a minimum quantity of milk or fat before they are finally placed in the permanent register and that males to be eligible for the register must be out of cows which have met production requirements higher than the minimum requirement which would entitle cows to entry. Sometimes the inspection by the committee takes place at the local fairs and sometimes the inspectors go from farm to farm systematically or upon request.

The countries where these selective registration systems are in operation are mostly countries where the breeds are native, and therefore the ancestry of the majority of the commercial stock differs only slightly from that of the pedigree stock. This makes more need for setting off, by some special requirement or inspections, those animals good enough to furnish future breeding stock than has been the case in the United States, where most of the pure breeds have been imported and at first differed greatly from the stock in the communities where they were introduced. In most of the countries where selective registration is in effect, the formal organization of the breed association occurred at a comparatively recent date. This contributes something to the feeling that there is no great gulf between the registered and the unregistered animals. Usually provision is made for admitting to registry high grades which are outstanding in their individual merit.

The following special items about selective registration in some other countries may suggest the special conditions encountered and the variety of ways adopted to meet them. They are by no means a complete account of the registration systems in use. The details of those are changed from time to time.[2]

GERMANY. German animal breeding practices and customs of registration were, of course, thrown into confusion by the war of 1914-18. The cow-testing associations and similar organizations were again operating on an extensive scale by 1924 or soon afterward. Since 1930 only cattle from herds which have been in cow-testing association work have been eligible to be shown at the national exposition. Since 1934 and especially since 1936 the compulsory features of cow testing have been greatly extended. In Hanover about 18 per cent of the cows were tested voluntarily in 1934, but under compulsory testing this rose to 83 per cent by the beginning of 1937. The German animal breeding law of 1936 brought about far-reaching control and compulsory inspection (Körung) of all classes of breeding stock. In many cases only one or two breeds could be kept in each district. This extended to sires used only in private herds, as well as to sires which were offered for public service. This system was not in operation long enough before the next war to demonstrate what effects it actually would have.

HOLLAND. The Netherlands Herdbook Association[3] at the Hague maintains four kinds of herdbooks: the calfbook, where service certificates and birth certificates are filed in order to keep the records of ancestry straight; the herdbook proper, to which animals are admitted on

[2] This is written in May, 1945.

[3] The Friesian Cattle Herdbook Association (with headquarters at Leeuwarden) has a slightly different procedure.

mature inspection; the advanced register for recording production; and the register of proved sires. The filing of service certificates soon after breeding is compulsory. The filing of the birth certificate must take place within a few weeks after the birth of the calf. Females are inspected and scored for entry into the herdbook proper soon after they drop their first calf. A score of 75 points out of a possible 100 is necessary for registration. Recording milk and fat production is not compulsory, but bulls are not accepted for registry unless they are from tested dams. Bulls are inspected and scored soon after they are 12 months old, before they go into service, but must be scored again after they are 24 months old. If a bull is accepted as a yearling but is rejected at the two-year-old scoring, the owner is allowed 30 days as a reasonable time in which to get another bull. All services made by the rejected bull more than thirty days after his rejection are treated as though the bull were a grade, but services made prior to that are accepted for purposes of registration. The inspection is so severe that about 55 per cent of all bulls offered for inspection are rejected. There is, of course, no way of judging how much selection the breeders have already practiced in deciding which bulls shall be offered for inspection.

SWITZERLAND. In Switzerland the physical conditions, especially the summer use of mountain pastures far from the home villages and the small size of many of the herds, compel more co-operation among breeders than is usual in the United States. Bulls must be shown at the large regional fairs for classification if the purchasers are to be eligible to receive the payments which the government provides for stimulating the use of improved sires. Such payments and other considerations are important enough that almost all bulls get shown at these large regional shows. Frequently as many as 1,300 Brown Swiss bulls are shown at one time at the largest fair. The cows are offered for inspection at the local or village fairs. Cows are scored during their first or second lactations, and those scores are printed along with their pedigrees in the printed herdbooks. The advanced registry records not only how much milk and fat the cow gave, but also how many hours she worked at draft purposes during the year. A considerable fraction of the Swiss cattle are used for milk, work, and beef, thus being in a sense triple-purpose cattle. Special stress is laid on longevity and regularity of breeding. Each cow which produces at least six living calves in the space of eight consecutive years is given a special distinguishing mark in the printed herdbook.

DENMARK. In Denmark there is widespread co-operation among farmers for many purposes, and the co-operative societies conduct the pedigree registration and other means of livestock improvement. Breed registry societies, such as those in the United States and Britain, do not exist. A committee of the union of co-operative societies directs the registration policies as the board of directors of a breed association would do in the United States. Of course, most members of this committee are breeders themselves or have in some other way made themselves thoroughly familiar with the practical problems which breeders encounter. Hence, the conduct of registration is not very different from that in the United States; but the committee members are responsible to the farmer co-operative societies, that is, to the customers who will use the improved breeding stock being produced. It has been agreed that 500 cows registered per year is a large enough number to supply the real needs of the breeders and farmers. The animal husbandry *consulents*[4] nominate something like twice that number of cows. Then the central committees consider the available data concerning these and discard those which seem least worthy until the number is reduced to 500. Cows rejected one year can be nominated again and perhaps accepted in a later year. No cow is considered until she has been in milk at least three consecutive years. No matter what her age when nominated, her actual uncorrected fat production must have averaged more than 400 pounds per year ever since she first freshened. The national committee in charge of swine breeding supervises the registration of swine. Registrations are accepted only for animals bred at the state-recognized swine breeding centers. Those are farms where the breeders have complied with certain regulations, including sending each year to the testing stations half as many test litters as they have sows in their herds.[5] At the testing stations these test litters (of four pigs each) are fed under standard conditions, and the rates and economy of gain are recorded. When each pig reaches 200 pounds, it is slaughtered at a nearby bacon factory and its dressing percentage and the type, conformation, and quality of its carcass are measured and scored. Also, each breeding center is inspected twice each year by a committee which scores it for: (1) management and general appearance of the farm, (2) conformation of the breeding animals, (3) fertility of the breeding animals, (4) efficiency in the use of feed by the test pigs from this center, and (5) slaughter quality of the test pigs from this center. The considerable costs of the swine improvement plans are mainly borne by the co-operative bacon factories. The plans are thus directed and financed indirectly by the customers who buy the improved breeding stock and not by the breeders who produce it. The government makes a small financial contribution

[4] Employees of the co-operative societies who provide technical advice on livestock matters, somewhat as county agricultural agents do in the United States, and also act to some extent as the business agents for their co-operatives.

[5] Lush, Jay L. 1936. Genetic aspects of the Danish system of progeny-testing swine. Iowa Agr. Exp. Sta., Res. Bul. 204.

to part of the work and extends legal authority where needed to the co-operative societies or to the co-operative bacon factories but does not itself take an active part in directing the work. Exhibiting at the fairs is not compulsory, but a large part of the breeders do that. The classification made at the fairs is entered in the herdbooks or other records pertaining to those animals, so that most pedigrees have in them considerable information about the type classification of the ancestors, especially of the sires. The central committees specify in detail the kind of records which each stockman, who aspires to be called a breeder, must keep. The committees or their representatives inspect those private herdbooks at certain intervals for accuracy and completeness. In some cases the owner is not permitted to make the entries himself but must keep the notes and papers concerning each event until the *consulent* comes on one of his regular visits, verifies them as well as he can, and makes the proper entries. This supervision and uniformity give the private herdbooks a semiofficial standing and make it unnecessary for the central committee to keep birth certificates or other records of that kind on animals not yet admitted to the national herdbooks.

These details will show something about how widely the pattern of selective registration may vary from country to country according to circumstances. If selective registration is adopted in the United States, still other devices will be needed to adapt it to the local needs and conditions.

AMERICAN APPROACHES TO SELECTIVE REGISTRATION

The Standardbred horse is an American breed which takes its name from the fact that in its early history animals must have been able to trot a mile in 2 minutes and 30 seconds or pace a mile in 2 minutes and 25 seconds before they could be registered. The Brahman Cattle Breeders Association (founded in 1924 and operating mostly in the coastal regions of Texas and Louisiana) requires inspection of animals before they are accepted for registry. The scores and other descriptive comments concerning each animal are filed with its registration in the secretary's office. No herdbook has yet been published. When the Jersey breeders established their Register of Merit in 1903, many of the cows were measured and scored as to type by some member of a group of about 19 authorized judges. These scores were not used, however, to determine eligibility for registry. They were intended to furnish evidence about the relation between type and production. Perhaps also some of the breeders then feared that unfortunate results would follow if the Register of Merit were allowed to operate with no attention at all to type.

Disqualifications. Many breeds have certain absolute disqualifications which are a bar to registry, even though the breeding is of undoubted purity. Thus, an Aberdeen-Angus which is red is not now eligible to registry; although red cows were eligible until about 1915. Red Aberdeen-Angus steers may still be registered to compete for prize money at the fairs. Neither is a red and white Holstein-Friesian nor one with black completely encircling the hoof-head eligible to registry. Most of these absolute disqualifications concern deviations from the peculiar type of that breed, which usually means details of conformation or color not directly affecting their usefulness as producers of meat, milk, eggs, wool, or power. No association will knowingly register an animal, either male or female, which is known to be a nonbreeder, except in the case of fat animals and geldings to be exhibited at shows. In none of the cattle breeds can a heifer born twin with a bull be registered until proof is furnished that she is fertile. This restriction is based upon the common experience that most such heifers ("free-martins") will be barren. Years ago some of the breeds had rules against registering offspring from dams which had formerly produced crossbred young to the service of grade sires or sires of other breeds. Only a few breeders actually professed a belief that this would affect the breeding value of subsequent purebred offspring ("telegony"), but the rule was placed in the by-laws of many associations on the long chance that there might be something to it. In 1924 it was removed from the by-laws of the last large association (a sheep one) which had had this rule. Breeders of dogs are reputed to believe more in telegony than breeders of other livestock.

Age Restrictions. Many American breed associations levy an extra charge on registrations not made until after the animal passes a certain minimum age. In some breeds the animals cannot be registered at all after they pass a certain age. One reason for this is to complete the registration before the breeder loses or forgets some of the facts in the case. The opportunity for fraud is also diminished by having the registration made promptly. More animals are registered than if registration could be postponed indefinitely without penalty. One unfortunate result is that some animals are registered young which would not be registered if the breeder could wait until he could see them as mature animals. A plan which promises to combine the advantages without the disadvantages of early registration is "deferred registration" such as that in effect with the Guernsey, Jersey, and Ayrshire breeds. Under this plan the breeder files a birth certificate soon after the calf is born. The fee for this is small. This birth certificate entitles the breeder to defer final registration until the animal is mature, without incurring any penalty for lateness.

Voluntary Plans. The voluntary Herd Classification plans of the dairy cattle registry associations in the United States embody some selective registration in the fact that the registration certificates must be surrendered on animals classified as poor and that bulls out of cows classified as fair are not eligible to registry. The Herd Classification plan also records the animals in several different classes of individual merit, so that one who does not know the animals may learn what the classification committee thought of their individuality. The type classification is printed, after the name and number of each classified animal, in the various printed herdbooks. This puts more meaning in pedigrees in which these classified animals appear.

A feature of selective registration in the Herd Improvement Registry of dairy cattle is that the breeder who surrenders the registration certificates for his poorest producers is entitled to have those omitted from the published average for the production of his herd. The extent to which this is actually being done may be seen from the fact that, in the first 13 volumes of the Holstein-Friesian *Herd Improvement Registry,* 14,733 certificates of registry were surrendered voluntarily. This was 10.3 per cent of the total number of cows on test. Of course, some of these surrendered certificates were for cows already dead or barren. Even so, the figures indicate some selective registration against the least productive animals even after the expense of registration has been met.

The Ayrshire Association registers bull calves under six months old for half price if the sire and dam, or the dam and paternal grandam, or all four grandparents are in the Ayrshire Advanced Registry.

Such individual registration of poultry as exists is mostly connected with records of performance in such a way that it is selective.

Abandoned Plans. In 1889-91 the Holstein-Friesian Association paid bounties of $5 each for the castration or vealing of bull calves which were eligible to registry. This was a systematic attempt to induce the breeders to discard their ordinary and inferior male calves. The plan was discontinued because nearly as many bull calves were registered while the plan was in effect as before, and it was a heavy drain on the association treasury, more than $20,000 being thus expended in the three years it was in effect.

The Hereford association in 1895 and 1896 had a rule that 10 per cent of all applications for registry of bulls from each herd should be rejected, but the Secretary had no guide as to which ones those should be. The ruling was repealed in 1897.

SUMMARY

Selective registration is a form of official mass selection which consists in preventing individuals deemed inferior from leaving any registered descendants. Such selection cannot do any more than the individual breeder *could;* it may do more than he *would* if left entirely to his own initiative. Collective wisdom, as imparted by the inspector, might help the less well informed and might be of considerable help to the customer. Probably it could not rise to the level of the ablest individual breeder's ability.

Selective registration is widely practiced in many foreign countries, but some of their conditions are distinctly different from those in the United States.

Some of the American associations are experimenting with devices embodying some voluntary principles of selective registration. Doubtless these will be carried farther if they are found satisfactory and if the problem of expense can be solved.

REFERENCES

American Jersey Cattle Club. 1941. Selective Registration.

Gardner, Malcolm H. 1932. Selective herd-book registration. Holstein-Friesian World, 29:411–12 and 428.

Van den Bosch, I. G. J. 1930. The scoring system in the Netherlands. Holstein-Friesian World, 27:337 et seq.

———. 1930. The Holstein industry in America. Holstein-Friesian World, 27: 283.

Wriedt, Christian. 1930. Heredity in livestock. (Especially pp. 151–67). New York: The Macmillan Company.

CHAPTER 17

Type and Production Records

Breeders pay attention to outward conformation in making their selections for two reasons. In the first place, they may want a certain type because it has a market value. If a market demand exists for a certain type, the breeder may care little whether that type really will furnish to his customers the maximum profit or other satisfaction. The fact that they want it and are willing to pay for it is the thing of immediate practical importance to him. In the second place, breeders may believe that type and productiveness are closely enough correlated that if one selects for type he will get at least part of the productiveness he wants.

In many cases, especially among meat animals ready for market, a certain conformation not merely indicates production but actually comes close to being production, since the desired production is largely a matter of sizes and proportions of various parts. At the other extreme are cases where the desired production depends far more on the quality and rates of physiological processes than it does on the sizes and shapes of organs or parts which can be judged on the live animal. The closeness of the correlation between type and production may be of any degree ranging from almost perfect in such a case as the width of the loin or thickness of the round on fat steers, through such relations as may possibly exist between width of head and width of body, to correlations which are practically zero. An example of a correlation which was once thought to be high but has since been found to be practically zero is the relation which a half century ago was widely supposed to exist between the escutcheon of a cow and her producing ability.

Reliance on type as a means of estimating productive ability may be necessary when reliable records of production are not available. Production records on most animals come slowly and expensively. Sometimes, as in poultry and dairy cattle, they are not available on both sexes. Even where production records can be fairly simple and complete, as in the case of cows and hens, it is still true that many purebred animals do not have their production recorded. The situation is still less satisfactory among meat and work animals, where productivity is not easily nor completely measured. A breeder often has an opportunity to buy an animal on which no production record has been made, or he may have to sell some of his young stock before they are old enough to have production records. Such a breeder, even though he has more faith in production records as indicating an animal's productivity than in conformation as a similar index, none the less wants to make as much use as he can of the animal's conformation in estimating its probable productivity.

Type has some sale value in all classes of livestock. In extreme cases beauty may be the main object. This is often encountered in "pet and fancy stock," such as rabbits, dogs, pigeons, and guinea pigs, and is a prominent feature of some of the larger livestock such as saddle and coach horses. If the breeder's customers center their demand on type, he is of course interested mainly in that, and in productiveness only in that his animals should remain healthy and fertile. To appear healthy is, in most cases, an important part of the breed ideal for type also. If his customers are looking for productiveness regardless of beauty, the breeder is interested in type only as it may help him get that productiveness more surely and quickly than if he did not pay attention to type.

The stockman usually wants lifelong productiveness in each animal rather than a maximum single record from each, although the advertising value of an extremely high record may sometimes mean more financially to a breeder than a higher average record which does not become phenomenal in any one year. A single production record is not a perfect index of such life productivity, as was emphasized in chapter 13. The question of how much attention to pay to type and how much to pay to production records in selecting for lifetime productivity is, therefore, a question of comparing and combining two indicators, neither of which is perfectly accurate. Yet it would be a rare coincidence if the usefulness of the two happened to be exactly equal. The principles of estimating an unknown quantity from two known quantities which are partly correlated with it are such that a slightly more accurate estimate may usually be made by using both the known quantities than by using the more accurate of them alone, although the best proportion in which to weight the two is much affected by their relation to each other.

Each intermediate step weakens the effectiveness of selection. Thus, if we want to select for quality x, and it is correlated imperfectly with w, we will not come so close to getting x by selecting w as we would if we could select x directly.[1] If x cannot be observed directly but is rather

[1] As a numerical example, Rasmusson has shown (1930, *Nordisk Jordbrugsforskning*, pp. 247–55) that even if the correlation between x and w is .8, one wishing to select a certain number of individuals which excel the population average in x by at least one standard deviation would need to examine 7.8 times as many if he selects them indirectly by looking for w as he would if he could select them directly by examining them for x.

closely correlated with *y* and not so closely with *w*, we would come nearer getting the value of *x* we want by selecting for *y* than by selecting for *w*. Unless *w*'s correlation with *x* is altogether due to *w*'s correlation with *y*, we would come still nearer to getting the desired value of *x* by selecting for both *y* and *w*; but the proper amount of attention to be given to *y* or to *w* would depend upon how closely they were correlated with *x* and with each other. The guiding principle on this subject is that every needless intermediate step in selection should be avoided as far as possible but that a little is usually gained by paying some attention to other things besides the one which is most closely correlated with productivity. The very real danger in that is that one will pay so much attention to these minor things that he cannot pay enough attention to more important things which are more closely correlated with lifelong productivity.

THE CORRELATION BETWEEN TYPE AND PRODUCTION

In most actual studies of the correlation between type and production the correlation between one estimate of type or conformation and one production record of each individual in the population studied was measured. A few samples of those are mentioned in the following paragraphs.

When official testing began in the Jersey breed, certain judges inspected and scored the cows admitted to the Register of Merit. Gowen studied these data to see what correlation existed between the scores and the production records. Most of the correlations between the scores for each individual point of conformation and the actual production records of those cows were of about the magnitude of — .07 to + .19. The correlation between the production record and the total score of the cow ran somewhat higher, since it took into account all of the points scored. When the study was confined to the scores turned in by the nine judges (out of the nineteen recognized ones) whose scores most closely agreed with the milk yield of the cows, the correlation between their total score and the milk yield was + .38. While this is a real correlation, it was obtained only after discarding the scores of half of these men who were believed by the association to be competent to score the cows.

In similar studies on early Holstein-Friesian records Gowen used measurements made on some of the first officially tested cows. Table 14 presents the correlation coefficients he found between the seven-day milk yield and various body measurements and weights.

The maximum correlation between yield and any body measurement was .36. The correlations found were real and of some use in selecting the high producers, but they were by no means as high as the correlations between different records made by the same cow. A considerable part of these correlations with measurements resulted from differences in general size. Within a breed the largest cows tend to be the heaviest producers and naturally tend also to have the largest measurements.

Engeler's study[2] of the yields of 455 Brown Swiss cows and their scores when they were inspected for registration showed a correlation of only + .04. In another study[3] of 138 cows in one herd he found a correlation of + .32 between milk yield and score.

In dairy cattle there have been several studies of show-ring placings and production records, where both were known. Because only a small

TABLE 14

CORRELATION COEFFICIENTS BETWEEN SEVEN-DAY MILK YIELDS AND CERTAIN ASPECTS OF CONFORMATION, AGE BEING CONSTANT (AFTER GOWEN)

Characters Correlated	Correlation Coefficient
365-day milk yields in different lactations	.66
7-day with 365-day milk yield (same lactation)	.60
7-day with 365-day milk yield (different lactation)	.46
Weight with 7-day milk yield	.42
Body length with 7-day milk yield	.36
Body girth with 7-day milk yield	.25
Body width with 7-day milk yield	.28
Hip height with 7-day milk yield	.24
Shoulder height with 7-day milk yield	.22
Rump length with 7-day milk yield	.18
Thurl width with 7-day milk yield	.01

range of types was included—no really poor types would be among those for which the placings were recorded—such studies throw little light on the correlation between type and production. They demonstrate that show animals can produce well, but so might many others if tested under the same circumstances.

A more suitable basis for studying the correlation between type and production is in such data as the Holstein-Friesian Herd Classification. Table 15 shows a summary of such data.[4] The figures show that type and production do tend to go together in this population. On the average the fat production increased 24.6 pounds with each grade the cow was higher in the type classification. The correlation between the two

[2] 1933. Die Ergebnisse statistischer Auswertung 40-jähriger Herdebuch-Aufzeichnungen beim schweizerischen Braunvieh. Bern: Verbandsdruckerei A. G.

[3] Schweizerische Landw. Monatshefte 19, No. 6, 1941.

[4] As of October, 1942. The records are on a thrice-a-day mature basis and include the first 365 days of the lactation.

is a little less than + .2. That leaves plenty of opportunity for very high producers occasionally to be of poor type and for some animals of high type to be poor producers. The correlation in the general population of dairy cattle is probably a little higher than this, since, in the herds submitted for classification, most of the cows thought by their owners to be "Fair" or "Poor" in type already would have been culled

TABLE 15
AVERAGE HERD TEST RECORDS OF HOLSTEIN-FRIESIAN COWS AS CLASSIFIED FOR TYPE UNDER THE HERD CLASSIFICATION PLAN

Type Classification	Number of Cows	Average Production	
		Milk	Fat
Excellent	261	17,215	601
Very good	1,377	15,988	554
Good plus	2,213	15,754	544
Good	2,138	14,960	514
Fair	426	14,316	488
Poor	25	12,612	431

unless their production was unusually good. On the other hand, part of the apparent correlation may have come about because large size gives an advantage, both in classification and in production. Also, the classification may have been influenced in some cases by knowledge which the classifying officer had of the cow's prior production.

Table 16 shows the summary of such data on Jersey cows to April, 1946 on a twice-a-day 305 day basis. The regression of fat production on type—the average increase in fat production with each increase of one grade in type classification—was 12.8 pounds which is only a little less than the 24.6 from the Holstein data when allowance is made for lactation length and times milked per day.

Engeler's conclusion from studies in Switzerland is: "Form, produc-

TABLE 16
AVERAGE HERD TEST RECORDS OF JERSEY COWS WHICH WERE ALSO CLASSIFIED OFFICIALLY FOR TYPE

Type Classification	Number of Cows	Average Fat Production
Excellent	801	483
Very good	4,213	460
Good plus	6,060	448
Good	2,700	434
Fair	369	420

tion, and health are not so closely related that they can be substituted for each other as bases for selection. These three characteristics are to a high degree independent of each other and to a high degree are transmitted independently to the offspring. The goal of selection consists in preferring those animals which to the fullest extent possess all three characteristics, phenotypically as well as genotypically."

On poultry several studies have been made of the relation between body measurements and production or of the success which a man actually attains in attempting to cull out the poorest producers from each lot, dividing them into two groups and noting their production immediately after the culling. In the culling demonstrations the apparently high success is largely the result of immediately preceding conditions, whereby the man doing the culling is able to identify those which are laying at that time. Since the test pens are not kept for a full year afterward, he seems to have been remarkably successful in picking out the poor producers. It is somewhat the same situation as if one were to go to a dairy farm and rank the cows according to his estimate of their milk-producing ability. If his rankings were then compared with the cows' actual productions in the following week or two weeks, he would seem to have been remarkably successful; but much of this success would be due to the fact that he could tell which cows were dry and which were recently fresh at that particular time! The available evidence makes it seem doubtful that the relation between productiveness and conformation is really any closer in poultry than it is in dairy cattle.

Studies of the relation between weight, conformation, and pulling ability in draft horses have shown (Rhoad) correlations of about the magnitude of + .23 to + .35 between different measurements and ability to pull and + .45 between weight and ability to pull. However, nearly all of the correlation between measurements and ability to pull seemed to be an indirect result of general size, because the relation between these measurements and pulling ability within groups of horses which were all of the same weight was practically zero. In a similar study Brandt found that weight, age, and five measurements had a multiple correlation of .47 with maximum pull. Most of this seemed to depend on weight and height.

Studies of individual beef cattle have been less frequent because there is no one measure which comes as near expressing their real productivity as the actual production records of dairy cattle, poultry, and draft horses do in those cases. The studies which have been made show moderate or low correlations except for anatomical traits, such as fullness of round, which can be estimated closely when judged in the live animal before slaughter. There have been many studies of commercial grades of groups of feeder steers as related to their subsequent performance in the feedlot. Usually the steers of the lower grades gain about as rapidly as those in the higher grades, but they sell at a lower price. Whether they are equally profitable to the feeder depends mostly on whether he can buy them at a low enough price. Because of the lower sale price, the lower commercial grades are less profitable to the man who breeds them.

Studies of type and production in sheep have been more numerous but have dealt mostly with the production of wool or with the feeding qualities of groups of lambs from the crossing of various breeds. Questions of type are often the subject of discussion among breeders of Merino and Rambouillet sheep[5] and Angora goats.

Several experiment stations have conducted swine type tests which have shown differences in the rates of gain and in the kind of meat produced by swine of types varying from "very chuffy" to "extremely rangy." Studies of the ability of men to predict which pigs would make maximum gains have generally shown correlations of something around + .4 to + .7 between the estimate and the actual individual gain. These high correlations (compared with much lower ones on steers) are generally reduced to somewhere near the level of + .15 to + .30 when corrected for differences in initial weight. Greve[6] studied eight different measurements on 205 sows of the Hoya breed near Hanover in Germany and concluded: "All the results show that it is not possible by using body measurement to find which sows have high breeding productivity."

Summarizing the actual evidence on the correlation between individual type and individual production, there is no complete analysis of the problem in any class of livestock; but the nearest approach to that has been in dairy cattle or poultry. In general, these studies have established the existence of correlations between type and production; but such correlations are generally much lower than the correlation between one production record and another production record of the same animal. The conclusion seems inevitable that if one is interested mainly in production he should pay much more attention to production records than to estimates of type, although it does not follow that he could afford to neglect type altogether. An actual production record is not quite a perfect indication of the lifelong production which is actually wanted. The emphasis placed by many men on outward evidences of health and "constitution" may have some justification in the relation of those to lifelong productivity. Many of the breeder's deci-

[5] See Texas Bul. 657. Also Michigan Quart. Bul. 26:31–33.

[6] Zeit. f. Zücht., Reihe B., 46:91, 1938.

sions must be made before he can know from actual experience what that animal's lifelong production will really be. Where type has any direct relation to lifelong productivity, the latter can be predicted more accurately by taking into account both type and production records than by using either alone, but the one which is least closely related to lifelong production should be given much less emphasis.

SUMMARY

A breeder may pay attention to type merely because it will add to the market value of the animals he expects to produce. Aesthetic considerations have much to do with the commercial value of some kinds of animals.

A breeder may also pay attention to type because he believes it to be a useful indicator of the lifelong productivity of the animal. Such indicators, even though not as reliable as production records, would be useful under a variety of circumstances actually encountered, especially among animals on which production records are lacking.

What is wanted is lifelong productivity. A single production record is not quite the same thing, although in most cases it is a more accurate indicator of lifelong productivity than is the individual's type, if either were to be considered alone.

Where type is related directly to productivity, a more accurate estimate of productivity can be made by taking type and the available production records both into account than by considering either alone, but there is danger of paying too much attention to the less accurate of the two indicators.

The practical problem confronting the breeder is to give the proper amount of attention to each. If he pays too much attention to type, the selection he can practice for production is automatically made less intense.

REFERENCES

There are many published studies which bear in part on this topic. The following ones will illustrate the methods of study and will indicate the kind of conclusions which are generally drawn. This list is not complete but is representative of the more recent ones written in English and readily accessible to most students in the United States.

For Dairy Cattle:

Aldrich, A. W., and Dana, J. W. 1917. The relation of the milk vein system to production. Vermont Agr. Exp. Sta., Bul. 202.

Brody, S., and Ragsdale, A. C. 1935. Evaluating the efficiency of dairy cattle. Missouri Agr. Exp. Sta., Bul. 351.

Copeland, Lynn. 1938. The old story of type and production. Jour. Dy. Sci. 21:295–303.

Gaines, W. L. 1931. Size of cow and efficiency of milk production. Jour. Dy. Sci. 14: 14–25.

Garner, F. H. 1932. A study of some points of conformation and milk yield in Friesian cows. Jour. Dy. Res., 4:1–10.

Gowen, John W. 1924. Milk secretion. pp. 32–49.

———. 1923. Conformation and milk yield in the light of the personal equation of the cattle judge. Maine Agr. Exp. Sta., Bul. 314.

———. 1925. The size of the cow in relation to the size of her milk production. Jour. Agr. Res., 30:865–69.

———. 1926. Genetics of breeding better dairy stock. Jour. Dy. Sci., 9:153–70.

———. 1931. Body pattern as related to mammary gland secretion. Proc. Nat. Acad. of Sci., 17:518–23.

———. 1933. Conformation of the cow as related to milk secretion. Jour. Agr. Sci., 23:485–518.

Swett, W. W.; Graves, R. R.; and Miller, F. W. 1928. Comparison of conformation, anatomy, and skeletal structure of a highly specialized dairy cow and a highly specialized beef cow. Jour. Agr. Res., 37:685–717.

For Poultry:

Bryant, R. L., and Stephenson, A. B. 1945. The relationship between egg production and body type and weight in Single Comb White Leghorn hens. Va. Agr. Exp. Sta., Tech. Bul. 96.

Jull, M. A.; Quinn, J. P.; and Godfrey, A. B. 1933. Is there an egg-laying type of the domestic fowl? Poul. Sci., 12:153–62.

Knox, C. W., and Bittenbender, H. A. 1927. Correlation of physical measurements with egg production in White Plymouth Rock hens. Iowa Agr. Exp. Sta., Res. Bul. 103.

Marble, D. R. 1932. The relationship of skull measurements to cycle and egg production. Poul. Sci., 11:272–78.

Miller, M. Wayne, and Carver, J. S. 1934. The relationship of anatomical measurements to egg production. Poul. Sci., 13:242–49.

Sherwood, Ross M., and Godbey, C. B. 1928. Construction of score card for judging for egg production. Poul. Sci., 7:263–74.

Waters, Nelson F. 1927. The relationship between body measurements and egg production in Single Comb White Leghorn fowls. Poul. Sci., 6:167–73.

For Horses:

Brandt, A. E. 1927. Relation between form and power in the horse. Trans. Amer. Soc. Agr. Eng., 21:PM—3–4.

Dawson, W. M. 1934. The pulling ability of horses as shown by dynamometer tests in Illinois. Proc. Amer. Soc. An. Prod. for 1933, pp. 117–21.

Dinsmore, Wayne. Learn to judge your horses and mules. Leaflet No. 196 of the Horse Association of America.

Laughlin, H. H. 1927. The Thoroughbred horse. Carnegie Institution of Washington. Yearbook No. 26, pp. 56–58. (See preceding and following yearbooks for other short notes on this.)

Rhoad, A. O. 1929. Relation between conformation and pulling ability of draft horses. Proc. Amer. Soc. An. Prod. for 1928, pp. 182–88.

For Swine:

Bull, S.; Olson, F. C.; Hunt, G. E.; and Carroll, W. E. 1935. Value of present-day swine types in meeting changed consumer demand. Illinois Agr. Exp. Sta., Bul. 415.

Nordby, J. E. 1932. Type in market swine and its influence on quality of pork. Idaho Agr. Exp. Sta., Bul. 190.

Scott, E. L. 1930. The influence of the growth and fattening processes on the quality and quantity of meat yielded by swine. Purdue Agr. Exp. Sta., Bul. 340.

For Beef Cattle:

Black, W. H.; Knapp, Bradford, Jr.; and Cook, A. C. 1938. Correlation of body measurements of slaughter steers with rate and efficiency of gain and with certain carcass characteristics. Jour. Agr. Res. 56:465–72.

Hultz, F. S., and Wheeler, S. S. 1927. Type in two-year-old beef steers. Wyoming Agr. Exp. Sta., Bul. 155.

Lush, Jay L. 1931. Predicting gains in feeder cattle and pigs. Jour. Agr. Res., 42: 853–81.

———. 1932. The relation of body shape of feeder steers to rate of gain, to dressing per cent, and to value of dressed carcass. Texas Agr. Exp. Sta., Bul. 471.

For Sheep:

Bosman, V. 1933. Skin folds in the Merino sheep. S. African Jour. Sci., 30:360–65.

Cole, C. L. 1943. Type selection vs. record of performance in sheep breeding. Quart. Bul. Michigan Agr. Exp. Sta. 26:31–33.

Hultz, Fred S. 1927. Wool studies with Rambouillet sheep. Wyoming Agr. Exp. Sta., Bul. 154.

———. 1935. A five-year study of Hampshire show sheep. Wyoming Agr. Exp. Sta., Bul. 207.

Jones, J. M.; Dameron, W. H.; Davis, S. P.; Warwick, B. L.; and Patterson, R. E. 1944. Influence of age, type and fertility in Rambouillet ewes on fineness of fiber, fleece weight, staple length, and body weight. Texas Agr. Exp. Sta., Bul. 657.

Joseph, W. E. 1931. Relation of size of grade fine-wool ewes to their production. Montana Agr. Exp. Sta., Bul. 242.

Spencer, D. A., and Hardy, J. I. 1928. Factors that influence wool production with range Rambouillet sheep. USDA, Tech. Bul. 85.

CHAPTER 18

Breed Type

Breed type means the complex of external characteristics which is typical of a breed or is considered ideal for that breed. The term is often used in distinguishing one breed from another breed used for much the same purpose. Many ingredients of breed type are conspicuous details of conformation and color which have no relation to the economic productivity of the animals. Examples are the shape of horns in cattle, the dish of face and size and shape of ear in swine, color of face and shape of ear in sheep, and color generally. It is those features of breed type which are the subject of this chapter.

Attention is paid to breed type mainly because it is a "trademark" which is some additional evidence that the animal really conforms to the ideals of the breed. Probably the men who breed purebred animals average somewhat higher in honesty than men in most other lines of work, since the foundation of the pedigreed livestock business is the honesty of the men who sign their names to the pedigrees. Without general honesty on this point purebred animals could not command the premium they do. Yet there will always be a few mistakes and frauds. The existence of a definite breed type, especially if that is a combination of characters hard to obtain without absolutely pure breeding, is one check upon errors in registration. If a breed has a type of this kind, an animal deviating markedly from that type will be regarded with suspicion. From this point of view the breed type which is the hardest to attain or which is the most easily upset in crosses is the most highly prized. This often goes to extremes in the case of "fancy stock," such as pigeons, rabbits, or dogs, the desired type or standard being kept just high enough or changed just often enough that only a small proportion of the breed attain it.

Of course, breed type also is a matter of beauty to the men who have long been breeding and admiring that breed. But beauty is very much a subjective matter. Most of us can bring ourselves to think that any particular type is beautiful if we work with it and study it long enough and find it profitable. Naturally the breeders of other breeds may not share our enthusiasm for the supposed beauty of our breed.

Part of the demand for breed type originates more or less unconsciously with breeders who are enthusiastically steeped in the tradition of the ancient purity of their breed. It is easy for such men to persuade themselves that "the best animals of the ———————— breed with the purest blood are always thus and so," and to believe that deviations from that description indicate impurity of breeding or that something went wrong with the hereditary process.[1]

Insistence upon conformity to breed type is actually harmful only as it weakens the intensity of selection for economically important points. This it must do to some extent.

An example of how insistence on breed type may change a breed is the occurrence of red spots on the faces and red rings around the eyes in the Hereford breed. Many of the Herefords imported to America carried these red markings. There was at first no prejudice against this; and, in fact, certain breeders rather preferred these, which they called "brown eyed." Eventually the tide of favor swung toward faces as white as possible. Today one sees few purebred Herefords which have complete red rings around the eyes. In the extreme southwest part of the United States, Herefords with pure white eyelids are more subject to cancer of the eyelid than are those with red eyelids. This is not an important matter, because only a small fraction of those with white eyelids develop cancer. Moreover, the cancer develops slowly and only upon the older animals. A ranchman usually has time to cull those affected and to ship them to market without suffering a complete loss. This is a minor disadvantage, but many ranchmen wish that the Hereford had kept its original high frequency of red-eyed cattle. Why did the Hereford breeders select the white-eyed type when there was no criticism of the utility of the red-eyed ones? The answer seems to be that among the very first things to appear in crosses of Herefords with other cattle were red spots on the face and red rings around the eyes. To many a cattleman, the presence of red spots on the face or red rings around the eyes of Hereford cattle indicated impurity of breeding. With this customer opinion confronting him, it was almost inevitable that the breeder of purebred Herefords should select for those which had the most nearly white faces and white eyelids.

Another striking case was the strong preference for yellow color in honeybees which arose soon after yellow Italian bees began to replace the black or German bees in the United States. Many beekeepers

[1] *See* W. Engeler's interesting account of the theory of "racial constancy" and selection for breed type in European animal breeding writings from 1800 to 1880. Pages 45–58 in "Neue Forschungen in Tierzucht," Bern, 1936.

inferred that the yellow color was itself *the cause* of the practical qualities—gentleness and superior honey-gathering ability—they wanted. They began to select for yellow color itself. Charles Dadant found in Italy some bees darker than most Italian stocks but more productive. These he could hardly sell to American beekeepers who had by that time come to believe that the yellower the bee the better. Who knows how many years the practical progress of beekeeping was retarded by that color craze!

Such examples are by no means confined to the United States. Even in Denmark, where such high emphasis is placed on practical utility in all livestock, the Red Danish cattle are not eligible for prizes at some of the important shows if they have a large amount of white or any roan color. The reason is that Shorthorn blood was used many years ago in some attempts to improve the Red Danish breed by crossing. In time that came to be regarded as a failure, and every effort was made to cull from the breed all animals carrying any traces of that Shorthorn outcross. Who knows how much of the Shorthorn dislike for a dark muzzle in Britain and the United States today is a similar hangover from the controversy about the use of the "Galloway alloy" in the days of the Collings? Or how much of the Guernsey dislike (in the United States—little attention is paid to it in Guernsey) for a dark muzzle stems from a desire to emphasize the distinction between them and the Jerseys? Similarly, in Denmark the insistence of the breeders of Landrace swine that their swine shall have very large and drooping ears seems logically explainable only on the ground that this is one of the few outward distinctions between the Landrace and the more or less competitive Yorkshire, both breeds being bacon hogs, long-bodied, and solid white. In the tropical parts of Brazil, breeders selecting among pure zebu cattle try to get them with ears as long as possible. Presumably this is an aftermath of the extreme competition which prevailed when the zebu cattle and their grades were first getting a foothold. The grades with the most zebu blood generally had the longest ears. Hence length of ear, originally preferred because it indicated a high percentage of zebu blood, came finally to be considered as itself a sign of higher merit.

Some of the things which constitute breed type cannot be fixed. Laboratory experiments with various piebald races of animals such as guinea pigs, have shown that a considerable amount of variation in extent of white spotting is not hereditary at all and would still exist in a homozygous strain or pure line. There is every reason to think that the same situation prevails in such spotted breeds as the Holstein-Friesian or Guernsey or Spotted Poland-China. No amount of selection, no matter how long continued or with what inbreeding system it was combined, could ever produce an absolutely uniform breed. This is in spite of the fact that in these breeds there are modifying genes which tend to restrict or extend the pigmented areas. Doubtless the same thing is true with spotting which takes a more definite form as, for example, the amount of white in the Hereford pattern or the amount and uniformity of the white belt on Hampshire swine.[2] Wherever this is the situation and the breed type is really unfixable, it is especially regrettable when otherwise desirable animals are discarded from the breeding herd on account of failing to conform closely to a rigid standard of color markings. Not only are their good qualities lost to the breed but, ironically enough, their discarding does not cause the breed to conform more closely to the standard breed type than if they had been kept for breeding purposes.

The matter of breed type is receiving less attention from breeders today than it has many times in the past. The practical breeder cannot afford to neglect it altogether wherever it still has some cash value in the market in which he must dispose of his surplus. He needs to satisfy his customers as much as he can without losing much real productivity from his stock. Any more efforts in selecting for breed type than his customers' demands absolutely force on him detract from his ability to select for things of practical utility. When a breeder hears that his customers very much want certain features of breed type, it behooves him to be skeptical about whether they really will pay him much more for the animals which have those things. Some statements of this kind are just sales talk or buyer's talk[3]; others are details in an almost endless and unbalanced catalogue of all characteristics which ever have been noticed. Many of these details will have little or no detectable effect upon the amount the buyer really will pay.

SUMMARY

Breed type serves as a trademark, which is to some extent additional evidence of purity of breeding over and above the printed pedigree. It is not as valid evidence on this point as is sometimes believed.

Some of the elements of breed type probably cannot be "fixed" so that they will be perfectly uniform. Effort spent in breeding for those not only weakens selection for more important things but even fails to improve breed type.

Breed type becomes positively harmful when so much attention is paid to it that animals above the average in real usefulness are dis-

---

[2] For a plausible genetic explanation of white belt in swine, see Olbrycht's article in *Annals of Eugenics* 11:80–88, 1941.

[3] *Proverbs* 20:14.

carded because they do not conform to breed type in matters which are of little or no economic importance. The more points considered in selection, the less effective can selection be for each of them. Breed type should be kept in a minor place, but the practical breeder cannot afford to ignore it altogether if his market places some value on it.

REFERENCES

Van Riper, Walter. 1932. Aesthetic notions in animal breeding. Quart. Rev. of Biology, 7:84–92.

Wentworth, E. N. 1926. Breed, show and market standards. Proc. of the Scottish Cattle Breeding Conference in 1925, pp. 212–18.

Wriedt, Christian. 1930. Heredity in livestock. See especially the chapter on "Fairs and Fancy Points."

CHAPTER 19

The Show Ring and Animal Breeding

The original purpose of fairs was to provide a place for trading. In some parts of Europe, even yet, the man who has a few animals for sale may take them to a weekly fair. If no buyer makes an acceptable bid, he takes them home again to wait for the next fair. It is a small step from exposing animals for sale to exposing them for having their merits appraised by other breeders. The exhibition of animals not intended for sale was a prominent part of animal breeding practices even as far back as the early history of the Shorthorn breed. Charles Colling's famous "Durham Ox" was started in 1801 on a tour of exhibition which lasted for six years. This tour was more like the sideshows than like the show rings of today. The Colling-bred "White Heifer That Traveled" started on a similar tour a little later. The Booths were famous for showing their breeding stock at many fairs. Thomas Bates on occasion exhibited his Shorthorns and prided himself on his ability to judge, although he was outspoken in his criticisms of the evils of overfitting and of keeping cattle just for showing. According to Wriedt, the first public show corresponding to our modern livestock shows was held in 1798 in Sussex, England. The first public show in Denmark for all kinds of livestock was held in 1810. The first public show in Wurtemberg was in 1817. Most of this early showing was for advertising purposes, and the premiums offered were small or consisted only of trophies. According to Jull the first poultry show in the United States was held in Boston in 1849. Poultry showing in England began at about the same time.

THE SHOW RING IN RELATION TO BREED IMPROVEMENT

There are two ways in which the show ring may affect breed improvement. First, it may keep the breeders informed about the ideals of the breed. If they use that information in their own selections, the show ring can be an important factor in guiding the direction in which the breed is to be changed. Secondly, the show ring might be used to find the best animals in the breed to such an extent that breeders could accept the show ring placings as guides in buying and selling their breeding animals. While this condition, of course, is never entirely reached in any land, yet the advertising and popularity which certain animals may acquire through their high winnings in the show ring may go far to get them or their sons or daughters used extensively in many herds which otherwise would never have sought them. This may have some effect upon the genetic composition of the breed if it is done year after year, since it constitutes a mild grading-up process toward the prize winners and the herds from which they came. This is a form of selection favoring the genes which are most frequent in the type of animals which the judges place highest. Even though the judge sees but a tiny fraction of the animals of the breed, his approval or disapproval may help determine which animals become paternal grandsires or great grandsires of the breed. Since one herd rarely is a prominent prize winner for more than a few years, this does not often emphasize any one animal or family in a way which could be called linebreeding. It is emphasis of an ideal, rather than emphasis of an animal.

In the shows in the United States and Britain, the main object is to emphasize the visible ideal which is held by the breeders of each breed. No attempt is made in American and British shows to judge pedigrees. The few attempts to give some weight to production records, in addition to what is visible to the judge, have met with only partial approval from those who tried the experiment. The shows, especially the larger ones, give tremendous emphasis and advertising value to championships and first places, out of all proportion to the usually small differences in real merit between first and second or third place animals. The main thing is to exalt before the public the most nearly ideal combination of visible characteristics which can be found and to give the breeders a clearer picture of the perfect animal to guide them in their own selections. It is only incidentally that attention is directed to the animal for its own sake. The major fairs in the United States usually perform well this function of exalting the ideal. The larger fairs do not give much help to the beginning breeder, since the placings he sees mostly concern small differences between animals nearly all of which are good ones, and the interest in judging is mostly centered on the placings of the top two or three animals.

In the 4-H Club shows in the United States and in many of the shows in continental European countries, there have been earnest attempts to remedy this and to make the shows a place of instruction in judging from top to bottom. This is intended to benefit the large part of the public and the considerable fraction of breeders who have not become experts in judging and perhaps never will, but who wish some information or training in the kind of judging they themselves can do in their own herds. In their applications of judging they may have an opportunity to cull perhaps the poorest third or fourth of their females; or perhaps they must select a sire from among a group of moderately good males, no one of which is really of championship caliber. The devices used to make these shows more instructive to the beginner and to the general public mostly concern: placing all animals shown, no matter how poor they are; having grades or descriptive terms which are kept as nearly constant as possible from show to show and from year to year so that the written record of the animal's show ring placings may have a standard meaning; stabling or tying the animals during the fair in the classes and in the order in which they were placed so that visitors may see and study the placing at any time during the fair; minimizing showmanship in the placings; and having as many breeders participate in the judging as is reasonably possible. It is probable that some of these practices can with advantage be adapted to American conditions for those shows where a majority of the animals come from nearby farms and where most of those who show are relatively inexperienced.

The show ring can do only a small amount toward ranking the animals of the breed in the order of their real breeding value. In the first place, only a tiny portion of the total number of animals of the breed are shown. Table 17 shows this in a general way, using Iowa as an example because its state fair is large and there are many purebred animals in Iowa. The figures in this table give only a rough idea of the extent of actual participation in showing, however. Besides the qualifications listed under the table and the fact that the years are not identical, many of the exhibitors and animals were from outside the state.[1] In the second place, differences in showmanship, grooming, and such things may affect the animal's show ring ranking, although they have no bearing on its breeding worth. Such practices may even prevent the judge from ranking the animals as nearly in the order of their breeding merit as he could if no preparation for showing were made. An example is length of wool, which is an important part of the practical merit of a sheep, especially of sheep of the Merino and Rambouillet breeds. Yet the length of fleece which the sheep wears when it enters the show ring may be so altered by shearing early, stubble-shearing, blocking and trimming, etc., that the judge cannot afford to pay as much attention to it in the show ring as he usually would if he were culling his own sheep where he knew that all had had substantially the same length of time in which to grow the fleeces he saw before him. In the third place, the

[1] In Denmark, where participation in the fairs is more general and each owner can show only in the district where he lives, about 6,000 bulls and 10,000 heifers and cows altogether are shown annually. This is about 7 or 8 per cent of the bulls but only about 0.4 per cent of the females among all the cattle of Denmark over one year old.

temporary condition of the animal must count for much because of the judge's duty to set before the public an animal which at that very moment comes nearer the visible ideal of the breed than any other animal in the ring. In the fourth place, the judge's ability to find the best breeding animals, even if these other obstacles could be overcome, is limited, of course, to the correlation which exists between outward appearance and real breeding value. In the fifth place, it is difficult to compare a placing in one show with a placing in another unless one was present at both shows and remembers what kind of animals were present at each. What comparison could one make from the information

TABLE 17

NUMBERS OF REGISTERED ANIMALS IN IOWA IN THE 1930 CENSUS AND NUMBERS EXHIBITED AT VARIOUS IOWA FAIRS FROM 1930 TO 1936

Kind of Animal	Farms or Breeders Concerned		Number of Animals		
	Farms With Registered Females in 1930	Exhibitors* at State Fair in 1936	Registered Females on Farms in 1930	Registered Animals Exhibited at State Fair in 1936	Average Number Exhibited at County and District Fairs Each Year, 1930 to 1935†
Draft horses	1,974	72	5,241	383	4,355‡
Beef cattle	5,556	113	60,685	713	12,809§ (beef, dairy and dual-purpose cattle)
Dairy cattle	5,297	112	33,272	719	
Dual-purpose cattle	437	19	4,003	200	
Sheep	1,352‖	73	15,677‖	830	4,551
Swine	7,446‖	226	66,189‖	2,238	15,118

* Including 4-H Club members who exhibited registered animals.

† The number of these fairs ranged from 76 to 82, averaging 78 in the six years. The number exhibited each year is the sum of the numbers exhibited at each of these fairs. Each fair was about two to four days in length, and they were scattered over a period of about nine weeks during August and September. Doubtless many animals were exhibited at more than one of these fairs and hence are counted more than once in these numbers, but no way was found to estimate how many of these duplications there were.

‡ Includes all horses without distinction between light and draft horses.

§ Includes all cattle.

‖ All registered animals, whether male or female.

that one sire won first as an aged bull at the Page County Fair in 1935, a second sire was third as a two-year-old at the Iowa State Fair in 1933, while a third sire stood tenth as a Junior Yearling at the National Dairy Show in 1931?

Probably the ideals of the show ring are usually those of a majority of the breeders, but it is not certain whether that is because the show ring leads the breeders or whether it merely reflects their current opinions after those have been formed by other circumstances, just as the driftwood on a river shows the course and speed with which the water moves but does not cause or guide that movement. Changes in ideals do sometimes occur; sometimes those get well started even against the disapproval of the current show ring ideals. A case of that was the marked change in ideals in the Poland-China and most other American breeds of swine between 1910 and 1920. When such changes are in process, the show ring may help them to spread more rapidly by giving the breeders occasion to meet and discuss the subject with examples before them.

In spite of its imperfections, no good substitute has yet been devised for the show ring as a means of indicating what kinds of individuals are best in the breeds of beef cattle, hogs, sheep, and draft horses. Even in animals such as dairy cattle and poultry, where there are reasonably simple and accurate tests for production, the show ring has not been displaced by these tests. Among all farm livestock, the Thoroughbred and Standardbred horses come nearest to relying upon production records with little use for shows. Among poultry breeders there is a rather wide gap between those who breed for production and those who breed for show. In dairy cattle there is a similar but less extreme divergence of opinion as to the usefulness of the show ring.

SPECIAL FEATURES OF SHOWS IN CONTINENTAL EUROPEAN COUNTRIES

Brief mention will be made here of some show ring practices in non-English-speaking countries. Some of these might be useful, especially in local fairs, if adapted to American conditions, or may be interesting because of their distinct contrast with the practices common here.

All animals exhibited in a class are usually placed from top to bottom, although it is permissible for the judges to indicate that two or more are equal. The animals are usually divided into at least three and not more than five classes. Prize money and usually the permanent records make no distinction between animals in the same class. Thus, in a single class of bulls there may be four "first prize" bulls, six "second prize" bulls, five "third prize" bulls, six "fourth prize" bulls, and three judged too poor to receive any prize. Those who are at the fair will know the individual ratings within each class. Those individual ratings are printed in the list of awards prepared immediately after the judging is completed; but they may not go into the permanent records and will not appear in the pedigrees of descendants of these animals. In those pedigrees it will be stated simply that this animal was "second prize" at a certain fair in a certain year. Every reasonable attempt is made to keep the standards of judging constant so that "second prize" will mean the same thing at all fairs in all years in any one country.

In the case of cattle and horses, classes which are judged together are tied together during the daytime in the order of their placing until the fair is over, so that a visitor can study at almost any time during the show the placing in any class in which he is interested. He does not need to be on hand to see it judged. Physical difficulties may prevent that with swine and sheep. Over each animal or pen is posted prominently its classification and often its score, perhaps accompanied by the judges' criticisms and commendations of certain things about it. The catalogues contain for each animal the production records and scores or classifications of its ancestors.

Sometimes pedigrees are classified or scored also. Practice varies about whether (1) the pedigree score and the individuality score are combined into a single net score for the animal, (2) separate prizes are given for pedigree classification and for individuality classification, or (3) the prizes are given only for individuality, while the pedigree scores are printed merely for information. Since 1930 at the German national show cows must have records of production to be eligible for showing.

Many breeders participate in the judging. Sometimes they work singly and sometimes in committees. Where the classes are large, there may be almost a separate committee for every class.

Much use is made of progeny groups of one kind or another. Those vary more in the rules as to numbers required, etc., than do the "get-of-sire" and "produce-of-dam" classes in American shows. At some of the Danish shows at least two-thirds of the progeny of an older bull must be exhibited if he is to gain a prize for type. Some of those progeny groups may be judged on farms before the show. That is often done with stallions.

Showmanship is minimized in many ways. Usually the attendant makes little effort to pose his animal. In the bull shows in Switzerland the judging is done behind closed gates and not even the owner is allowed to be present. When the judges finally get a class placed in what they believe is the correct order, the bulls are allowed to stand for a while tied to the rail. Then the judges come back to look for defects which may become evident after the animals have stood for a time, and to make sure that the placing is satisfactory. The judging takes place in the first half day in these Swiss shows, but the animals must remain on exhibition during the daytime for three or four days.

In many countries there is more selling of exhibited animals than is usual in the United States. More than half of the bulls at these Swiss shows may change owners before the show is ended. In the Argentine shows nearly all prize-winning animals are auctioned afterward. The owner may withhold his animal from the sale if he wishes, but this is not often done. In case he does that, the owner pays the management a fee to cover expenses and to pay the auctioneer what he would have received if this animal had been sold.

BUSINESS ASPECTS OF THE SHOW RING

The show ring is one of the best channels for advertising surplus stock. Many potential customers will not inquire whether the animals which won the prizes are closely related to those which the breeder is offering to sell them. Because of this, breeders who have many animals for sale can sometimes pay large prices for good show animals owned by some one who is not intending to exhibit them. The money thus spent is an advertising expense, just as surely as if it had been spent for newspaper space. It may be profitable if it helps keep the name of the breeder before his potential customers in a favorable light and if he has many animals to sell.

The show is an excellent place to meet other breeders and exchange ideas and experiences which may be of considerable practical value. In this way one can do much to keep informed on matters of concern to the breed and can learn of events or changes of reputation which he would not otherwise learn so soon. Sometimes the members of the breed association present will hold a meeting some evening after the judging is over to discuss matters which can be handled only in a co-operative manner.

The show ring provides an opportunity to learn judging, at least among the better animals, or to keep up to date the judging knowledge one already has. One will learn much by standing at the ringside, making his placing before the judge does, and then trying to see why the judge's placing was different from his own.

The practice of fitting out a show herd and going from one show to another on a long circuit insures that the average show ring merit of the individuals exhibited at each fair shall be higher than if there were no circuit system. Thus the show ring will more nearly achieve its object of showing the public the ideal which is held by each breed. Against this must be balanced the fact that it tends to destroy local participation in the fair. Many a breeder, who might take his animals to the nearest fair if he did not have to compete with professional showmen, will leave his animals at home and go to the fair as a ringside spectator under the present system. The long circuits promote professionalism in showing because the skill of the showman has a chance to be rewarded in many shows, not merely in one. The high rewards for success in professional showmanship are an incentive toward such practices as surgical operations to correct defects and toward all manner of deception in showing the animal. The show tends to become more of a contest between showmen and less of a court of inquiry as to which animal really would be most useful for further improving the breed. In doing so, it may acquire some of the sporting interest which attaches to a horse race, but it becomes less useful to agriculture. The long show circuit may keep the herd away from home for a long time when the animals should be used for breeding.

Professional showmen sometimes take advantage of the circuit system by exhibiting several different breeds of the same species of livestock, especially at fairs where the competition will be light. This is particularly common with exhibitors of sheep where prize money is offered for so many different breeds. Such show herds or flocks are sometimes called "carnival outfits" or "gypsy herds." Few men are successful breeders of more than one breed of each kind of livestock. The most which can be said in economic justification of the carnival outfit is that it will advertise a breed in a region where that breed is little known. This may lead to some increase in sales by breeders of that breed and at rare intervals may be the means of introducing to a region a breed which has some real usefulness there but would not otherwise find a foothold so soon. The managers and directors of fairs are usually reluctant to reduce the prize money offered for the rare breeds to quite as low a level as would be proportional to the number of them which are bred in that region.

Many animals which played a prominent part in breed history were themselves prize winners, but there have also been some which were not shown or did not place high and yet did have more influence on the breed than any of their contemporary prize winners. Champion of England, who, more than any other one animal, was responsible for the "Scotch type" of Shorthorn, was of doubtful individual merit as a calf and was nearly discarded without being tried as a sire. In the Hereford breed Anxiety 4th was not shown, although it is said that he was an excellent individual. His owners regarded him as too valuable to risk fitting for showing. His sire, Anxiety, had been lost in just that way. Many of the sons and grandsons of Anxiety 4th were shown, but Beau Brummel, the grandson who became the most influential animal of the whole breed, was shown only once and placed fifth in his class that time. It is related that he would have ranked higher if he had been especially fitted for showing. The noted Shorthorn sire, Avondale, was fourth in his class at the 1908 International but later, as a sire, excelled the three which placed above him. It does not seem that any valid general conclusion can be drawn from such individual cases. They demonstrate that show ring ranking is not an infallible guide to future success as a breeding animal, but not even the most enthusiastic admirer of the show ring would maintain that it is.

Whether an animal has an important influence on a breed depends on the opportunity it has and on chance circumstances, as well as on the kind of heredity it really has. As a rule those animals which stand high in the show ring are given a better opportunity as breeding animals than those which do not stand so high, but there is much variation in this. The financial circumstances of their owners and other incidental circumstances, which have no relation to an animal's breeding worth, are often the controlling factors in determining what influence an animal shall have on the breed. The Hereford bull, Beau Brummel, and the Aberdeen-Angus bulls, Black Woodlawn and Earl Marshall, were offered for sale to grade herds while yet young; but, fortunately for their breeds, the sales were not completed and the bulls were used for many years in purebred herds. The Percheron stallion, Brilliant 1899, was used for nearly a dozen years in France and then was sold to America, where he was used but one year on purebred mares.[2] After that he was used for nearly 15 years on grade mares only. The colts he sired in France and the kind of grade colts he sired in the latter half of his life indicate that the history of the Percheron breed might have been materially different if he had stood at the head of a stud of purebred mares during the latter half of his life. It is said[3] that Mr. Gentry once offered to take $25 for Longfellow, the Berkshire boar who afterward became the most famous sire of his breed. In fact, the buyer, who was merely looking for a boar to ship to a tenant, was given his choice of two pigs at that price. Mr. Gentry, trying to help him out, suggested Longfellow and the suspicious buyer promptly chose the other pig!

Since differences in visible merit are partly determined by the genes the animals have, it is to be expected that, if all animals were given equal opportunity, prize winners would usually have a higher proportion of prize-winning offspring than would breeding animals which were not prize winners themselves. But the animals' opportunities to be shown and to be used for breeding are different and little is known definitely about the heritability of differences in show ring merit. Rice found .21 for the regression of daughter on dam in the official type classification of Holstein-Friesians but herd differences in environment might have been responsible for more of this than heritability was. Proportion, symmetry, and balance are emphasized so much in the

[2] See pp. 237–39 in *A History of the Percheron Horse,* by A. H. Sanders and Wayne Dinsmore, Chicago, 1917.

[3] Page 6 of *The Breeders Gazette* for February, 1932.

show ring that epistatic gene interactions seem likely to be important causes of differences in show ring placings. For these reasons it is impossible to say how many more of the offspring of prize winners will be capable of winning prizes themselves than will be the case among the offspring of those which did not win prizes. There have been many studies of the pedigrees of groups of prize winners, but only a few of these have included comparisons with the pedigrees of a representative sample of the whole breed. Those few have indicated that the pedigrees of the prize winners are substantially the same as the average pedigrees of the breed, so far as concerns ancestors much more than two or three generations back in the pedigrees.[4] Some of the animals prominent as parents, grandparents, or great grandparents of the prize winners have not been so prominent in the average pedigrees. It seems possible to interpret this either as meaning that the prize winners come largely from only a few contemporary families to which the breed will later be graded up, or as resulting incidentally from the fact that only a few owners make a regular practice of showing at the large fairs, and any sires used extensively in their herds will almost inevitably be prominent in show ring winnings a few years later. In either event, it can hardly be maintained that the prize winners constitute very distinct families or strains within the breed.

SUMMARY

The show ring is a means of emphasizing the current ideals of the breed. If breeders are guided much in making their own selections by noting the types of animals which are placed high in the show ring, the show ring can have an important part in guiding the direction in which the breed average shall move.

It is not certain whether the show ring really causes the changes in the ideals of the breed or merely reflects the ideals currently held by a large portion of the breeders. There have been times when the breed ideal changed, even against opposition from the show ring. Probably the show ring cannot lead the whole breed far in a direction contrary to the ideals of the commercial breeder.

To a limited extent the show ring may help in rating the animals of the breed according to their breeding value. It is not very effective in this because: (1) the correlation between outward appearance and real productiveness is low for many characteristics; (2) so few of all purebred animals are shown; (3) considerable attention is paid to fitting, to temporary conditions, and to showmanship; and (4) many important things about which the breeder may know, such as amount of milk and

[4] *Jour. of Heredity*, 22:245–49, and 27:61–72.

fat produced by dairy cattle, number of pigs weaned by sows, length of fleece on sheep, etc., must for practical reasons be given only a little attention by the judge since he cannot know exactly what those were.

In spite of these limitations there is not yet any good substitute for the show ring in measuring the general merit of the meat animals. Even breeders of animals which, like dairy cattle, have reasonably complete measures of productiveness not connected with the show ring continue to make extensive use of shows.

In a business way the show ring is an important means of advertising. The fairs offer opportunities to make sales and to exchange news and ideas with other breeders.

REFERENCES

Eckles, C. H. 1933. We can improve our dairy shows. Successful Farming, 30 (No. 3) : 20.

Lush, Jay L. 1935. Observations of European livestock shows. The Cattleman, 21 (No. 11) :21–28.

The Agricultural Council. 1935. Denmark agriculture. Copenhagen. 324 pp.

Wentworth, E. N. 1926. Character correlations, livestock judging and selection for type. Proc. of the Scottish Cattle Breeding Conference for 1925, pp. 195–236.

CHAPTER 20

Likeness Between Relatives—Degrees of Relationship

The idea of relationship is familiar to all. Proverbs such as "Like father like son," "A chip off the old block," "What's bred in the blood will out in the bone," and "Blood will tell" are found in every language. Their antiquity attests the fact that people have always known in a general way that offspring tend to resemble their parents, and that brothers and sisters show many of the same "family characteristics," and that more distant relatives are usually less like each other than close relatives are. Genetics itself is defined as "the science which seeks to account for the resemblances and differences exhibited among organisms related by descent."

The first scientific attempts to measure the degrees of resemblance between different kinds of relatives were made late in the last century by Sir Francis Galton and his associates. In fact, correlation coefficients and many of the modern statistical methods now used for other purposes, too, were devised by them primarily for this purpose. With the rediscovery of Mendelism, interest in heredity shifted from the biometrical method to studies of the transmission of individual genes and, for a time, it was even supposed that the two points of view were antagonistic. Within recent years many of the statistical consequences of the Mendelian nature of inheritance have been explored, and the two fields of knowledge have been unified, each complementing the other.

THE BASIS OF RELATIONSHIP

Relatives resemble each other in various degrees because each offspring gets a sample half of the genes which its parent had. Relationship between two individuals is simply probability that, because they are related by descent, they will be alike in more of their genes than unrelated members of the same population would be. Closer relationship merely means higher probability of genetic likeness.

The parent-offspring relationship is the simplest one. It is fundamental in the sense that all other relationships are combinations of chains of parent-offspring relationships. In populations where there is no inbreeding, the parent-offspring relationship is 50 per cent, simply because each offspring has received half of its genes from each parent.[1]

Half of the genes of each offspring are identical with half of the genes of each parent, since the offspring received them from that parent. The rest of their genes (those which the parent did not transmit to this offspring and those which the offspring received from the other parent) may or may not be alike, just as two individuals of the same population may have some of the same genes merely because those genes are common in that population. Where there is some inbreeding these other genes will have some extra probability of being alike also. The extra relationship which inbreeding may cause will be discussed in the next chapter.

Half brothers are 25 per cent related because, on the average, one-fourth of their genes are duplicates which both received from the common parent, another fourth also came from that parent but are opposite members of the pair it had, while the remaining half came to each of them from the parent the other one did not have. This half of the genes are no more and no less apt to be alike than if the half brothers were unrelated members of the same population.

The most probable situation among a pair of full brothers is that one quarter of their genes will be duplicates received from the sire, another one quarter will be duplicates received from their dam, another quarter will have come to both from their sire but in each locus the genes will be opposite members of the pair he had, while the other fourth will have come to both from their dam but will be opposite members of the pairs she had.

This fact that pedigree and heredity are not identical was known before Mendelism but was then regarded as a mystery. Now we know that it is a natural—indeed an inevitable—consequence of the segregation of genes in parents which are not completely homozygous. However, anachronistic traces of the older view still persist in our speech and writings. Wonder is still sometimes expressed when two brothers are noticeably unlike, or such unlikeness is inferred to be evidence that the characteristic in question is not hereditary. In our everyday speech and among persons unfamiliar with genetics it is not yet generally appreciated that even for a perfectly hereditary trait (one unaffected by environment, dominance, or epistasis), full brothers or parent and offspring will usually differ in half as many of their genes as will unrelated members of the same population.

The probabilities stated for half and full brothers are averages; that is, they are more likely to happen than any other one result. Yet the

[1] Modifications of this for sex-linked genes will be discussed later in a separate section.

laws of chance cause some pairs of paternal brothers to receive more than one quarter of their genes as duplicates from their sire, while other pairs of brothers get less than one quarter. Although the average or most probable result remains at one quarter, it is theoretically possible for paternal brothers to have received anywhere from none to 50 per cent of their genes as duplicates from their sire. But if the number of genes is large, either of these extreme happenings would be very rare. The standard deviation of individual cases around the expected average of 25 per cent is $25/\sqrt{n}$ per cent where n is the number of independent pairs of genes involved. With $n = 25$, for example, about two-thirds of all pairs of paternal brothers would have received at least 20 and not more than 30 per cent of their genes as duplicates from their sire.[2] That still leaves room for some individuals actually to be noticeably more alike in their genes than others which have the same expected relationship. If we are interested in only one or a very few pairs of genes, such as the pair for the black-red contrast or the pair for the horned-polled contrast in cattle, relationship will mean little for a single pair of animals. However, even for a single pair of genes, the relationship figure will become dependable if we want to describe the average situation in a large group of pairs thus related.

Each additional generation, which intervenes in the line of descent through which two individuals are related, halves the fraction of their genes which are likely to be exact duplicates received from the ancestor which they have in common. That is why any one line of relationship between two animals gives an amount of relationship which is 1/2 raised to the nth power, where n is the number of generations (Mendelian segregations) intervening between the two animals in that line or path of relationship. If they are related through more than one line of descent, each such bit of relationship must be evaluated separately. Then these are added to obtain the total relationship.

Two individuals chosen at random from the population which is used as the basis for computing the relationship would have many genes alike, merely because those genes are widespread in that population. Among pairs of allelic genes chosen at random, q^2 will be AA and $(1 - q)^2$ will be aa, leaving only $2q(1 - q)$ of such pairs to be unlike. Relationship between two individuals is the *extra* likeness due to common ancestry. It shows what fraction of those genes which would be

[2] Linkage makes the n of this formula something more than the number of different linkage groups but less than the total number of genes involved. With 20 to 30 pairs of chromosomes in most farm animals, an effective n of something like 25 to 100 appears reasonable for considering the reliability of individual relationship coefficients when the animals' whole genotypes or their heredity for complicated characteristics are being considered.

unlike in pairs of individuals chosen at random from this same population are probably alike in the related pair. The average genetic likeness between random animals of this population is the zero point on the scale on which relationship is measured. Zero relationship does not mean absolute unlikeness in every gene any more than zero on the thermometer needs to mean the coldest temperature possible, or sea level means the lowest altitude possible.

The question of what population should be used as the base or zero point for measuring relationship in any particular case thus has some importance. In considering evolutionary questions, the population might logically be the whole species or even a larger group of some extremely remote date. This is what the taxonomist means when he says, for example, that sheep are more closely related to goats than they are to cattle but are more closely related to cattle than they are to horses. But in applying the idea of relationship to individual animals or herds, we never carry it back to such a remote base, partly because pedigrees necessary for computing relationship are not known that far back, partly because chance variations from the most probable distribution of genes will in some cases have been in the same direction in successive generations and can have become large, and also because the time involved in evolutionary questions is so enormous that selection and even mutation have had opportunity to produce important changes which would not show in the pedigrees. The most remote bases we actually use in animal husbandry are in connection with the history of different breeds where we may, for example, group the Jersey and Guernsey together as "Channel Island breeds," or group together the black and white lowland cattle of the regions along the shores of the North and Baltic seas as a group of breeds more closely related to each other than they are to the Channel Island breeds or to the mountain breeds of central Europe.

The most convenient population to use for a base in animal breeding problems with known individual pedigrees is usually the breed at a date not often more than four to six generations in the past. For example, two Shorthorns might be "unrelated" relative to the Shorthorn breed in 1910. This is the same thing as saying that they are probably no more and no less alike genetically than the average pair of Shorthorns chosen at random from among those born in 1910. Yet, if their pedigrees were traced further, they might be found to be related 20 per cent relative to the Shorthorn breed in 1870 and 50 per cent relative to the foundation animals entered in the very first volumes of the Coates Herd Book. If their pedigrees could be traced back to the time when they had ancestors in common with other bovine breeds or races, it might possibly be found for example, that they are probably alike in 70 per cent of the genes which would be different in random pairs of cattle from a population which included all cattle now living in Europe. No practical purpose would be served by tracing pedigrees that far; but the example may explain the apparent inconsistencies which occur when we compare relationships between members of the same breed, between animals of mixed breeding, or between animals belonging to different breeds or even to different species. The apparent inconsistencies arise because the populations chosen as bases for computing the relationships are not the same. The inconsistency is no more a real one than if we say that a certain mountain peak is 2,500 feet above the plane at its base but 12,000 feet above sea level. The height of the peak is the same in either case—the two figures differ merely because the base from which the height is measured is different in the two cases.

The processes of computing relationship do not allow for changes which mutation and selection may have caused in gene frequency. The errors caused by neglecting mutation are not serious unless the base for relationship was hundreds of generations in the past. Those caused by neglecting selection might be important for genes with major effects and under intense selection even when the base date is as recent as six or eight generations in the past. This is additional reason for not computing relationships to a distant base date. Instead one considers what the breed or race average was at a fairly recent date, how it differed from the average of other breeds, etc., and then considers the two related individuals in terms of how like each other they probably are in genes which would have been different in animals descended from the same breed or race at the base date but without ancestors in common since.

If two animals are related to an extent which is worth knowing for practical purposes (i.e., in addition to knowing whether they are members of the same breed), that relationship will usually come from ancestors not more than four or five generations back in the pedigrees of either. Even where there is some reason to express relationship relative to a more distant base, it is usually sufficient to trace the pedigree to a date about four or five generations back and then to assume that the ancestors at that time were a random sample of the breed. For example, in a study[3] of Holstein-Friesians born in 1909 these were found to be related to each other about 2.6 per cent relative to the foundation stock of about 1883. If a present-day Holstein-Friesian is related 40 per cent to another, both pedigrees being traced back only to 1909, we are not

[3] Lush, Jay L., Holbert, J. C., and Willham, O. S. 1936. Genetic history of the Holstein-Friesian cattle in the United States. *Jour. of Heredity*, 27:61–72.

apt to be seriously in error if we assume that the relationship found if both were traced back to 1883 would be about 41.6 per cent (40 per cent plus 2.6 per cent of the remaining 60 per cent). In other words, 40 per cent relative to 1909 is about the same as 41.6 per cent relative to 1883 in this breed.

In human relationships it is usually convenient to assume that the foundation ancestors in the two pedigrees being compared were random samples from the same population. This may lead to some discrepancies in a population like that of the United States, where some individuals are descended entirely from ancestors coming from one race while others are descended from crosses between two or more rather distinct races. People of the same race might consider themselves unrelated and yet on the average be alike in more of their genes than two first cousins who come from a racial mixture. If pedigrees were known as far back as the time when the races originally diverged from each other, the figures for human relationship would be reasonably consistent also, except for changes produced by mutation and natural selection and accumulated chance variations since the races ceased to intermarry freely. But human pedigrees are not known that far. Many people would have difficulty in even naming all eight of their great grandparents.

THE MEASUREMENT OF RELATIONSHIP

Measuring relationship is evaluating the probability that the two related individuals will have duplicate genes because they are related by descent. Each line or path of relationship is evaluated separately. The results are then added to get the total probability of likeness in their genes. It is often convenient to separate direct relationship and collateral relationship in the computations, although a given percentage of relationship represents the same probability of genetic likeness, regardless of whether it is collateral or direct. Direct relationship is that which comes about because one animal is the ancestor of the other, as parent and offspring or grandsire and grandson. Collateral relationship is that which comes about because both animals are descended in part from some of the same ancestors, as half and full brothers, uncle and niece, cousins, etc.

The first thing to do is to examine the pedigrees and find all the paths or lines of descent by which the two animals are related. To evaluate the closest paths first is usually more convenient and less likely to lead to duplication or omissions. Usually two individuals are not connected by many different paths of relationship unless there has been some inbreeding.

Direct relationship is measured by what animal breeders commonly call "percentage of blood." By "blood" is meant inheritance in general or the genes considered collectively. The physical substance, blood, is not actually transmitted from parent to offspring at all. The young embryo makes its own blood. Figure 29 shows the percentages of "blood" arranged in the form of a pedigree. The fractions come naturally out of the halving nature of Mendelian segregation.

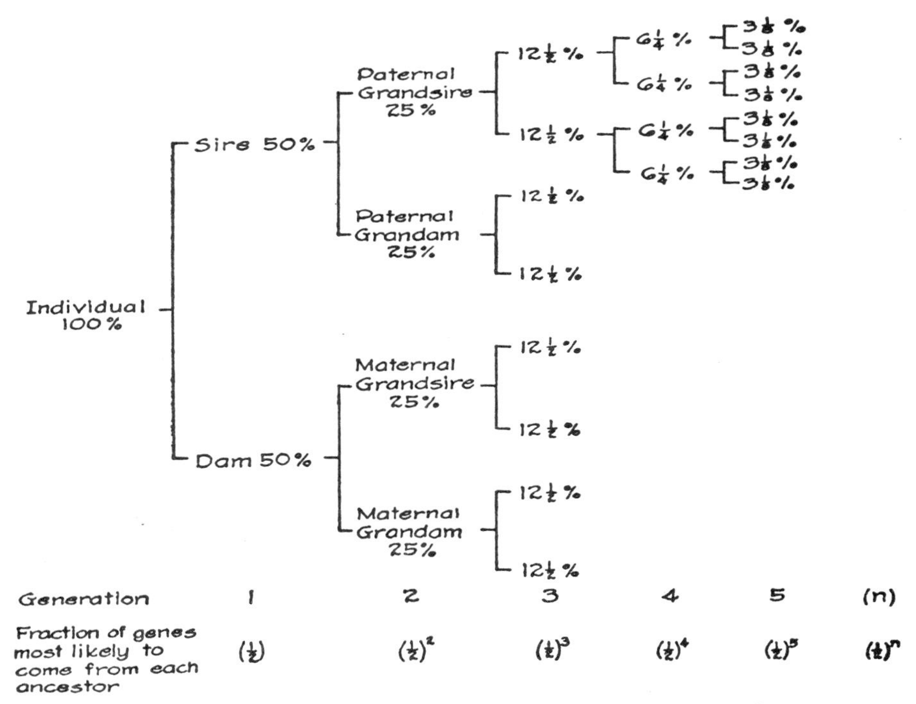

FIG. 29. The fraction of an individual's genes most likely to come from each ancestor.

The most probable proportion of an individual's genes to come from each of its ancestors is 50 per cent from a parent, 25 per cent from a grandparent, 12½ per cent from a great grandparent, and so on, the percentage being halved with each additional generation the ancestor is farther back in the pedigree. Because of the part chance plays in Mendelian inheritance, these percentages need not be exact for an ancestor farther back than a parent. So far as concerns any one great, great, great grandparent, the most probable expectation is that the individual will have inherited 1/32 of all its genes from that ancestor. Even in animals having 30 pairs of chromosomes, this does not average quite two chromosomes from each ancestor in that generation. Since there is some crossing-over, it is probable that the individual will still have at least a few genes from every ancestor in that generation. However, it could happen that certain ancestors in that generation would have contributed nothing at all to this individual's inheritance. So far as concerns the inheritance which the descendant has, such ancestors might as well never have existed except that they were a part of the living machinery without which this individual would not have been produced. As one goes farther back in the pedigree there are more and more opportunities for chance to have altered the percentage of blood. Somewhere not very far from the fifth or sixth generation, one must come to the place where it is probable that at least a few of the ancestors have not contributed anything to the inheritance which the individual has. However, when some of the ancestors in any one generation have contributed less, others have contributed more. All of the genes which the individual has, unless a recent mutation has occurred, were somewhere in every generation of its pedigree but were scattered among a larger number of ancestors with each additional generation back in the pedigree.

Where two ancestors are related only as ancestor and descendant—that is, where there is not also some collateral relationship between them—their relationship is simply the sum of the percentages of the ancestor's blood which the descendant carries. A correction for inbreeding (see next chapter) is necessary where ancestor and descendant differ in their inbreeding. If the ancestor is more highly inbred than the descendant, the relationship will be larger than percentage of blood; while, if the ancestor is less highly inbred than the descendant, the relationship between the two will be a little less, but in most actual pedigrees the correction for inbreeding is very small.

Figure 30 shows the pedigree of an Aberdeen-Angus heifer sold in 1931 in the Strathmore sale. This heifer "carries 81¼ per cent of the blood of" Earl Marshall. This figure is the sum of: 25 per cent for each of Earl Marshall's two appearances as grandsire, 12½ per cent for each of his two appearances as great grandsire, and 6¼ per cent for his one appearance as great great grandsire in this pedigree. By similar computations it will be seen that Earl Marshall 50th carries 87½ per cent of the blood of Earl Marshall and that Blackcap Empress 74th carries 75 per cent of the blood of Earl Marshall. The percentage of Earl Marshall blood in the daughter is naturally the average of that in her parents. These figures for percentage of blood need only a small correction for inbreeding (larger than usual in this case) to become the coefficients of relationship of Earl Marshall to these three descendants of his.

Collateral relationship between two animals is computed separately for each line of descent by which it is possible to go from one of them back to the common ancestor and then down to the other. Each generation in this line of descent is another Mendelian segregation halving the fraction of genes which are likely to be duplicates in the two animals because of their common descent. If there are many more than four or

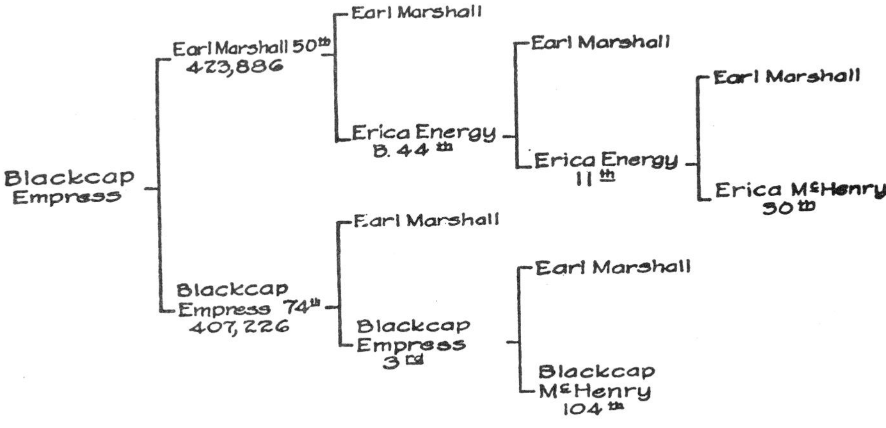

FIG. 30. A pedigree showing high relationship to the Aberdeen-Angus bull, Earl Marshall.

five intervening segregations, the amount of relationship through any one such line will be insignificantly small; but if there are many such lines, their total may be large enough to be of some importance.

The probability that cousins will have the same genes may be computed by extending the same process used for computing the relationship between half brothers. For each pair of genes the chance is one-half that a grandparent will give the same gene to both its offspring. Only one-half the times when this does happen will this same gene be transmitted to the one cousin. In only one-half of those cases will the other offspring transmit the same gene to the other cousin. There is, therefore, one chance in eight that an identical gene shall reach two cousins from a common grandparent. Even this concerns only half of their inheritance, since the other half comes from their other parent. Hence the probable genetic likeness between cousins on account of common descent from one grandparent is that 1/16 of their genes will be identical because of this. If, as is usually the case with human first cousins, they have two grandparents in common, this adds an equal probability of their having the same genes through descent from that other grandparent. This makes a total probability that 1/8 of their genes will be alike because of the common ancestry, while the rest of their genes are no more and no less apt to be alike than if they were unrelated members of the same freely interbreeding population.

Breeders sometimes measure collateral relationship by "percentage of common blood," but this can be very misleading. Full brothers have 100 per cent common blood but are only 50 per cent related; double first cousins (all four grandparents the same) have 100 per cent common blood but are only 25 per cent related. Quadruple second cousins (that is, individuals having no parents or grandparents in common but the same set of 8 great grandparents) would have 100 per cent common blood but would be related only 12½ per cent. Percentage of common blood is almost useless as a measure of relationship because the probability of the animals being genetically alike depends so much on how far back one must go to find that common blood. Percentage of common blood always sounds like closer relationship than it really is.

In Figure 30 the relationship between Earl Marshall 50th and Blackcap Empress 74th is entirely collateral since neither is an ancestor of the other and both have Earl Marshall for an ancestor. There are three different paths through which Earl Marshall 50th may have received genes from Earl Marshall and two paths through which Earl Marshall may have transmitted genes to Blackcap Empress 74th. Matching each of the three ways with each of the two ways makes six different ways in which these two descendants of Earl Marshall might have received duplicates of any gene which was in him. The fact that Earl Marshall is the sire of both contributes 25 per cent to their relationship. The two different ways in which he is the sire of one and the grandsire of the other contribute 12½ per cent each. Descent from him as grandsire on both sides contributes 6¼ per cent. Descent from him as sire of one and great grandsire of the other contributes another 6¼ per cent. Finally the small probability that these two animals would get identical genes by the long route from Earl Marshall as great grandsire of one and grandsire of the other contributes another 3⅛ per cent to their relationship. This makes a total of 65⅝ per cent, all of it coming through their descent from Earl Marshall, but that is still to be corrected for their inbreeding. A somewhat simpler way of figuring their relationship in this case, where they have only one ancestor in common, is to find that one of them has 75 per cent of the blood of Earl Marshall, while the other has 87½ per cent, and to *multiply* those two percentages together. This method gives the same answer; but if the two animals had more than one ancestor in common, the computations would have to be made separately for each such ancestor. This method will lead to difficulties if the common ancestors are related to each other.

These rules for computing relationships are nothing but counting the number of Mendelian segregations which have intervened in each line of descent connecting two individuals, and using $(1/2)^n$ as the fraction of their genes which are likely to be identical because the two animals received duplicate genes in that way from their common ancestor.

When one animal is an ancestor of the other and they are also related collaterally because both are descended from a third animal, it is usually more convenient to compute the direct relationship first. An example of this can be had from Figure 30 if we wish to learn how closely Earl Marshall 50th is related to his daughter, Blackcap Empress. They are directly related as sire and daughter, and in addition he may have received from Earl Marshall some genes for which she received duplicates through her dam. He traces to Earl Marshall in three different lines, the number of generations in each being 1, 2, 3, respectively, while she traces through her dam to Earl Marshall in two lines, the number of generations being 2 and 3 in those. Combination of each of the two with each of the three makes six different ways in which these two animals may have received duplicate genes by descent from Earl Marshall. The sum of the separate values for these six paths equals 32 13/16 per cent collateral relationship to be added to the 50 per cent direct relationship. That total is still to be corrected for the inbreeding, which in this case is intense enough to make that correction rather large.

Figure 31 shows another example. The relationship of X to A is both direct and collateral, while the relationship between X and Z is entirely collateral, neither being an ancestor of the other. The arrow diagram of the pedigree is often very convenient for showing at a glance the nature of the relationship. If the pedigrees are complicated by much inbreeding, the arrow diagram is almost necessary in order not to omit some lines of relationship nor count some a second time without being aware of having done so. The relationship of X to Z in Figure 31 will illustrate why percentage of common blood is not dependable as a measure of collateral relationship. It is not legitimate to say that 75 per cent of X's blood is the same as 100 per cent of Z's blood and to estimate their relationship through manipulating these figures of "common blood" in any way. It is legitimate to do this separately for each common ancestor and to say that Z has 50 per cent of the blood of D and X has 25 per cent of that blood and that Z and X are therefore related 50 per cent of 25 per cent, or 12½ per cent, through D. Also, it is legitimate to say that both Z and X have 50 per cent of the blood of C and that 50 per cent of 50 per cent equals 25 per cent of relationship between Z and X through descent from C. This 25 per cent may then be added to the 12½ per cent of relationship through D to give the total relationship of 37½ per cent.

Relationship between two individuals cannot be higher than 50 per cent unless some inbreeding has been practiced. The only exception to this concerns identical twins, and members of the same "clone" in plants and animals which can be propagated asexually. Relationship between members of such isogenic lines is 100 per cent, since they are

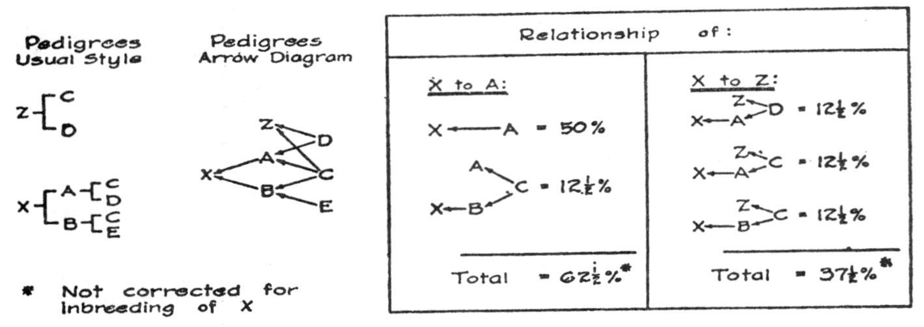

FIG. 31. Showing how relationship is computed.

really duplicates of the same zygote and there has been no intervening Mendelian segregation or recombination to permit them to have unlike genes. Since not much inbreeding occurs in most animal species or in man, we rarely have a chance to see the resemblance between animals related much more closely than 50 per cent. It is largely for this reason that identical twins are such interesting evidence about the importance of heredity. An increase from 50 to 100 per cent in genetic likeness makes a marked difference in the variation to be expected between individuals, especially in characteristics which are highly hereditary.

Two animals cannot be very closely related if all their common ancestors are distant ones. Rarely is much gained by going back farther than three or four generations in the pedigrees of two animals to see whether they are related. In an extreme case two animals might have the very same 16 great great grandparents, and yet their relationship to each other would be only 6¼ per cent if they had no parents, grandparents, or great grandparents in common and if there were no inbreeding involved.

Relationship is sometimes measured in "degrees," especially for legal purposes. In civil law a degree of relationship corresponds to a generation of Mendelian segregation. Thus parent and offspring are "related in the first degree," grandparent and offspring are "related in the second degree," uncle and nephew are "related in the third degree," etc. When the individuals are related through more than one line of descent, each line is computed and stated separately without combination into a single figure. In canon law and common law only the number of generations in the longer line of descent from the common ancestor is counted. Thus uncle and nephew are related in the second degree by canon law and common law.

THE EFFECTS OF SEX-LINKAGE

In the mammals the male is the heterogametic sex. His sex-linked inheritance comes entirely from his mother. Her sex-linked inheritance is present in duplicate, and among the pairs of sex-linked genes which are heterozygous in her, chance at segregation determines what sex-linked inheritance a son shall receive. The result is that 100 per cent of his sex-linked genes are like 50 per cent of hers. The square root of the product of these is the relationship (71 per cent) of son and dam for sex-linked genes. The sex-linked inheritance of a female comes equally from both parents; but since her sire does not have a duplicate set of sex-linked genes, chance plays no part in determining what sex-linked inheritance she shall receive from him. So far as his daughters are concerned, the situation is the same as if the sire were homozygous for all his sex-linked genes. The relationship between sire and daughter is also 71 per cent for sex-linked genes, since 100 per cent of those in the sire are like 50 per cent of those in his daughter. Daughter and dam are related 50 per cent for sex-linked genes, just as they are for autosomal genes. Sire and son are not related at all for sex-linked genes, since the son cannot receive any of those from his sire. The practical consequence of sex-linkage is to make the parent-offspring relationship higher between opposite sexes than it is between parent and offspring of the same sex. In birds, moths, and some fishes the female is heterogametic, but that leaves the practical situation nearly the same: namely, that males are a bit more closely related to their dams and females to their sires than either is to the parent which is the same sex. The lowest relationship for sex-linked genes is that between sire and son in the mammals, but between dam and daughter in birds.

This difference in relationship to ancestors of different sexes is partially, but not entirely, equalized as one goes further back in the pedigree, since a dam may transmit either the sex-linked genes she got from her sire or the sex-linked genes she got from her dam.

Since the farm animals have about 20 to 30 pairs of chromosomes and only one pair carries the sex-linked genes, it is improbable that the total amount of sex-linked inheritance is much larger than 3 to 5 per cent of the total inheritance. Hence, for practical purposes no material error is introduced by neglecting sex-linkage when computing relationships. (See table 20.) Occasionally there may be individual matings in which sex-linkage will play a noticeable part.

PRACTICAL USES OF THE RELATIONSHIP COEFFICIENT

The most important practical use of the relationship coefficient is to predict the merit of relatives of animals whose merit is known. This is the whole basis for using the merit of relatives to aid in making selection more effective. If nothing at all is known about an animal or its relatives, the only prediction we can make is that it will be an average animal of its breed. Neglecting for the moment the effects of environment, dominance, epistasis, and selection, if we know an animal related 40 per cent to the unknown one, the most probable prediction is that the unknown animal will deviate from the breed average in the same direction as the known relative does, but only 40 per cent as far. If a cow's genotype for fat production is 100 pounds above the breed average the most probable genotype of her daughter, if nothing is known about the sire except that he belonged to the same breed, is that the daughter's genotype will be 50 pounds above the breed average. Numerous practical difficulties and necessary precautions beset this use of relationship as a measure of how much weight to put on each relative in estimating the breeding worth of an animal. Most of those hinge around the inescapable fact that we are not sure of the genotypes of the relatives but only know their apparent merit. The two may be widely different if outward merit is much affected by environment, dominance, or epistasis. Other difficulties concern the weight to give each relative in combining information about several related ones, not all known equally well, into the best single estimate of what the unknown animal will be. The practical aspects of this were discussed in chapter 14. They put serious limitations on the practical usefulness of the principle—which is true when comparing equally well-known relatives of which one is to be used alone in such estimates—that the attention to be given such relatives is in proportion to their relationship to the animal to be estimated. The relationship between an individual and the average genotype of a group of its relatives may be higher than its relationship to any one of them. The simplest and most important example of this is the case of half sibs. In the extreme case of a characteristic perfectly hereditary in the simple additive way, and unselected half sibs from a random breeding population, the equation for predicting with the least error what an individual will be from knowing p of its paternal half sibs and m of its maternal half sibs is:

$$\text{Animal} = \text{breed average} + \frac{p}{p+3}(\text{paternal half sibs minus breed average}) + \frac{m}{m+3}(\text{maternal half sibs minus breed average})$$

As p and m increase from one to at least four or five, the usefulness of this equation goes up rapidly and it soon comes close to the limit set by perfect knowledge of the genotypes of sire and dam of the individual being predicted. Of course, in actual practice there are effects of environment and dominance and epistasis and possible previous selection of half sibs which will usually make it wise to put less weight than this on what the half sibs indicate. The above equation shows quantitatively what we are recommending if we advise choosing a son of a proved sire and a proved dam.

To the extent that the trait being measured is affected by environment, dominance, and epistasis, actually observed likeness will be less than the genetic likeness unless each pair of related individuals tended to be exposed to an environment alike for them but different from the environment of other pairs. In that case the observed may be higher than the calculated. Thus, in a random bred population half sisters are genetically as much alike as grandam and granddaughter; but each pair of half sisters is more apt to have been exposed to the same more or less unusual environment than is each grandam and granddaughter pair. Hence, in such data as official dairy records, it is to be expected that the observed likeness between half sisters will be larger than that between grandam and granddaughter. For the same reason paternal half sisters, which are usually contemporaries, may be expected actually to resemble each other a little more closely than maternal half sisters. The latter may have been born and reared several years apart.

Dominance and epistasis make observed likenesses generally lower than calculated ones, but in certain relationships—of which the most important is that between full sibs—dominance produces a little extra likeness on its own account. This partly offsets its general tendency to lower likenesses.

If breeders' ideals diverge enough that they have been selecting for distinctly different types within the same breed, then animals in the same herd are apt to resemble each other, relative to the whole breed, more than their relationships indicate.

GALTON'S LAW

Two distinct although related ideas are sometimes confused under the name "Galton's Law." The first is the observed fact that the correlation between parent and offspring is nearly $+.50$ in populations where there is not much inbreeding and where the trait being measured is highly hereditary. The correlation between an animal and a more remote ancestor is halved for each additional generation which separates them. This is precisely what was shown in Figure 29 and was discussed under "percentage of blood." It was an observed fact in Galton's time and had been used by practical breeders in England from at least as long ago as 1815.[4] The only material difference now is that we understand why it is a natural consequence of the laws of inheritance, provided the population is mating at random and the trait being studied is entirely hereditary in the simple additive manner. Naturally the extent to which these conditions are fulfilled varies in different populations and even for different traits in the same population and often is known only within rather wide limits.

The other part of Galton's law—which unfortunately is the part most frequently quoted—is true only in a specialized statistical sense. The square of the correlation coefficient measures the portion of the variance in one variable which disappears in data in which the other variable is constant. In this special statistical sense the individual's inheritance is ¼ determined by its sire and ¼ by its dam. That instantly raises the question: What determines the remaining ½? Galton reasoned that this same process would be repeated with the sire and with the dam, each of them being ¼ determined by its sire and another ¼ by its dam, and that if this process were pursued far enough the fractions would ultimately add up to one. Consequently he proposed as a general "law" that the individual's inheritance was ¼ determined by its sire and ¼ by its dam, 1/16 by each of the four grandparents, 1/64 by each of the 8 great grandparents, and so on, each ancestor contributing just ¼ as much to the total inheritance as did the one a generation nearer to the individual. This is usually pictured as in A, of Figure 32. Because this seemed a logical scheme, it was accepted rather widely and even today is sometimes quoted with approval. Unfortunately, the obvious inference from this diagram (that if one knew all about all of the ancestors, he would know all about the heredity of this individual) is not true.

Had Galton used multiple correlation technique (as Yule pointed out at the third International Genetics Congress in 1906) he would have found that the partial correlation between grandparent and grandson or granddaughter, the intervening parent being held constant, is zero in any population in which the correlation between parent and offspring is .50 and the correlation between grandparent and grandson or granddaughter is .25. If the parent's heredity is not known, it is still true in the technical statistical sense of the word that the individual is "determined" 1/16 by each of its grandparents. That is shown for the paternal grandam by D in Figure 32. But if the parent's heredity is known, knowledge of the grandparents adds nothing to the

[4] According to the encyclopedia of J. G. Krünitz.

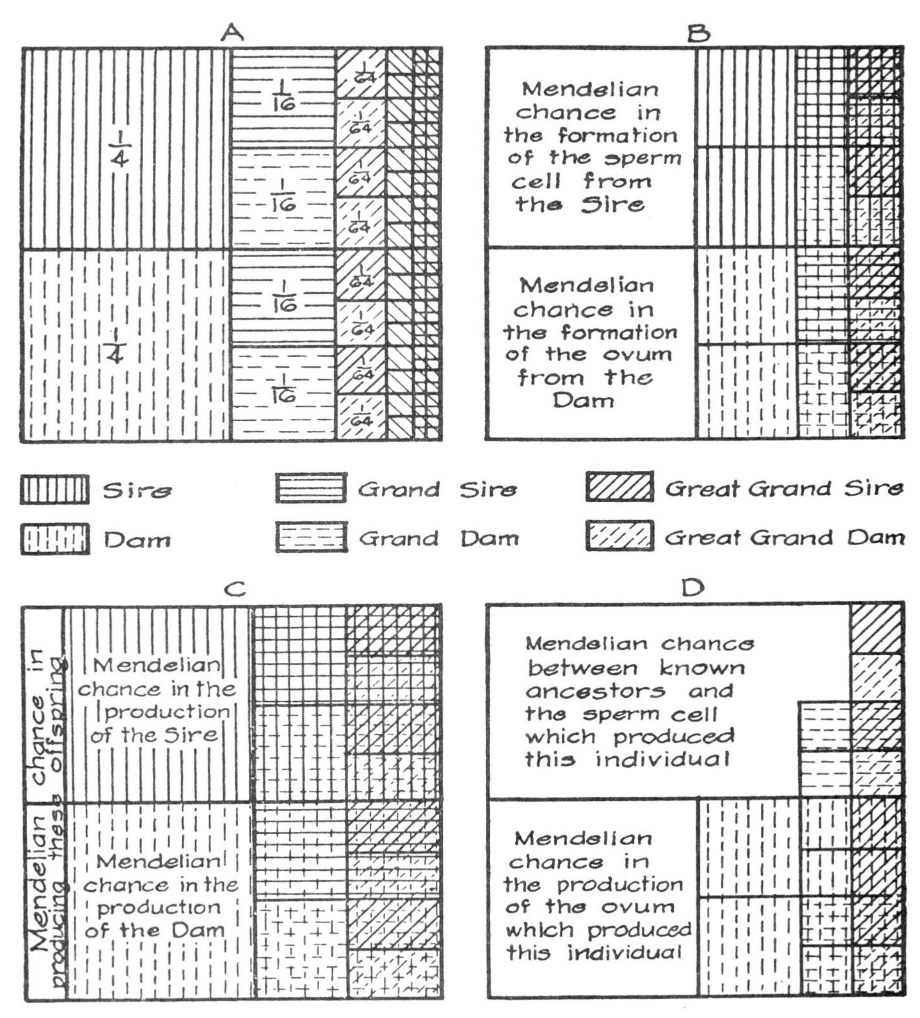

FIG. 32. Galton's "law" of the relative importance of various ancestors, pictured for a characteristic perfectly hereditary in the simple additive way. *A:* As formulated by Galton, each ancestor being considered singly. *B:* Correctly pictured for an individual animal in a random breeding population. That which can be predicted from the more remote ancestors is included in what can be predicted from the intervening ancestors. Mendelian chance at segregation accounts for half of the determination in the statistical sense of that word. *C:* Same as *B* except that this concerns the average inheritance of nine unselected offspring from one pair of parents. The effects of chance in the nine segregations which produced the sperm cells and in the nine segregations which produced the ova tend to cancel each other to such an extent that Mendelian chance now determines only one-tenth of the average inheritance of these nine instead of the one-half which it determines for each individual. *D:* Determination of the individual by known characteristics of its ancestors in the case where males cannot express the characteristic themselves and neither the sire nor the two grandsires were progeny-tested.

knowledge of the individual. That which could be predicted from a knowledge of the grandparent is included in that which can be predicted from a complete knowledge of the parent. In actual practice the correlation between parent and offspring is usually less than .50, and something is still to be gained by studying the more remote ancestors, as was mentioned in chapter 14.

There remains the question: If, under the simple conditions, the individual's inheritance is half determined by its parents and not at all by more remote ancestors when the merits of the parents are known, then what does determine the other half? Remember that (as always in breeding problems) we mean variations, not absolute magnitudes; i.e., we mean what causes the differences between it and the other individuals in the population to which it belongs. The answer in this same statistical sense is that the individual is half determined by chance at segregation of the genes in its parents. This is pictured in Figure 32 by B, which is correct for a characteristic which is entirely hereditary in the simple manner and for an individual animal in a random-bred population. The individual certainly cannot have any genes which its parents did not possess, no matter whether its more remote ancestors had them or not.

The average heredity of many offspring from a particular pair of parents presents a different problem. The results of chance at Mendelian segregation will be different from one offspring to another and hence will tend to cancel each other. In the simplest case, where there is no epistasis or dominance, where the trait is not affected by environment, and where mating is at random, the average of n full sibs is $\frac{n}{n+1}$ determined by their parents and $\frac{1}{n+1}$ by chance at segregation of the genes in their parents. Determination of the progeny average by the genotypes of the two parents starts at ½, or 50 percent, when there is only a single offspring but becomes 80 per cent for four offspring, 90 per cent for nine, and approaches close to 100 per cent as the numbers of offspring become very large. This is pictured in C of Figure 32 for the case where $n = 9$. This is why complete knowledge of the genotypes of the parents would permit predicting with almost perfect accuracy what the average of their progeny (if numerous) will be, even though it would not permit highly accurate prediction of what each individual offspring will be. D of Figure 32 shows Galton's law when the merits of some of the near ancestors on one side of the pedigree are not known.

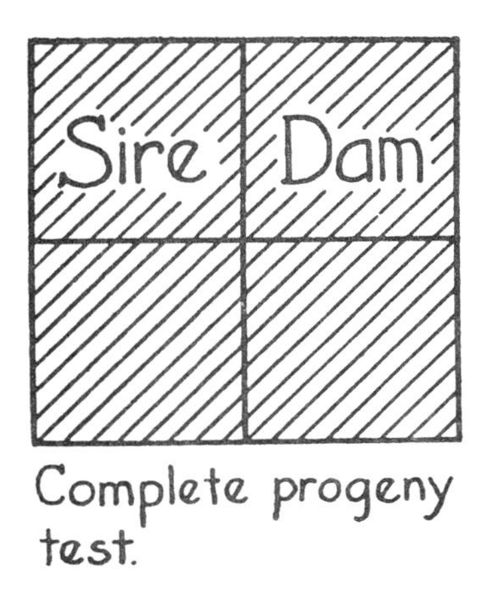

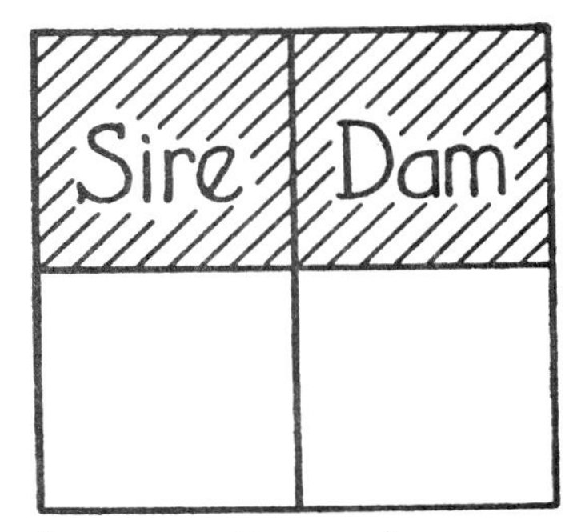

Limit of pedigree study.

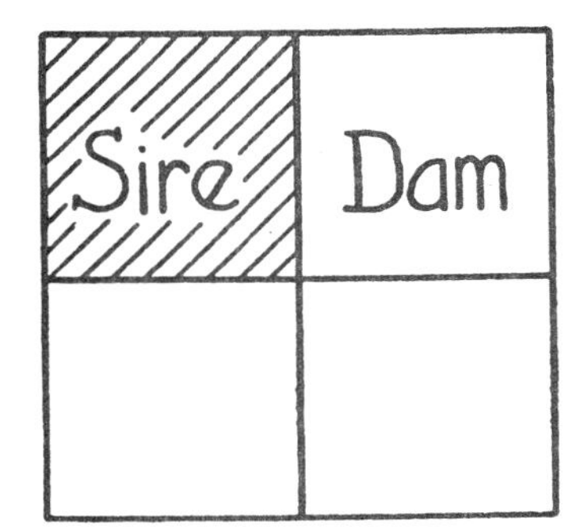

Son of a completely proved sire and a dam of unknown merit.

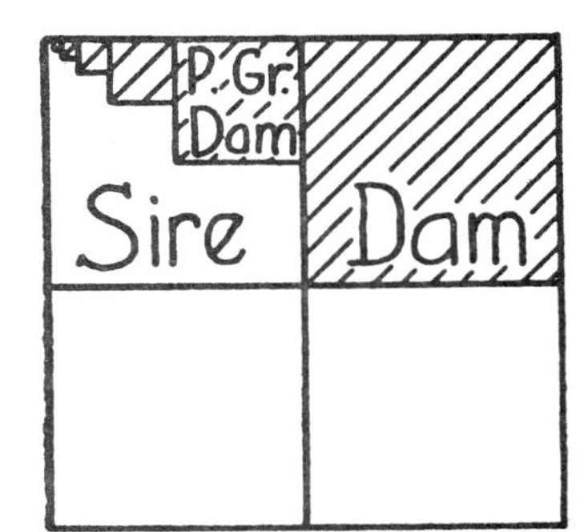

Complete knowledge of female ancestors only.

Fig. 33. Relative accuracy of a progeny test and of various kinds of pedigrees when the information in each is as complete as is theoretically possible under the conditions named. Shaded areas represent portions of the individual's inheritance determined (in the statistical sense) by complete information of the kind pictured. The correlation between the actual and the predicted breeding value of the individual would be the square root of the shaded fraction of the area; i.e., 1.0 in the upper lefthand corner, .71 in the upper right, .50 in the lower left, and .58 in the lower right. The unshaded areas represent possibilities for Mendelian segregation to have made the individual's actual breeding value higher or lower than its expected breeding value.

SUMMARY

The genetic resemblance between individuals is based on the probability that they have received identical genes from animals which are ancestors of them both.

The "coefficient of relationship" is the percentage of genes which are probably identical in the two related individuals because of their relationship by descent. The rest of the genes in those two individuals are no more and no less apt to be alike than if the two individuals were chosen at random from the base population. The relationship coefficients are therefore relative to the population chosen as a base. For most practical purposes that population is the breed at a date not much more than four or five generations back in the pedigrees being traced.

The familiar figure for "percentage of blood" is the coefficient of relationship of ancestor to descendant, provided (1) that both ancestor and descendant are equally inbred or not inbred at all and (2) that they are not related in any other way than as ancestor and descendant. "Percentage of common blood" cannot be used to measure collateral relationship without modifying it intricately to allow for the number of ancestors from which that "common blood" comes and to allow for the number of generations each such ancestor is removed from each of the two animals whose relationship is being measured.

The probable resemblance between collateral relatives must be figured separately for each line of descent through which the relatives may have received the same genes. The contribution of each such line is the fraction ½ to the nth power where n is the number of Mendelian segregations intervening in the line of descent from the one animal back to the common ancestor and down to the other animal again. This requires some correction in case the related animals or the common ancestors are not equally inbred.

Total relationship is computed by adding the relationships from each separate line of descent if two individuals are related through more than one line.

"Galton's law" is correct in the sense that the relationship between ancestor and descendant is halved with each additional generation which intervenes between them. It is not correct in the sense that the individual's heredity is completely determined by the heredity of its ancestors. In that sense in a random-bred population the individual is one-fourth determined by each parent and one-half determined by chance in Mendelian segregation. Determination by more remote ancestors is included in the determination by the parent.

The existence of sex-linkage causes sons to resemble dams a little more than they do their sires and makes daughters resemble their sires a bit more closely than they do their dams. This effect must be small in most cases.

Relationship measures the probability that individuals will be alike *in their genes.* Their actual outward likeness depends also upon how much the traits being measured are affected by environment, dominance, and epistasis, upon the extent to which their environments were correlated, and upon whether their ancestors were mated like to like.

The most important practical use for relationships is in predicting the most probable merit of unknown or perhaps even unborn individuals from the merit of their known relatives.

REFERENCES

Davenport, E. D. 1907. Principles of breeding, pp. 525–34.

Fairchild, David. 1921. A genetic portrait chart. Jour. of Heredity, 12:213–19.

Wright, Sewall. 1922. Coefficients of inbreeding and relationship. Amer. Nat., 56: 330–38.

———. 1923. Mendelian analysis of the pure breeds of livestock. I. The measurement of inbreeding and relationship. Jour. of Heredity, 14:339–48.

CHAPTER 21

The Consequences and Measurement of Inbreeding

Inbreeding is the mating of closely related animals. Everyone agrees to that general definition, but there is much diversity of usage about how closely related the mates must be before the mating should be called inbreeding. Many practical breeders restrict the word inbreeding to the mating of full brother and sister or of parent and offspring. Others would call the mating of half brother and sister, or the mating of grandparent to grandson or granddaughter inbreeding. The broad scientific definition is that inbreeding is the mating of animals more closely related to each other than the average relationship within the population concerned. Such matings tend to make the offspring more homozygous than if their parents were of average relationship to each other. Mating of animals less closely related than the average of the population concerned is outbreeding.

The population concerned would usually be the whole breed when this definition is applied to animal breeding, but might be the race or the whole species when considering the part which inbreeding may play in evolution. The intensity of the inbreeding is very slight, however, unless the mates are quite closely related or the inbreeding is continued for many generations. This leads to the convenient situation that the great majority of all breedings which take place within a pure breed are practically neutral so far as any inbreeding or outbreeding effect is concerned and may be classed as random breeding, even though one does not know the average relationship within the breed. The practical use of the definition that inbreeding is the mating of closely related animals merely requires agreement as to how close that relationship must be.

It is impossible to define inbreeding simply as the mating of related animals. All animals that can be mated at all are related, at least slightly. Each individual has two parents, four grandparents, eight great grandparents, and so on, the number of ancestors doubling each generation. In the tenth generation of its pedigree an animal will have more than a thousand ancestors if there has been no inbreeding. If two animals are unrelated in the nearest ten generations of their pedigrees, the thousand ancestors of the one cannot include any of the nearly contemporary thousand ancestors of the other. If there has been no inbreeding, each animal has more than a million ancestors in the twentieth generation of its pedigree. If the pedigree is followed much further, these numbers become greater than the number of animals alive at that time could have been. For example, in man there are about three generations to the century. The number of ancestors each person had living at the time of William the Conqueror would be about 2^{26}, or roughly a little more than 67 millions, if there had been no mating of relatives. Now there never were that many people in Great Britain at any one time. Anyone descended entirely from British ancestors must have had an enormous amount of repetition of ancestors that far back in his pedigree, especially since many individuals living at any one time leave no descendants. One whose ancestors came from several nations has only to follow his pedigree a few centuries further back (no further than to 900 A.D. at the outside) to find that, if there had been no mating of relatives, he would have had more ancestors alive at that time than there were people on earth!

This situation is more extreme in livestock breeding. The Brown Swiss breed in the United States is descended entirely from 129 cows and 21 bulls which were imported into this country. In American Rambouillet pedigrees about 45 per cent of the lines traced back at random end in sheep from the von Homeyer flock in Germany. Over half of the random pedigree lines of the Shorthorn breed go to one bull, Favourite. Similar things are true of other breeds, although few breeds are yet explored in detail from this point of view. Moreover, this includes only what has happened since pedigree recording began. That is a comparatively short time—about 150 years in the Shorthorns and only about 50 years in the other two breeds mentioned.

The definition of inbreeding *must* be relative to some group or population. Pure breeding, for instance, is inbreeding relative to the whole species but need not be and rarely is inbreeding of noticeable intensity relative to the breed. Figure 34 illustrates the situation diagrammatically from closest inbreeding to widest outbreeding. To the left of random mating are the inbreeding matings while the outbreeding ones are to the right. The closeness of the two lines to each other in Figure 34 represents how closely the mates are related to each other. That ranges from complete identity in the case of self-fertilization (possible in most plants but impossible in any farm animal) through close relatives, members of the same breed, members of different breeds but the same species, members of different species within the same genus, and perhaps even members of different genera. The offspring of species crosses usually exhibit some degree of sterility. Generic crosses are very rare. Presumably the genotypes of the individuals are so unlike that the union, even if possible at all, produces no living offspring.

THE CONSEQUENCES OF INBREEDING

The primary effect of inbreeding is to increase the probability that the offspring will inherit the same thing from sire and dam. This tends

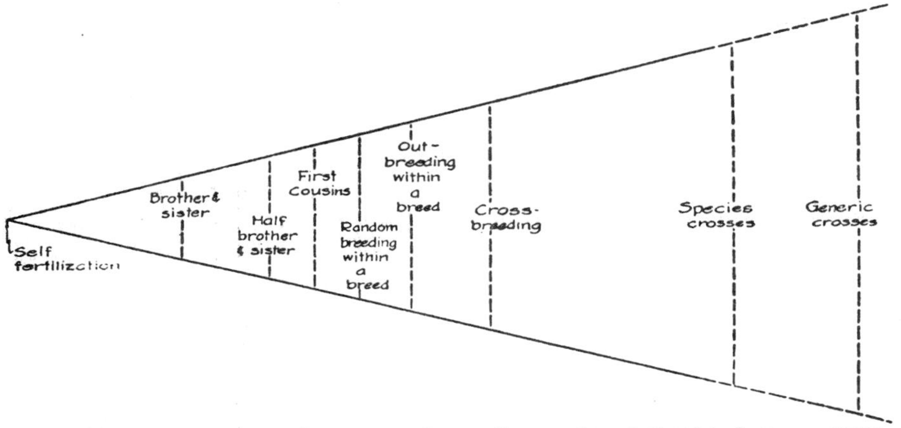

FIG. 34. Degrees of inbreeding arranged according to the relationship between mates.

to lower the percentage of heterozygosis in the population and to produce relationships higher than 50 per cent. All the other effects of inbreeding result from those. In each inbred line the genes which are to be in the next generation are a sample of those which were in the preceding generation. Because the sample is small the gene frequency in it will often by chance differ considerably from the gene frequency in the generation from which the sample came. Thus the gene frequency can wander back and forth until it reaches either zero or one. Then the line is homozygous for that particular gene or its allel. This homozygosis cannot be lost, except by mutation, as long as the inbreeding is continued. The genes which are heterozygous are still subject to the possibility of becoming homozygous each generation the inbreeding continues. The fewer animals there are in the inbred line the smaller is the sample of gametes which are needed to constitute the next generation and the farther the frequency of each gene can drift up or down in any one generation.

Because this change in gene frequency is random (equally likely to be up or down), inbreeding is in conflict with selection whenever selection is tending to keep a gene at some equilibrium frequency, as when the heterozygote is preferred. Inbreeding is continually causing the gene frequency to drift away from that equilibrium point in either direction and selection is continually tending to take it back there. When the inbreeding is very mild and selection is intense, the gene frequency cannot get far from its equilibrium value. But when the inbreeding is intense it may overwhelm selection and carry gene frequencies far away from their equilibrium values, some above and some below.

If the inbreeding is intense and continues long enough, and if there are no mutations, the ultimate condition approached in each inbred line is complete homozygosis in all pairs of genes. In some pairs it will be the less desirable and in other pairs the more desirable gene which becomes homozygous. Each inbred line is likely to differ from every other inbred line in regard to which particular combination of genes becomes homozygous in it, if many different pairs of genes are involved. Inbreeding in animals almost never comes close to complete homozygosis in actual practice. Even self-fertilized plants reach an equilibrium point where the further loss of the remaining small amount of heterozygosis just equals the new heterozygosis which mutations produce in each generation. Mutations are so rare that for practical purposes they can be neglected in considering inbreeding but they are mentioned here for completeness.

The Mendelian basis of the inbreeding effect can be illustrated most simply by the extreme case of self-fertilization. Each generation in each inbred line consists of one individual in which, of course, each gene is either homozygous or heterozygous. That is, gene frequency can have only the values: .0, .5, or 1.0. Genes which are already homozygous in the whole line must remain so as long as the inbreeding continues, unless a mutation occurs. Pairs of genes which are heterozygous may be considered as a population of the two kinds of genes in equal numbers, from which two genes are to constitute the population in the next generation. The probabilities that those two will be *AA*, *Aa*, or *aa*, are in the ratio 1 : 2 : 1. Thus half of all heterozygous genes in a self-fertilized individual may be expected to become homozygous in its offspring. Selection favoring either homozygote will tend to hasten the rate of approach to homozygosis, while selection favoring the heterozygote will tend to retard it. If there are many heterozygous pairs of genes, several will become homozygous in every individual; and selection can have only trivial effects in delaying any such rapid rate of approach toward homozygosis. For example, if a self-fertilized individual is heterozygous for 10 pairs of genes, only 1 offspring in 1,024 would be as heterozygous as its parent. Even in species with reproductive rates which would permit intense selection, it would be impossible to recognize genotypes accurately enough to find such a rare heterozygous offspring every time without mistake. If the inbreeding rate were lower or the number of heterozygous genes were less, selection would have more chance to alter the consequences of inbreeding. Actually, with each animal heterozygous for scores of genes, many of which have only minor effects which may be blurred by the effects of environment, the difference between intense selection and no selection under self-fertilization has little more effect on the outcome than the fate of a man dropped into the Niagara River just a few yards above the falls would be affected by whether he is a good swimmer or a poor swimmer! The difference in swimming ability might be of tremendous importance if he were in comparatively still water, as is roughly analogous to the situation in a population in which the inbreeding is very mild, but would only rarely make a detectable difference in the results in the presence of the much more powerful force of the swiftly moving water.

Self-fertilization is impossible in the higher animals, but the Mendelian basis of the inbreeding effect may be illustrated with the continued inbreeding of full brother and sister. For each allelic set of genes this constitutes a population of four genes—two in the sire and two in the dam—from which four are to come to constitute the next generation. Gene frequency can have only the values: .00, .25, .50, .75, or 1.00. Although only one of them will occur in any one generation in any one line, six different types of matings are possible. Those types are:

Type	Mating	Frequency of A in this inbred line
1	$AA \times AA$	1.00
2	$AA \times Aa$	.75
3	$AA \times aa$	.50
4	$Aa \times Aa$	.50
5	$Aa \times aa$	.25
6	$aa \times aa$	.00

If either the first or the last of these prevails, that line will remain homozygous for that gene indefinitely, unless a mutation occurs. If either the second or the fifth type prevails in one generation, there is one chance in four that this line will become homozygous in the next generation, two chances in four that it will remain the same, and one chance in four that the next generation will be like the fourth type of mating. When the line is of the third type, it will change to the fourth type in the next generation. When it is of the fourth type, there is one chance in eight that it will become homozygous in the next generation, four chances in eight that it will change to types two or five, one chance in eight that it will change to type three, and two chances in eight that it will remain the same. Once the inbred line becomes type one or type six, it remains that way. It can shift back and forth among the other four types; but from them it will occasionally drift into types one or six, from which it cannot return. Hence the ultimate end of all such lines would be type one or type six if the inbreeding were continued long enough. The drift from one type to another is so rapid in a population as small as a line inbred full brother by sister that, after the inbreeding gets well under way, nearly one-fifth of all the genes which are still heterozygous in the line at any given time become homozygous in the next generation. In larger populations the gene frequency would fluctuate less extremely, but in any finite population it would do some shifting. Inbreeding is only an extreme form of a process which exists to some degree in all populations. The fact that the number of breeding individuals in the poplation is finite permits gene frequency to vary because of the sampling which takes place when the genes of one generation are replaced by the genes of the next.

This Mendelian proof of the nature of the inbreeding process was studied as long ago as 1914 by Fish, Jennings, and Pearl,[1] but it becomes extraordinarily difficult to follow even for regular inbreeding systems milder than full brother by sister and practically impossible to follow in the irregular inbreeding which occurs with farm livestock. Wright in 1921 published a generalized explanation of the consequences of milder forms of regular inbreeding and of irregular inbreeding. In 1931 he generalized this still further to establish the identity of the inbreeding effect and the general consequences of finite population size, even in cases which we would not ordinarily call inbreeding.[2]

By changing heterozygotes to homozygotes, inbreeding brings to light many of the recessive genes which would otherwise remain hidden. Most recessive genes have less desirable effects than their alleles. Inbreeding, therefore, usually lowers the average outward merit of the inbred animals. Inbreeding permits more rapid improvement of the breed by getting the recessive genes out from the shelter of their dominant alleles so that they can be found more readily. For example, if 1 per cent of the calves in the Aberdeen-Angus breed are now born red and the breeders were all to begin suddenly to inbreed as intensely as mating parent and offspring or full brother and sister, the percentage of red calves would in the first generation of inbred calves go up more than threefold. The distribution of the genotypes before and after the inbreeding would be as follows:

[1] *Amer. Nat.*, 48:57–62, 491–94, 693–96, and 759–61.

[2] *Genetics*, 6:124–43 and 16:106–29.

Before inbreeding:	81 % *BB*	18 % *Bb*	1 % *bb*
After one generation of inbreeding:	83¼% *BB*	13½% *Bb*	3¼% *bb*

The percentage of heterozygosis would decrease one-fourth, the genes which would have been in heterozygous individuals now being equally divided among the two kinds of homozygotes. The gene frequency has not changed, but only the zygotic ratio. On account of dominance, the desirable increase in *BB* and decrease in *Bb* individuals would not be apparent outwardly. The only readily apparent change would be the threefold increase in the proportion of red calves.

For ages men have observed this general fact that inbreeding tends to produce a certain amount of degeneration or decline in individual merit. Formerly it was thought that the inbreeding in some mysterious way actually damaged the inheritance and thus directly caused this degeneration. Now it is understood that the inbreeding merely acts somewhat as a detective does in uncovering crime—not in creating it. The undesired recessive genes are there all the time, but homozygous recessive individuals appear more frequently when inbreeding begins.

More than any other breeding method, inbreeding tends quickly to separate the population into many distinct families, each of which is uniform within itself but distinct from others. This splitting of the population into distinct families offers more opportunity for selection between families than could take place in a random-bred population. Inbreeding without selection leads toward a variable population composed of unlike lines. If inbreeding were carried to completeness without selection, the composition of the resulting population would be the same as if the gametes from the initial population had suddenly been transformed into zygotes by doubling all their chromosomes. If all the genes acted only in an additive way, the genetic variance in the new entirely homozygous population would be exactly twice what it was in the original random-bred population, but would be entirely between families, with no genetic variance within families. If many of the lines were discarded, the population as a whole might then become more uniform than it was before the inbreeding began; but that would be primarily the result of discarding the poorer inbred lines rather than a result of the inbreeding itself.

Inbreeding is especially powerful in forming families because the genetic likeness of the mates does not depend upon the breeder's ability to recognize which animals have similar genes when he mates them together, as is the case with selection and with assortive mating. The effects of inbreeding are not limited by the breeder's mistaking the effects of environment, dominance, or epistasis for the effects of genes, as the effects of those other breeding methods are. That is why inbreeding is particularly powerful and useful in breeding for characteristics which are only slightly hereditary. This difference also leads to the general situation that under an inbreeding system the mates are more alike in their heredity than they are in their appearance, whereas under assortive mating they appear more like each other than they really are. Where the correlation between genotype and phenotype is low, this difference may be extreme.

Inbreeding may also produce some degeneration of individuals by scattering the genes which in certain combinations produce a desirable effect but which separately produce neutral or undesirable effects. If there really is much of this epistasis, and if a breed has been under strict selection for some time, there will have been a tendency to bring together, in combinations where they will show good effects, those genes which nick together well. Since most of those combinations will still be heterozygous, at least for some of the genes concerned, any inbreeding will tend to separate those combinations again and thus will lead to some apparent loss of individual merit in addition to that coming entirely from uncovering recessives.

Also inbreeding will lower the average merit of the population directly by reducing heterozygosis among those pairs of genes in which the heterozygote is more desirable than either homozygote.

MEASURING THE INTENSITY OF INBREEDING

Since the primary effect of inbreeding is to increase the probability of receiving duplicate genes from sire and from dam, the measure of inbreeding ought to be one which shows how much decrease in heterozygosis is to be expected from that particular kind of inbreeding. Such a coefficient will mean little so far as concerns one pair of genes in one animal, but it may tell much about the average condition of one pair of genes in a whole herd or breed, and it will tell much about the average heterozygosity of the whole group of genes in an individual animal.

There seems to be no possibility that we shall ever be able to count the actual number of heterozygous genes in each animal. Each animal must be compared with the average condition in the population chosen for a base. The most convenient population to use for a base in animal breeding is usually the breed at the date to which the pedigrees are traced. It is then assumed that those foundation ancestors were a random sample of the breed at that date. Occasionally that assumption may lead to some error, when a pedigree if traced another generation or two would have revealed that the foundation ancestors were highly related to each other.

The inbreeding coefficient (devised by Wright[3]) starts at zero for random mating, in which case the probable proportion of heterozygosis is $2q\ (1 - q)$, and increases toward 100 per cent as the probable proportion of heterozygosis goes toward zero. The inbreeding coefficient has to be expressed relatively rather than in terms of the actual number of heterozygous genes. For example, if the average Shorthorn of 1910 was heterozygous for 200 pairs of genes, then a Shorthorn, so bred that its inbreeding coefficient is 25 per cent relative to 1910 as a base date, will probably be heterozygous for only 150 pairs of genes. On the other hand, if the average Shorthorn in 1910 was heterozygous for a thousand pairs of genes, a Shorthorn showing an inbreeding coefficient of 25 per cent would probably still be heterozygous for about 750 pairs. An inbreeding coefficient of 25 per cent in one breed may not mean the same number of heterozygous genes as an inbreeding coefficient of 25 per cent in another breed. This need not bother us, since we really take that into account at the start when we recognize that the one breed is more uniform than the other.

The inbreeding coefficient of an individual is exactly one-half the relationship between its parents unless those parents are themselves inbred, in which case some correction (usually small) for that is to be made. The formula is as follows:

$$F_X = \tfrac{1}{2}\ \Sigma[\ (\tfrac{1}{2})^n (1 + F_A)\ ]$$

where F_X is the inbreeding coefficient of animal X; n is the number of generations (the intervening segregations) in a line by which sire and dam are related; F_A is the inbreeding coefficient of the common ancestor (A) out of whom that line of descent divides; and Σ is the summation sign meaning that each such path of relation between sire and dam is to be evaluated separately and then all the results are to be added together. The $\Sigma[\ (\tfrac{1}{2})^n]$ part is exactly the formula discussed in the preceding chapter for relationship when no inbreeding is involved. Thus, when the parents are not inbred the inbreeding of the offspring is exactly half of the relationship of its parents to each other. The factor $(1 + F_A)$ corrects for the fact that, because an inbred ancestor (A) is probably homozygous for more pairs of genes than a random-bred ancestor, two gametes coming out of it will usually be alike in more of their genes than two gametes from a non-inbred ancestor. As a numerical example,

[3] *Amer. Nat.*, 56:330–38 or *Jour. Heredity*, 14:339–48. A different measure of inbreeding had been proposed earlier by Pearl (*Amer. Nat.*, 48:513–23, 1914). It was based on a ratio between the number of different ancestors an animal actually had and the number it would have had if there had been no inbreeding. In general, it was a useful standard for comparing different intensities of inbreeding but in many cases gave inconsistent results. It is now of historical interest only, although references to it occasionally are made in current writings.

if a non-inbred ancestor is heterozygous for 200 pairs of genes, the average situation concerning two gametes from it is that they will have unlike genes in about 100 of those loci. From the same population an ancestor inbred 25 per cent would probably be heterozygous for only about 150 pairs of genes. Two gametes from it would be unlike in only about 75 loci. If the ancestor were complete inbred ($F_A = 1.0$, which can only be approached but not actually reached in farm animals), all the gametes from it would be exactly alike. This would be equivalent to eliminating one Mendelian segregation, and hence one 50 : 50 chance of unlikeness, from each line of relationship between sire and dam through this ancestor.

As a Mendelian example of why the formula for the inbreeding coefficient takes the form it does, consider the case of *X*, a double grandson of *A*, with pedigree as follows:

$$X \begin{cases} W \begin{cases} A \\ B \end{cases} \\ \\ Y \begin{cases} A \\ C \end{cases} \end{cases}$$

W and *Y* are related to each other 25 per cent, and $F_X = 12.5$ per cent or 1/8. The one line of descent connecting *W* and *Y* is $W \leftarrow A \rightarrow Y$. To see what will probably happen to the genetic composition of *X*, we will consider a pair of genes, *Rr*, for which *A* is heterozygous. What is the probability that *X* is homozygous (*RR* or *rr*) through having received duplicate genes in this locus from *A*? We are not interested in genes from *B* or *C* since, as far as the pedigree shows, they are unrelated to each other or to *A*, and our inbreeding coefficient measures only how many of the genes heterozygous in the foundation animals (those at the base date to which the inbreeding was computed) have probably become homozygous in *X* because of its inbreeding. *W* and *Y* are each equally likely to have received *R* or *r*. They each are equally likely to transmit whichever gene (*R* or *r*) they did receive from *A*, or the allel to it which each received from its other parent. The probabilities with respect to the genes of the *Rr* pair in *X* are as follows:

4 chances in 16 that neither gene came from *A*.
4 chances in 16 that *R* came from *A* but the other allel came from a grandam.
4 chances in 16 that *r* came from *A* but the other allel came from a grandam.
2 chances in 16 that both genes came from *A* and *X* is *Rr*.
1 chance in 16 that both genes came from *A* and *X* is *rr*.
1 chance in 16 that both genes came from *A* and *X* is *RR*.

If either of the last two events happened, X is homozygous for a gene for which A was heterozygous. Together the last two events have a probability of one-eighth of happening. When we say that the inbreeding coefficient of X is ⅛, we are saying that X is probably homozygous for ⅛ of the genes which were heterozygous in the ancestors at the foundation or base date to which the pedigree of X was traced. If we combine the probabilities of each of the above events happening, and include also what would have happened if A had been RR or rr, we find that if many are bred like X (i.e., with the inbreeding that results from being double grandsons) from a population of grandparents whose zygotic frequencies are: $q^2RR : 2q\,(1-q)\,Rr : (1-q)^2rr$, then the most probable zygotic ratio among those bred like X is:
$[q^2+Fq\,(1-q)]\,RR : 2q\,(1-q)\,(1-F)\,Rr : [(1-q)^2+Fq\,(1-q)]\,rr$. In other words this amount of inbreeding will probably transform ⅛ of the heterozygous gene pairs into homozygous ones, half of that ⅛ being added to each of the two kinds of homozygotes. Thus the inbreeding is impartial between the two homozygotes, tending on the average to add to each of them one-half of the decrease it causes in the frequency of heterozygotes. Of course chance fluctuations concerning that fraction are large in the necessarily small population which is any one inbred line.

The measurement of inbreeding, even in the most complicated pedigrees, is simply computing the amount of heterozygosis probably lost because of the inbreeding. In the simpler cases, where sire and dam are related through only one or two lines, the computations are easy. It is only necessary to find the common ancestor, count and add the number of segregations between sire and ancestor and between dam and ancestor and compute ½ to one higher power than that. If much of this is to be done, it is convenient to memorize or have handy a table of $(\frac{1}{2})^n$ for values of n from 1 to 7 or 9. When n is more than 6, this fraction is less than 1 per cent. Little is gained by investigating any one relationship too distant to contribute even this much, although if there are very many such lines, their total might be important.

In the more complicated cases it may be necessary and is convenient to draw the pedigrees in the arrow style shown in the middle and bottom of Figure 35. In this form of pedigree each ancestor is shown only once. An arrow leads from it to each descendant. Unless it had more than one descendant it did not provide any inbreeding itself, but merely transmitted to its one offspring some of the genes received from its parents. When pedigrees are drawn in arrow style it is usually easy to see at a glance what kind of a breeding system had been used and toward which ancestors the inbreeding had been directed. Printing difficulties are an obstacle to using the arrow style widely, as well as the fact that in most pedigrees in sale catalogues and advertisements there is little or no inbreeding visible, and in many cases the owner does not wish to call attention to that small amount. At the bottom of Figure 35 are shown the computations for the amount of inbreeding coming from each line through which sire and dam are related.

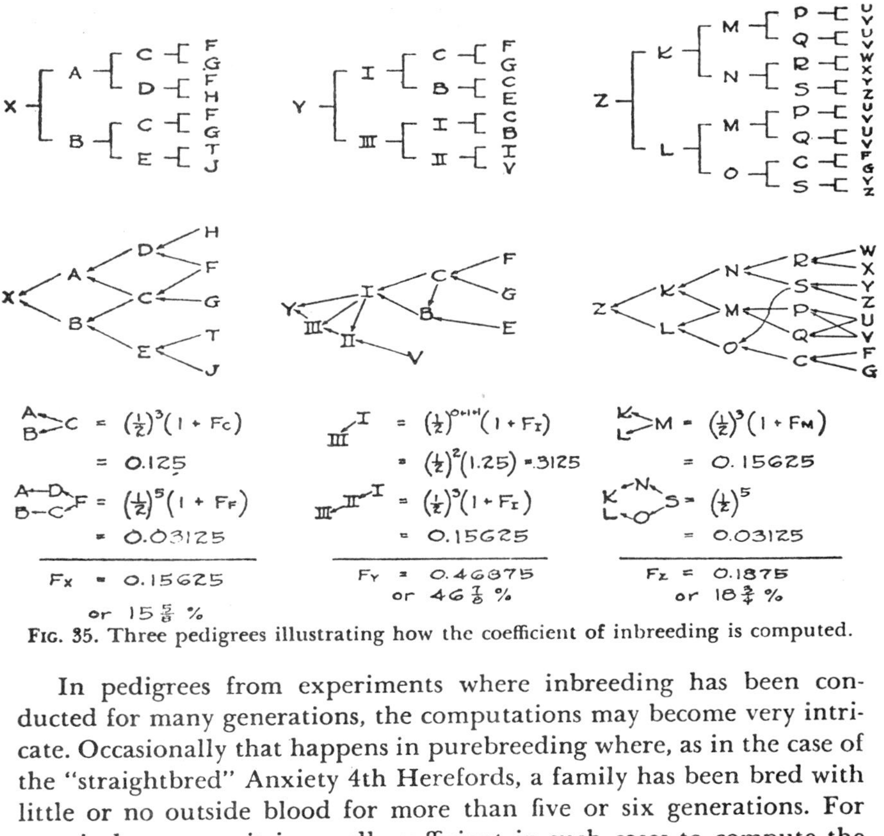

A–C, B–C: $= (\frac{1}{2})^3(1 + F_C) = 0.125$

A–D–F, B–C–F: $= (\frac{1}{2})^5(1 + F_F) = 0.03125$

$F_X = 0.15625$ or $15\frac{5}{8}\%$

III–I: $= (\frac{1}{2})^{0+1+1}(1 + F_I) = (\frac{1}{2})^2(1.25) = .3125$

III–II–I: $= (\frac{1}{2})^3(1 + F_I) = 0.15625$

$F_Y = 0.46875$ or $46\frac{7}{8}\%$

K–M, L–M: $= (\frac{1}{2})^3(1 + F_M) = 0.15625$

K–N–S, L–O–S: $= (\frac{1}{2})^5 = 0.03125$

$F_Z = 0.1875$ or $18\frac{3}{4}\%$

FIG. 35. Three pedigrees illustrating how the coefficient of inbreeding is computed.

In pedigrees from experiments where inbreeding has been conducted for many generations, the computations may become very intricate. Occasionally that happens in purebreeding where, as in the case of the "straightbred" Anxiety 4th Herefords, a family has been bred with little or no outside blood for more than five or six generations. For practical purposes it is usually sufficient in such cases to compute the inbreeding for only the last four or five generations and assume that the ancestors at that date were typical of this family, thus making the coefficient relative to this family at that date instead of making it relative to the whole breed.

Figure 36 shows for some regular inbreeding systems the sharp differences between the most intense systems theoretically possible and some of the milder ones which might more readily be practiced with farm animals. The milder inbreeding systems are much less intense at the start but if continued long enough in an entirely closed population can bring the population to a high degree of homozygosis. How very long that would be in terms of one breeder's lifetime can be seen by multiplying the number of generations in Figure 36 by two and one-half years in the case of swine, four or five years in the case of cattle and sheep, and ten or more years in the case of horses.

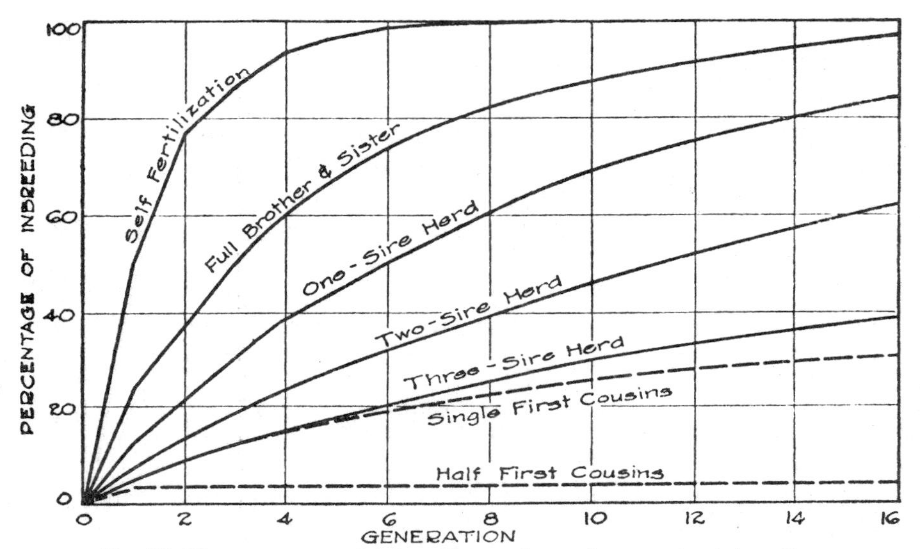

FIG. 36. The percentage of inbreeding under various regular inbreeding systems. (After Wright in *Genetics,* 6:172.)

Figure 37 shows similarly what happens to the relationship between full brothers under the same inbreeding systems. Unrelated families are apt to drift farther apart under inbreeding than they would under random mating, but each tends to become uniform within itself. A one-sire herd where no females are ever purchased but each new sire is unrelated to the herd will approach but never rise above the level of an average relationship of 33 1/3 per cent between herd mates. By contrast a one-sire herd in which neither sires nor dams are purchased and there is no overlapping of generations will even in the first inbred generation reach an average relationship of 39 per cent between herd mates and in the next generation will pass 50 per cent. The whole herd will then be more uniform than if all members were full sibs.

THE RATE OF INBREEDING IN ISOLATED POPULATIONS

The complete formula for the inbreeding coefficient is unwieldy in estimating the consequences of any breeding plan which is to extend for more than a few generations. For example, one might have a herd of cattle just big enough to justify keeping two sires at a time; and he might plan to raise his own sires without ever introducing any new stock from other herds. Soon his sires would be related to all the females on which they were to be used, but this relationship would vary. Many would be half sisters, some would be cousins, some would be less closely

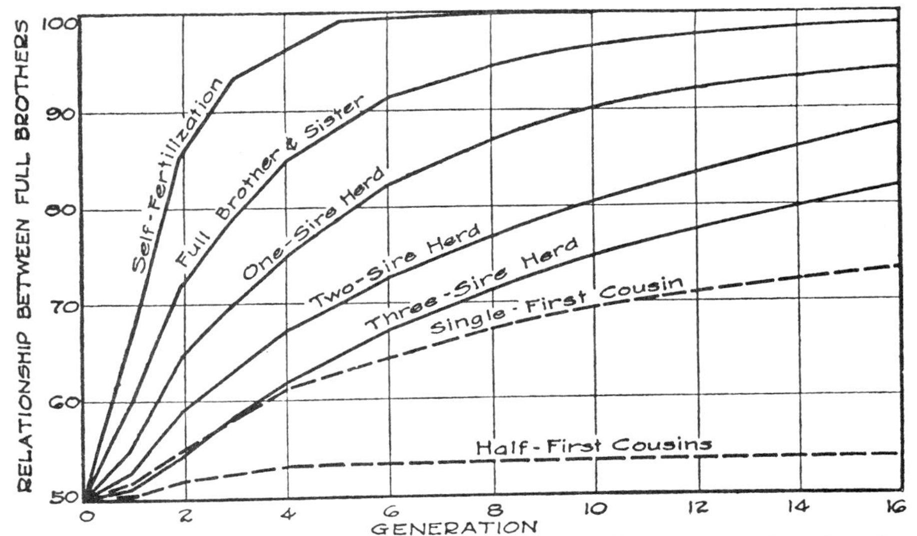

FIG. 37. The relationship between full brothers under various regular inbreeding systems. (After Wright in *Genetics*, 6:170.)

related older cows from the preceding generations, a few might be full sisters, dams, three-quarter sisters, etc. Figure 38 shows an actual example of this from a Shorthorn herd where only one sire and no females had been bought in 20 years.[4] To compute the average inbreeding in such an actual herd after it has been produced is a tedious job and not very practical, since the animals are already produced whether one likes them or not. For practical purposes one wants to estimate and compare the consequences of various possible plans, so that the one thought most favorable can be adopted and the less favorable ones left untried.

If a population is kept entirely closed to outside blood, about $\frac{1}{8M} + \frac{1}{8L}$ of the remaining heterozygosis will be lost per generation, where M is the number of males and L is the number of females reach-

[4] *Jour. of Heredity*, 25:208–16.

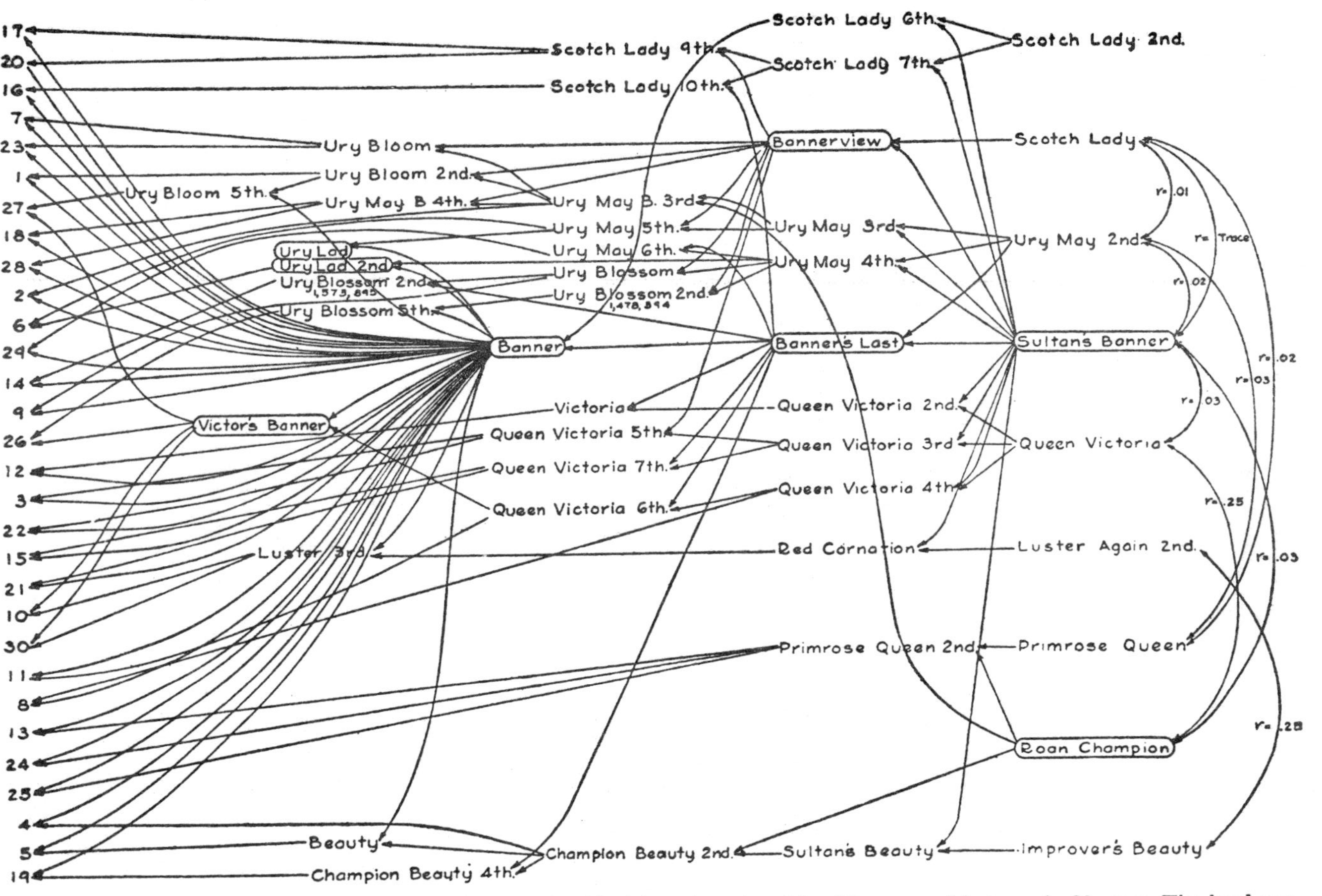

FIG. 38. Pedigree of a Shorthorn herd in which no animal had been introduced for 17 years and but one in 20 years. The herd was built largely around Sultan's Banner with some secondary linebreeding to his sons, Bannerview and Banner's Last and quite a bit of secondary linebreeding to his double grandson, Banner. (From *Jour. of Heredity*, 25:208.)

ing breeding age in each generation.[5] In a herd where there are 2 sires and 40 females in active use, this would be 1/16 + 1/320, or about 6.6 per cent of the remaining heterozygosis. In animal breeding, L will usually be so much larger than M that the term $\frac{1}{8L}$ can be neglected without much error. Then the formula becomes simply $\frac{1}{8M}$ giving inbreeding rates of 12 per cent, 6 per cent, 4 per cent, and 3 per cent per generation, respectively, for one-sire herds, two-sire herds, three-sire herds, and four-sire herds, closed to all outside blood. These rates can be reduced somewhat by avoiding inbreeding as far as possible under those conditions. No reduction at all could be produced in a one-sire herd, only a small reduction in a two-sire herd, more in a three-sire herd, etc. The maximum effect of avoiding all inbreeding as far as possible within a closed population tends toward halving the rate given by the formula.

By this formula we can calculate where purebred systems with rigidly closed herdbooks are drifting, so far as concerns any inbreeding which is inevitable because the members of a pure breed are becoming more closely related to each other. Even in a small breed with 200 sires and 2,000 dams used per generation, the formula shows only .069 per cent of the heterozygosis lost per animal generation. In other words it would take about 15 animal generations to lose 1 per cent of the remaining heterozygosis in such a small breed. There need be no fear that a closed herdbook will automatically lead to dangerously high inbreeding, even though the herdbook remains closed and the breed remains moderately small for centuries. However, show rings, advertising, and other sales efforts make some males more popular than others. These males have many sons and grandsons which go to head other registered herds. Less popular sires have no sons which see service in purebred herds and perhaps few daughters and no grandsons. This has the effect of making M in the formula much smaller than the actual number of males which have registered daughters in each generation. In several pure breeds so far studied, the decrease in heterozygosis per generation on account of the inbreeding practiced is not very far from 0.4 to 0.6 per cent. Even so, only about 10 to 12 per cent of the present amount of heterozygosis would disappear in another century of pure breeding like

[5] For the derivation of this formula see: *Genetics*, 16:107–11. Strictly speaking, M and L are the "effective numbers." They would equal the actual census numbers in the simple case in which all males and all females had equal chance to leave offspring. Many conditions can cause the effective numbers to be smaller than the actual numbers. At least a few of these will occur in practice. Hence the formula will underestimate the amount of inbreeding in closed populations. See *Amer. Nat.*, 74: 241–47. 1940.

that of the last 30 years. An occasional undetected fraudulent registration of a grade would further reduce this rate.

COMPLETE FORMULA FOR THE MEASUREMENT OF RELATIONSHIP

The measures of relationship can now be corrected for the effects of inbreeding. The complete formula for the coefficient of relationship between animals X and Y is:

$$R_{XY} = \frac{\Sigma[\frac{1}{2}^n (1 + F_A)]}{\sqrt{1 + F_X} \quad \sqrt{1 + F_Y}}$$

where n is the total number of Mendelian segregations in the path of descent through which X and Y are related. This differs from the approximate formula used in the last chapter only by having terms for the inbreeding of X and Y and of their common ancestor. The $(1+F_A)$ in the numerator allows for the fact that an inbred ancestor (A) is homozygous for more pairs of genes than a non-inbred one. Its use is illustrated in Figure 35 where the inbreeding of I adds to the relationship betwen itself and III, or the inbreeding of M increases the relationship between K and L. The terms in the denominator are to correct for the fact that inbreeding makes the population more variable, the inbred lines tending to drift apart from each other. Relationship is a fraction which has for its numerator the number of genes in which the two related animals are probably alike but which would be unlike in two random animals from the base population, and for its denominator the average number of genes in which two animals probably would be unlike if they were unrelated but descended from the base population by the same kind of breeding system as the two actually are. The denominator grows larger with their inbreeding because the inbreeding tends to decrease the proportion of heterozygosis and to throw the population toward the two homozygous extremes in each pair. In a highly inbred population two unrelated individuals (necessarily from different inbred lines if they are unrelated) have a considerably larger chance of being *AA* and *aa* than if they were from a non-inbred population, because in the inbred population there are more *AA* and *aa* and fewer *Aa* individuals present. This increase in the denominator naturally lowers the relationship unless there is a corresponding or greater increase in the numerator. Such an increase in the numerator can and usually does occur if the two animals are members of the same inbred line but not if they belong to lines which have separated and no longer interbreed with each other.

The complete formula for relationship between an animal and its ancestor where there is no collateral relationship may be written

$$R_{XA} = \Sigma[\tfrac{1}{2}^n] \sqrt{\frac{1+F_A}{1+F_X}} = \text{"Percentage of blood" times} \sqrt{\frac{1+F_A}{1+F_X}}.$$

If A and X are equally inbred, the term under the square root sign equals 1.0 and the percentage of blood is exactly equal to the coefficient of relationship. If the ancestor is more highly inbred, the figure for percentage of blood is not quite large enough. This is because there will be more than the average number of cases in which the ancestor is homozygous and therefore *both* its genes in that pair will be like the one its descendant receives from it. If the descendant is more highly inbred, the figure for percentage of blood is a little too large. This is because there will be cases in which the descendant is homozygous for a gene heterozygous in the ancestor. In such gene pairs the descendant gets 100 per cent of its genes from that ancestor and yet is not 100 per cent like the ancestor. For an example we may return to the pedigree of X (page 273) which was a double grandson of A. In one-fourth of the cases for any one pair of genes, X gets both its genes from the grandams and none from A; in one-half the cases, X gets one gene of the pair from A and the other from a grandam; in the remaining one-fourth of the cases, X gets both genes from A. On the average, therefore, X gets 50 per cent of its inheritance from A, just as the percentage of blood figure shows. However, in the one-eighth of the cases when X is *rr* or *RR*, it will get both genes from A but yet will not be exactly like him, because he was *Rr*. The corrections in the denominator thus keep the relationship coefficient a measure of probable genetic likeness and not merely a measure of source of the genes, as is percentage of blood. Not often is the difference in the inbreeding of A and X large enough to be important. A much more serious general defect of percentage of blood as a measure of relationship is that it does not include collateral relationship and that collateral relationship cannot readily be measured by any way of manipulating percentage of common blood.

In the pedigrees ordinarily encountered in actual animal breeding, the denominator of the relationship coefficient does not get very much larger than 1.0. Neglecting it altogether is not apt to lead to a serious mistake. However, it must be included if the formula is to be entirely correct. It is not often worthwhile to carry the computation of either R or F for individual animals farther than to the nearest 1 per cent. Often the nearest 5 per cent is close enough. The sampling variations inherent in Mendelism prevent one from being sure that the computed coefficient describes with extreme accuracy the actual situation in individual cases, even if one could make practical use of small differences in these coefficients.

The similarity between the formula for inbreeding and the complete formula for relationship shows how it is that an individual's inbreeding depends upon the relationship between its sire and its dam. The equations connecting the two, where S represents the sire, D the dam, and O the offspring, are:

$$F_O = \frac{R_{SD}}{2} \times \sqrt{(1 + F_S)\,(1 + F_D)}, \text{ and } R_{SD} = \frac{2F_O}{\sqrt{(1 + F_S)\,(1 + F_D)}}$$

Unless the sire and dam are highly inbred, the term under the square root sign will not differ much from 1.0; and it will be approximately correct to say that the inbreeding of the offspring is one-half of the relationship between its parents. This rule is useful when estimating the amount of inbreeding danger in a mating before that mating is made.

OTHER PROCESSES WHICH CAN AFFECT HOMOZYGOSIS

The inbreeding coefficient expresses probable changes in homozygosis based on no other assumptions than that inheritance is Mendelian and is equal from the two parents. It neglects sex-linkage and the small changes in homozygosis which may be made by mutation and selection and by assortive mating which is not inbreeding.

So far as concerns sex-linked genes, the inbreeding coefficient for the pedigree of a heterogametic animal has no meaning. The heterogametic parent behaves as if it were entirely homozygous for sex-linked genes in transmitting to its homogametic offspring, and transmits no sex-linked genes at all to its heterogametic offspring. Referring for convenience to the female as the homogametic sex (as is correct for the mammals but not for the birds and a few other animals), the inbreeding computed for a female's pedigree is not true for her sex-linked genes wherever the line of descent is from sire to son. On the other hand, wherever the line of descent is from sire to daughter, the homozygosity of females for sex-linked genes will be higher than the coefficient indicates. These two effects tend partly to cancel each other so that the inbreeding coefficient will generally measure the extra homozygosity of females, even for sex-linked genes. There will be cases where the coefficient will be systematically in error for the sex-linked genes; e.g., a double granddaughter of a male would not tend to have her sex-linked genes any more homozygous than if her parents had been unrelated, while a double granddaughter of a female would have her sex-linked genes 25 per cent less heterozygous instead of the expected 12½ per cent.[6]

---

[6] A more detailed explanation of the consequences of inbreeding on the distribution of sex-linked genes was presented by Wright in 1933. *Proc. Nat. Acad. Sci.*, 19:411–20.

Mutation is so rare an event that its neglect in the formula introduces no error of importance in practical breeding, although mutation needs to be included in evolutionary considerations where the time involved extends over an enormous number of generations. The formulas for including it are rather intricate.[7]

Selection affects homozygosis only incidentally through changing q and thereby the value of $2q\ (1 - q)$. Selection will usually require many generations to make large changes in q, except in those cases where intense selection is directed at a characteristic the main variations of which are highly hereditary and are determined by a very few genes. The effects of selection on homozygosis certainly need to be considered in connection with problems of evolution, but in general they are probably too small to make inbreeding coefficients much in error as measures of the change in homozygosis which has occurred within the last four to six generations. Assortive mating (as will be explained in chapters 27 and 28) has almost no effect on homozygosis except when (1) the assortive mating is almost perfect, (2) the characteristics concerned are highly hereditary, and (3) the number of genes producing the observed variation is small. Very rarely would all these conditions be met in actual practice.

PRACTICAL USES OF INBREEDING

Most breeders inbreed only when they must to attain some other object, such as linebreeding or forming and preserving family distinctness. The commonest reason for inbreeding is that a breeder must do some of it if he is to keep his animals closely related to individual animals he admires. Relationship between two animals cannot go higher than 50 per cent unless there has been some inbreeding. If a breeder continually uses unrelated sires, the relationship of each succeeding generation to the animals he has had in his herd will be halved. In a short time his animals will be very little related to the best ones of two or more generations ago. If he still has a good herd, that will be because he has been successful in selecting his subsequent sires. If, instead of using unrelated sires each generation, he uses sires closely related to the best animals he has had, then he may keep the future generations as closely related to those good ancestors as his present animals are. But, since both sires and dams of his stock are related to these good animals, he will be practicing some inbreeding. This is the essence of "linebreeding," which is the subject of Chapter 23. In this practice most breeders regard inbreeding as a necessary evil which must be endured if they are to keep their herd closely related to some noted animal, but which

[7] *Genetics*, 16:116–37.

should be kept as low as can be done and still accomplish the linebreeding program.

If the homozygous recessives can all be discarded, inbreeding can be a powerful help to selection against rare recessives. There may be so many recessives in the stock that the inbreeding will bring them to light in more animals than the breeder can afford to discard, or he may not recognize some of those with the less conspicuous effects in time to discard them before they become fixed on his whole herd. This is what breeders have in mind if they say that their breed is not yet perfect enough to make even linebreeding wise as a general policy. The price which must be paid for using inbreeding thus to improve the breed is the occurrence at first of a larger proportion of undesirable individuals. That price may be so high that the individual breeder will not find it economically possible for him to practice extremely close inbreeding as a steady policy.

Inbreeding may be used to test whether animals are good enough to justify deliberate and long-continued attempts to keep the herd closely related to them. Inbreeding is the severest test of the hereditary worth of an individual that can be made. Wriedt goes so far as to recommend that every dairy bull thought worth using in the first place should be bred as soon as possible to enough of his daughters to insure that there will be at least 15 or 20 of his daughters out of his own daughters to prove his breeding value. For several reasons this proposal seems too extreme for general adoption. However, it does rest on the truth that inbreeding is the severest and quickest test to find whether an animal has any undesired genes.

Inbreeding can be used to promote homozygosity. Homozygosity has little commercial value in the sale ring as yet. Hence it seems unlikely that many breeders could afford to inbreed intensely just for this object. Nevertheless, homozygosity is the most important element in prepotency, and the building of this into his herd is worth some effort by the breeder, provided he can maintain or improve the average phenotypic merit at the same time.

A breeder often practices inbreeding merely for economy. This is an unwise policy whenever the animals are of only average merit. He could then expect only offspring of average merit, even if there were no inbreeding. With the added probability of some phenotypic degeneration from inbreeding, he can expect the offspring to be below the average merit of their breed as individuals, although not necessarily below the breed average in their merit as breeding animals. When the animal to which the inbreeding is being directed is of superior merit, this reason of economy lends additional weight to the argument for inbreeding.

One of the important general reasons for practicing inbreeding is that it tends to form distinct families within the breed and thus permits more selection between families than would be possible under random breeding. (This will be discussed in more detail in Chapter 24). Selection between families can be much more accurate than selection between individuals, especially for characteristics which are only slightly hereditary; but the families must be rather distinct from each other if that is to be the case. The family averages of non-inbred families do not deviate from each other as much as individual genotypes do.

The producer of market animals has little reason to practice inbreeding. His market will not pay him any extra for prepotency. Improvements he may make in the average merit of his own herd by using inbreeding along with selection will be about halved next time he buys an unrelated sire from some other source, as most commercial producers do. Economy can be a valid reason with him, especially when he thinks the sire he has is better than the next one is likely to be. Partly offsetting this superiority of his present sire is the probability that some harmful results of inbreeding will occur in his herd.

THE DANGERS OF INBREEDING

Inbreeding makes desirable and undesirable genes homozygous impartially. If the rate of this is too rapid, every individual produced will be homozygous for some undesired genes as well as for some desired ones. If the inbreeding is too mild, many generations will be needed to accomplish much with it. The problem of the best rate at which to inbreed is one of keeping the inbreeding mild enough that the man in charge can avoid fixing the genes with undesirable effects and can fix as many as possible of the genes which have desirable effects. How mild or how intense such inbreeding can be depends upon several things, the most important of which are the skill of the man doing the selecting, the abundance of undesired genes in the stock with which he begins, the amount of linkage between desired and undesired genes in the initial stock, the amount of epistasis, dominance, or environmental effects which may deceive the breeder when he makes his selections, and whether he is breeding a group all by himself. If other men are breeding closely related lines, he can correct his mistakes by mild outcrosses to some of their herds without having to use totally unrelated animals.

Fragmentary evidence of various kinds indicates that inbreeding rates as high as 6 per cent per generation[8] under favorable circumstances may be pursued for many generations without noticeably harmful consequences. It is unlikely that inbreeding rates as high as 3 or 4

[8] As, for example, in a two-sire herd where no outside blood is introduced.

per cent can go on forever without harm, but certainly they can be continued for many generations. Many breeders when in possession of unusually good animals have had favorable results from mating half brother to sister or grandsire to granddaughters, but not many have continued to do that for more than two or three successive generations. Occasional matings of parent and offspring or, more rarely, full brother and sister have turned out well; but general experience is that those should be risked only when the stock is unusually good.

Among human beings inbreeding as intense as the marriage of first cousins has enough probability of undesired consequences that in some places it is forbidden by law or religious rule.[9] Inbreeding more intense than that is regarded as "incest" in nearly all human societies although there have been some exceptions, as among the Pharaohs in ancient Egypt. No doubt the biological principles are the same in man as in other organisms, although the abundance of undesired recessives may be higher or lower. An extremely important practical difference is that in farm animals, if inbreeding brings to light a few more defectives than would occur without it, they may be culled with only the small economic loss that their defectiveness entails; whereas in most civilized communities of man the codes of ethics and morals do not permit such drastic action with defective human beings. The care and support of each one too defective to take care of itself is a serious burden, whether it is kept in a private home or in a special institution. Rigid prohibitions of marriages of a certain degree, such as between first cousins, do not allow for the fact that such marriages may be desirable when the common ancestry is of unusual merit. Prohibiting the marriage of relatives does not improve the average heredity of a population. Neither inbreeding nor outbreeding makes the undesired genes systematically either rarer or more abundant, but inbreeding does bring them together so that more of them can show their effects and be culled—if selection is being practiced. Perhaps the general experience of man over centuries may be considered as indicating that around 6 per cent is in general the "stop, look, and listen" level of danger from inbreeding? If the common ancestry is of sound stock, the children of such marriages may be above average in merit. If the common ancestry has any serious defect, even a rare one, the probability of that defect reappearing in the children who have a chance to inherit it from both sides is much higher than if their parents were unrelated. The rarer the defect in the general population, the more extreme is this difference.

[9] See page 52 of *Time* for August 19, 1940, for a list of marriages forbidden by the Church of England in 1560. In 1945 prohibitions against ten of these categories of in-law marriages were removed. This was the first change in those rules in nearly four hundred years.

The inbreeding coefficient may be used to estimate the danger involved in any particular mating, if one considers also the merit of the ancestors from which the inbreeding comes. Inbreeding of 25 per cent coming from an outstanding ancestor might be safer than inbreeding of 10 per cent coming from a mediocre ancestor. Setting a definite percentage of inbreeding as the point where danger begins is much like setting a certain speed in automobile driving as the speed beyond which danger begins. In the case of the automobile much depends upon the condition of the highway, the field of vision, the mechanical condition of the car and the skill of the driver. Similarly with inbreeding, much depends upon the clearness of the goal, the accuracy of the tests and measures of merit, the initial scarcity of undesirable genes in the stock, the amount of culling which the reproductive rates permit, and the breeder's ability to recognize and discard genes which are on the verge of becoming fixed in his stock.

POSSIBILITIES OF PRODUCING INBRED LINES FOR COMMERCIAL CROSSING

Corn breeders have made a distinct success of producing inbred lines by self-fertilization and then crossing those lines which produce the most desirable crosses. The crossbred seed is sold for the production of commercial corn. Although the fundamental principles are the same, there are several differences in their application which make the success of such a breeding system appear less likely for animals than for plants, although modified systems based on the same principles may perhaps prove successful. In the first place, the closest possible inbreeding in animals is less than half as intense as self-fertilization. It would take many more generations with animals to reach the same degree of homozygosity. In the second place, the fertility of animals is lower than that of plants, so that not nearly as large a percentage of the individuals produced in each generation could be discarded. In the third place, the interval between generations in farm animals is longer than with annual plants; and the amount of time required to reach an equal degree of inbreeding would be longer for that reason also. In the fourth place, the individual animal is worth more money than the individual plant. Culling the undesired individuals which appear during the inbreeding will be more expensive than the same process applied to plants. In the fifth place, and partially offsetting these others, is the fact that the merits and faults of individual animals are usually better known than is the case with individual plants. Therefore, the individual selection which accompanies the inbreeding would be more accurate in animals. In the sixth place, the lower fertility of animals would make it economically difficult to sell the commercial producer as many as he would need of the crossbred stock from two successfully inbred lines, if such should ever be produced, or to sell him inbred females of one line and inbred males of another line so that he could make his own crosses. Poultry and swine seem more nearly suited to the economics of that than the other farm animals.

For those reasons commercial animal breeding will never practice such intense inbreeding alternating with such extreme outbreeding as is already practiced in corn breeding. On the other hand, a mild form of this is already happening in the crossbreeding which is practiced for producing commercial meat animals and in the practice, among breeders of purebred stock, of making outcrosses between distinctly unrelated lines within a breed, hoping thereby to produce excellent individuals. Perhaps it may be commercially possible to produce highly inbred sires to be used on practically random-bred high grade or purebred females. At present stockmen set so much store by individuality in their sires that few of them would use inbred sires unless these were also good individuals, but that would change quickly if it were demonstrated clearly that such use would be profitable.

EXPERIMENTS ON INBREEDING

Thorough and extensive experiments on inbreeding have been more numerous with plants than with animals. Many of the facts about inbreeding were discovered in experiments with corn. More come from the contrasting behavior of naturally self-fertilized and naturally cross-fertilized plants when those are experimentally inbred or are used in various crosses. Conspicuous examples of plants which in nature have a high percentage of self-fertilization include wheat, cotton, sorghum and oats. Corn and beets are examples of naturally cross-fertilized species on which extensive experiments with inbreeding have led to the production of inbred lines and the sale of crossbred seed on a commercially important scale. Strawberries and raspberries show much the same results as corn and beets, but the application is different because vegetative multiplication of the former is practical.

Among animals, laboratory experiments have been extensive on the inbreeding of rats, mice, and guinea pigs. Dr. King inbred white rats full brother and sister for more than 70 generations without finding degeneration. Mice have been inbred full brother and sister in many experiments. In at least one case this has been carried further than the 55th generation.[10] In the United States Department of Agriculture experiments on inbreeding guinea pigs, some lines have been inbred brother by sister for more than 30 generations. There have been several

[10] *Jour. of Heredity*, 27:21–24, 1936.

short experiments on inbreeding chickens and swine at a number of experiment stations. In 1945 the 38 lines being studied in the Regional Swine Breeding Laboratory ranged from about 10 to 70 per cent in inbreeding. Twelve of them were already inbred more intensely than three generations of full brother by sister and ten more were nearly that far along. Inbreeding experiments with poultry at the Iowa Station have reached a more intense stage than that of nine generations of full brother by sister mating, although the breeding system actually used was not that regular.

In farm animals other than chickens and swine, the small number of full sibs and the variations in the sex ratio prevent the long-continued use of such regular inbreeding systems as full brother by sister. Even in chickens and swine these are serious difficulties and reduce tremendously the amount of selection which can be practiced while the inbreeding is being done. For the other farm animals the most intense inbreeding plan which can be followed long is the use of a sire on his daughters as long as he lives, he to be followed by one of his inbred sons, which in turn would be followed by one of his inbred sons, and so on. There would be a few full brother by sister matings and some of the females would live longer than others, some, perhaps, even outliving two generations of sires. Hence such a system of inbreeding would be far from regular, and there would be comparatively few pedigrees which were exactly alike in the kinds of inbreeding they showed over a period of three or four generations. Before an inbreeding coefficient was devised for measuring the intensity of the irregular inbreeding shown in these many kinds of pedigrees, it was natural that experimenters should think that inbreeding experiments of that kind could not be interpreted in any unmistakable way and therefore would not return scientific information worth the money and effort they would cost. Now that Wright's coefficient of inbreeding, which was first proposed in 1922, largely removes this objection, it is probable that more experimental study will be made of irregular systems of inbreeding.

Some of what we know about the results of inbreeding in animals comes from the scattered and irregularly reported experience of breeders. It is difficult to be at all sure that such evidence is a typical sample of the results of inbreeding. There is the question of whether the animals inbred were typical of purebred animals in general. There is also the question of whether one hears of the typical results of such cases or only of the exceptional results. Any bad result which does appear is apt to be blamed on the inbreeding in spite of the fact that equally bad results sometimes occur when no inbreeding is practiced. There is usually an absence of adequate control; that is, of non-inbred animals kept under the same conditions. However, the results agree in general with those expected on theoretical grounds and with those actually obtained in laboratory experiments. The usual consequences of inbreeding in breeders' experiences is a degeneration which, however, is slight and irregular, affecting some characteristics in one animal and other characteristics in another and not affecting some individual animals at all. Even in Bakewell's time it was known and stressed that inbred animals are more apt to be prepotent and effective when used in outcrosses than are animals of equal individuality but not inbred.

The breeders who have practiced intense inbreeding for a long time have nearly always encountered enough degeneration that a cross with unrelated animals produced beneficial results. So universal has this experience been that breeders are rather generally convinced of the necessity of introducing "fresh blood" from time to time to "rejuvenate" a strain or herd. It is not always understood that this rejuvenating effect rarely occurs unless there has been some prior inbreeding. The explanation of these cases is that the herd becomes homozygous for undesirable genes which produce such small effects that the breeder scarcely notices them as they become fixed a few at a time, but instead just sees a gradual decline in vigor, fertility, size, etc. Since undesirable genes tend to be recessive, a cross with an animal from an unrelated herd often appears to remedy these defects at once.

SUMMARY

Inbreeding is the mating of animals which have a closer relationship to each other than the average relationship within the population concerned. Its measure is relative to some population, just as the measure of relationship is. Pure breeding is inbreeding relative to the whole species, but not many purebred animals are closely inbred relative to their breed.

The primary effect of inbreeding is to make more pairs of genes homozygous and to lower the percentage of heterozygosis correspondingly. Because this uncovers many recessive genes which would otherwise remain concealed by their dominant alleles, and because recessives generally have less desirable effects than dominants do, there is usually some degeneration in average individual merit when inbreeding is practiced.

Inbreeding does not of itself change gene frequency but does permit it to drift rapidly at random in each subgroup of the population.

Inbreeding is the most powerful tool the breeder has for establishing uniform families or strains which are distinct from each other. This it does by permitting gene frequencies to drift in different ways in different subgroups, by making the parents more homozygous, and by providing more and more ways in which members of the same family are likely to inherit the same genes because their parents are related to each other.

Some inbreeding is almost essential if selection is to have much success in those cases where a highly desirable effect is produced by a combination of genes which individually have undesirable effects; that is, for getting a population out of some of the lower "peaks of desirability" shown in Figures 20 and 21 and into higher nearby peaks.

The coefficient of inbreeding measures the percentage of genes which were heterozygous in the basic population but have probably become homozygous because of the inbreeding. It is subject to the sampling errors of Mendelian inheritance and hence means almost nothing for one pair of genes in one individual, but its sampling error may be small when it is applied to the average percentage of heterozygosis of one pair of genes in a whole population or to the average heterozygosis of the entire group of genes in one individual.

Among the more important reasons for practicing inbreeding are: (1) It is necessary if relationship to a desirable ancestor is to be kept high; (2) it helps uncover rare recessives so that they may be culled from the breed; (3) it forms uniform and distinct families so that interfamily selection may be possible in a more effective way than if inbreeding were not practiced; (4) it increases prepotency; and (5) it is sometimes economical, especially if the present sire is of such high merit that it will be difficult to find as good a one for a successor.

The danger of intense inbreeding is that it will make undesired genes homozygous at so rapid a rate that it will be impossible to discard all individuals homozygous for them. Some of the undesired genes will therefore become "fixed" in the whole herd. The lowered sale value of the defectives uncovered by the inbreeding will cause some loss. From the standpoint of breed improvement, that loss is balanced by the increased prepotency of those which are not defective; but the man who is breeding animals for the commercial market will not receive that compensation.

It seems reasonably certain that more opportunities for breed progress are lost by not inbreeding when inbreeding would be advisable than are lost by too much inbreeding. When inbreeding is too intense, the individual breeder may lose by that; but the progress of the breed is not apt to suffer. The best of the inbred animals are likely to give good results in outcrosses.

REFERENCES

The general subject of inbreeding was treated comprehensively in the following book, which, even a quarter of a century later, is obsolete in little except its treatment of the measurement of inbreeding intensity:

East, E. M., and Jones, D. F. 1919. Inbreeding and outbreeding. Philadelphia The J. B. Lippincott Co. 285 pp.

For explanation of the Mendelian basis of inbreeding and of what happens when various rates of inbreeding, mutation, and selection are balanced against each other, see:

Wright, Sewall. 1921. Systems of mating. Genetics, 6:111–78.

———. 1922. Coefficients of inbreeding and relationship. Amer. Nat., 56:330–38.

———. 1923. Mendelian analysis of the pure breeds of livestock. I. The measurement of inbreeding and relationship. Jour. of Heredity, 14:339–48.

———. 1923. Mendelian analysis of pure breeds of livestock. II. The Duchess family of Shorthorns as bred by Thomas Bates. Jour. of Heredity, 14:405–22.

———. 1931. Evolution in Mendelian populations. Genetics, 16:97–159.

———. 1940. Breeding structure of populations in relation to speciation. Amer. Nat., 74:232–48.

For reports of actual experiments on inbreeding animals, see:

Castle, W. E. 1930. Genetics and Eugenics, pp. 286–304. Cambridge: Harvard University Press.

Hodgson, R. E. 1935. An eight-generation experiment in inbreeding swine. Jour. Heredity, 26:209–17.

Hughes, E. H. 1933. Inbreeding Berkshire swine. Jour. Heredity, 24:199–203.

King, Helen Dean. 1918 and 1919. Studies of inbreeding. Jour. of Exp. Zoology, 26:1–54, 26:335–78, 27:1–36, and 29:71–112.

Strong, Leonell C. 1936. The establishment of the "A" strain of inbred mice. Jour. Heredity, 27:21–24.

USDA Yearbook for 1936.

Waters, N. F., and Lambert, W. V. 1936. Inbreeding in the White Leghorn fowl. Iowa Agr. Exp. Sta., Res. Bul. 202.

Willham, O. S., and Craft, W. A. 1939. An experimental study of inbreeding and outbreeding in swine. Okla. Agr. Exp. Sta., Tech. Bul. 7.

Woodward, T. E., and Graves, R. R. 1933. Some results of inbreeding grade Guernsey and grade Holstein-Friesian cattle. USDA, Tech. Bul. 339.

Wright, Sewall. 1922. The effects of inbreeding and crossbreeding on guinea pigs. I. Decline in vigor. II. Differentiation among inbred families. USDA, Department Bul. 1090. III. Crosses between highly inbred families, USDA, Department Bul. 1121.

———, and Lewis, Paul A. 1921. Factors in the resistance of guinea pigs to tuberculosis, with special regard to inbreeding and heredity. Amer. Nat., 55:20–50.

For brief statements of breeders' experience with regard to inbreeding, see:

Mumford, F. B. 1921. Breeding farm animals. pp. 217–42.

Wriedt, Christian. 1930. Heredity in livestock. pp. 68–113.

For studies of the amount and kind of inbreeding which has occurred in various breeds of livestock, see:

Berge, S. 1930. Inbreeding in Telemark cattle. (Translated title). Nordisk Jordbrugsforskning 204–16.

Brockelbank, E. E., and Winters, L. M. 1931. A study of the methods of breeding the best Shorthorns. Jour. Heredity, 22:245–49.

Calder, A. 1927. The role of inbreeding in the development of the Clydesdale breed of horses. Proc. Royal Soc. of Edinburgh, 47, Part 2, No. 8, pp. 118–40.

Carter, Robert C. 1940. A genetic history of Hampshire sheep. Jour. Heredity, 31:89–93.

Dickson, W. F., and Lush, Jay L. 1933. Inbreeding and the genetic history of the Rambouillet sheep in America. Jour. Heredity, 24:19–33.

Fletcher, J. Lane. 1945. A genetic analysis of the American Quarter Horse. Jour. Heredity, 36:346–352.

———. 1946. A study of the first fifty years of Tennessee Walking Horse breeding. Jour. Heredity, 37:369–373.

Fowler, A. B. 1932. The Ayrshire breed: A genetic study. Jour. of Dy. Res., 4:11–27.

Lörtscher, H. 1945. Inbreeding and relationship among the Jura horses today. (translated title). Landw. Jahrb. d. Schweiz, pp. 1–16.

Lush, Jay L. 1946. Chance as a cause of changes in gene frequency within pure breeds of livestock. Amer. Nat. 80:318–342.

———, and Anderson, A. L. 1939. A genetic history of Poland-China swine. Jour. Heredity, 30:149–56 and 219–24.

———; Holbert, J. C.; and Willham, O. S. 1936. Genetic history of the Holstein-Friesian cattle in the United States. Jour. Heredity, 27:61–71.

McPhee, Hugh C., and Wright, Sewall. 1925. Mendelian analysis of the pure breeds of livestock. III. The Shorthorns. Jour. Heredity, 16:205–15.

———. 1926. Mendelian analysis of the pure breeds of livestock. IV. The British Dairy Shorthorns. Jour. Heredity, 17:397–401.

Rottensten, Knud. 1937. Inbreeding in Danish Landrace swine. (Translated title). Nordisk Jordbrugsforskning, Hefte 3–4A, pp. 94–114.

Sciuchetti, A. 1935. Ein Beitrag zur genetischen Analyse der schweizerischen Braunviehrasse. Julius Klaus-Stiftung f. Vererb. Sozialanthr. u. Rassenh., 10:85–99.

Smith, A. D. B. 1926. Inbreeding in cattle and horses. Eugenics Review, 18:189–204.

Steele, Dewey. 1944. A genetic analysis of recent Thoroughbreds, Standardbreds, and American Saddle Horses. Kentucky Agr. Exp. Sta., Bul. 462.

Stonaker, H. H. 1943. The breeding structure of the Aberdeen-Angus breed. Jour. Heredity 34:322–28.

Willham, O. S. 1937. A genetic history of Hereford cattle in the United States. Jour. Heredity, 28:283–94.

Yoder, Dorsa M., and Lush, Jay L. 1937. A genetic history of the Brown Swiss cattle in the United States. Jour. Heredity, 28:154–60.

CHAPTER 22

Prepotency

Prepotency is the ability of an animal to impress characteristics upon its offspring to such an extent that they resemble that parent or each other more closely than is usual. If the offspring are all unusually like this parent, they will naturally tend to be unusually like each other. Writings on animal breeding are full of references to prepotency. Many of those no doubt exaggerate the supposed amount of prepotency beyond the actual facts. Differences in prepotency do exist, however, and are sometimes large enough to be practically important.[1] Purebreds are usually more prepotent than crossbreds or grades. An animal may be prepotent for undesirable as well as for desirable characteristics, but naturally in breeders' discussions prepotency for desirable characteristics is mentioned more often.

"Potent" and "impotent" in animal breeding usage refer to the ability or inability of the animal to reproduce or even to copulate normally. These terms do not refer to the merit of the offspring.

GENETIC BASIS OF PREPOTENCY

Differences in prepotency depend mainly upon dominance and homozygosis. In some cases a part may be played by linkage and epistasis.

The most important cause of differences in prepotency is the degree of homozygosis in the animals concerned. A perfectly homozygous animal could produce only one kind of gamete. All its offspring would receive exactly the same genes from it. Any genetic differences between those offspring would depend entirely on their having received different things from their other parents. On the other hand, an animal heterozygous for n pairs of genes could produce 2^n different kinds of gametes. This permits its offspring to differ genetically, not only in what they received from their other parent, but also in what they received from the common parent.

Dominance is the other important cause of differences in prepo-

[1] Wentworth, E. N. 1926. Prepotence in character transmission. Proc. Scottish Cattle Breeding Conference for 1925, pp. 146–63.

tency. Every offspring which receives a dominant gene will show the effect of that gene. If the gene is completely dominant and the parent is homozygous for it, then all of the offspring will appear exactly alike for the effect of that gene, regardless of the inheritance they received from the other parent. When a parent having many dominant genes is also highly homozygous, its prepotency is maximum.

A breed which has several conspicuous dominant traits will appear to be prepotent in crosses with other breeds. This does not mean that it will be prepotent in other characteristics. For example, crossing of Herefords with Aberdeen-Angus ordinarily produces white-faced, black animals without horns. In this case the Angus breed is prepotent for body color and for the absence of horns, but the Hereford breed is prepotent for the white face. Body color and the presence or absence of horns are conspicuous characteristics. One who does not examine the animals carefully might infer that prepotency is a general characteristic of the animal as a whole. Probably most statements that a certain animal transmitted all of its qualities uniformly to all of its offspring are based on careless observation of an animal which was homozygous for one or a few conspicuous dominant traits. To the man unfamiliar with Aberdeen-Angus cattle the mere fact that a group of cattle are hornless and black would make them seem impressively alike to him. But the man familiar with black-polled cattle would be looking for other things and would not be much impressed by this.

Linkage has the general effect of making most of the offspring of an individual fall into a smaller number of classes than if there were no linkage. If there has been considerable selection among the offspring of that animal, many having been discarded before we see them, linkage may thus give us the impression of more prepotency than we would have observed if all the genes had been segregating independently.

Epistasis may sometimes add something to apparent prepotency. Occasionally a sire will be homozygous for one or more genes which, when brought together with genes which many of his mates have, will produce conspicuous results, although the genes may have little apparent effect when the full combination is not present. As a result the offspring from such a sire will be unusually like each other and yet distinctly different from either their sire or their dam. It is not certain whether this happens often, but it is a possibility and is the most plausible genetic explanation for some cases reported.

THE MEASUREMENT OF PREPOTENCY

After the offspring are produced, there are two ways of measuring prepotency. The first is to measure directly the resemblance of this animal and its offspring, as compared with the usual resemblance of parent and offspring. The second is to measure how closely the offspring of this sire resemble each other, as compared with the usual resemblance between half brothers and sisters. In the first method any permanent effect produced in the parent by environment, dominance, or epistasis will appear in every comparison of that parent with each of its offspring. Therefore, the second method is generally to be preferred. Moreover, the second method is the only one available in cases such as measuring a dairy bull's prepotency for milk and fat production, since he cannot himself express those traits. A weakness of the second method is that the offspring are more likely to have been exposed to the same peculiarities of environment than parent and offspring are. Thus if one bull's daughters freshened in a poorly managed herd and a second bull's daughters all freshened in a well-managed herd, a breeder knowing only the records and not about the difference in management is likely to conclude mistakenly that the first bull was prepotent for low production and the second bull was prepotent for high production. Also, the second method will give a high figure for prepotency in those cases where the offspring resemble each other closely but are distinctly different from either parent, as might sometimes happen if there were much epistasis in a particular mating. Some breeders would not like to call such a sire prepotent, since the offspring do not resemble him even though they are unusually uniform.

Prepotency has its limits. In the absence of dominance and epistasis, the most prepotent sire in the world when mated to random-bred females cannot do more than make his offspring, which are half sibs, resemble each other as closely as ordinary full brothers and sisters from random-bred parents would. The relationship between half brothers which are not themselves inbred is $\frac{1 + F}{4}$, where F is the inbreeding coefficient of their common parent. This relationship is only one-fourth if the common parent is not inbred, but approaches one-half as F approaches 1.0. Now, if the genes of the common sire were all dominant in addition to being perfectly homozygous, we might have the appearance of still greater prepotency than this. The general effect of epistasis would be lower prepotency, since not all of the dams to which the sire was mated would have the genes necessary to nick well with those the sire carried. In exceptional cases epistasis might increase rather than decrease prepotency.

THE BREEDER'S CONTROL OVER PREPOTENCY

Dominance and epistasis result from the physiology and chemistry of the genes in their reactions with each other and with the environment in the growth of the individual. The breeder can do little or nothing to change them. Linkage is likewise not subject to the breeder's control.

The breeder's control over prepotency is limited to changes he can make in the homozygosity of his stock. For all practical purposes he changes homozygosity little except by inbreeding. The more highly inbred an individual is, the more apt it is to be homozygous for an unusual number of genes. The inbreeding coefficient is the best estimate which can be made of an individual's prepotency before that individual has actual offspring, by which its prepotency can be measured. Prepotency can be increased only a very little by the practice of mating like to like without inbreeding. The resemblance between parents and offspring is much increased by mating like to like, but when animals bred in this way are mated to unrelated or random individuals, they show only a little more prepotency than if they themselves had been random bred.

The increased prepotency of inbred animals has been known at least since Bakewell's time, but breeders do not generally pay much money for it in the sale ring. The inbred animal is usually less apt to be prepotent for its poor traits than for its good ones, but that is not always recognized. The undesired traits are more often recessive than dominant. Therefore, they are not apt to appear in the offspring of the inbred individual when it is mated to unrelated animals which do not show those undesired traits.

MYTHS ABOUT PREPOTENCY

Prepotency is not transmissible from parent to offspring as other characteristics may be, except insofar as it depends on dominance. No matter how homozygous a parent is, it cannot transmit that homozygosis to its offspring. Its offspring can be homozygous only if they receive the same genes from both parents. A high degree of homozygosis can be attained only by many generations of inbreeding but can be destroyed by a single generation of outbreeding. We sometimes read in animal breeding history of cases where there was an unbroken line of succession of noted sires from father to son and to grandson, and so on. The Baron's Pride line in Clydesdales is an example. Often the history of the case is accompanied by the inference or perhaps the outright statement that this sire transmitted prepotency to his sons and his grandsons. What really happens in such cases is that there is much selection in each generation, and to some extent each sire's mates are selected to be like him. As we look back on the breeding history, we note that one sire was more popular and successful than any other of his generation, just as his sire and grandsire were; but we do not notice the large number of half brothers which were discarded in each generation while finding the leading sire of that generation. If a sire is thought to be better than any of his contemporaries, he is likely to be bred to better mates than they are, and more of his sons are likely to be tried as sires in prominent herds. It is not necessary to invoke prepotency to explain why the most prominent sire of one generation is sometimes the son of the most prominent sire of the preceding generation.

Nothing which is known of the mechanism of heredity justifies the belief that masculinity in a male or femininity in a female indicates prepotency. Those traits are desirable to the extent that they indicate normal sex instincts and normal health of the sex glands but there is no reason for thinking that they indicate prepotency.

SUMMARY

Prepotency is the ability of an animal to make its offspring resemble that parent and each other more closely than is usual.

The genetic basis for prepotency is the degree of homozygosity of the animal and whether its genes are prevailingly dominant or recessive. To a small extent linkage and epistasis may play some part.

Almost the only control the breeder has over prepotency is the extent to which he builds homozygosis into his animals by inbreeding.

Prepotency is not transmissible from parent to offspring except insofar as it depends on dominance. Masculinity or femininity in appearance probably has nothing to do with prepotency, although it may be desirable as an indication of normal ability to reproduce.

CHAPTER 23

Linebreeding

The word "linebreeding" is in common use among breeders of pure-bred stock. It bears a good reputation and in that respect is in marked contrast with "inbreeding." Linebreeding is mating animals so that their descendants will be kept closely related to some animal regarded as unusually desirable. It is accomplished by using for parents animals which are both closely related to the admired ancestor but are little if at all related to each other through any other ancestors. If both parents are descended from the animal toward which the linebreeding is being directed, they are related to each other and their mating is a form of inbreeding in the broad sense of the word. If a man says an animal is linebred, this instantly calls forth the question: "Linebred to what?" In fact, he will not often make such an incomplete statement as that an animal "is linebred." He will say that this bull is "a linebred Domino" or these "are linebred Anxiety cattle" or "this bull is linebred to Prize-mere 9th." The use of the term linebred almost carries with it the necessity of specifying the animal or group of closely related animals toward which the breeding is directed.

Linebreeding thus differs from other forms of inbreeding primarily in that it is directed toward maintaining a high relationship to some chosen ancestor and secondarily, in that it is usually less intense than the most extreme inbreeding which might be practiced. Relationship to the admired ancestor rather than intensity of inbreeding is dominating the breeder's thought when he uses the term linebreeding, even though this same breeder if he were asked for a formal definition of linebreeding might give one which would mention nothing but the intensity of the inbreeding.

The pedigrees below show the difference between linebreeding and some other forms of inbreeding. The parents of *X* are double first cousins, having the same four grandparents. The parents of *Y* are half brother and sister. *Z* is produced by mating a male to his own granddaughter. *W* is produced by mating a sire to his daughter out of one of his own daughters. The intensity of the inbreeding is the same for *X*, *Y*, and *Z*. Yet *X* would rarely if ever be called linebred. Its sire and its dam are related through four different ancestors which, so far as the pedigree shows, may belong to four unrelated strains. If a breeder were to

- *X*
  - *A*
    - *C*: *G*, *H*
    - *D*: *I*, *J*
  - *B*
    - *E*: *G*, *H*
    - *F*: *I*, *J*

- *Y*
  - *K*: *M*, *N*
  - *L*: *M*, *O*

- *Z*
  - *M*
  - *P*
    - *Q*
    - *L*: *M*, *O*

- *W*
  - *M*
  - *S*
    - *M*
    - *L*: *M*, *O*

call *X* linebred, he would have to say that it was linebred to four different lines at once, which is something of a contradiction in terms. He would call *Y* linebred to *M* because *K* and *L* are related only through *M*, and *Y* has been kept almost as closely related to *M* as its parents were. *Z* is even more clearly a case of linebreeding because it is more closely related to *M* than *Y* is, although no more intensely inbred. Many breeders would call *W* inbred instead of linebred because the intensity of its inbreeding is so high. Others would call it "intensely linebred to *M*," since all of its inbreeding is focused on *M* and it contains 87½ per cent of the blood of *M*—a relationship of 75 per cent after allowing for *W*'s inbreeding.[1]

WHY LINEBREEDING IS PRACTICED

Animals do not live long enough for the breeder to get all the sons and daughters he wants from the best ones. Often an animal is old or even dead before its real superiority is recognized. If its sons and daughters are mated to unrelated individuals, the offspring will get only about one-fourth of their inheritance from this outstanding grandparent. If these in turn are mated to unrelated individuals, the influence of the outstanding ancestor is again halved. Unless some form of linebreeding is practiced, it is only a matter of three or four generations until even the most outstanding animal's influence is so scattered and diluted that no one descendant is very much like it. Linebreeding takes advantage of the laws of probability as they affect Mendelian inheritance to hold the expected amount of inheritance from an admired

[1] For other illustrations see Iowa Agr. Exp. Sta., Bul. 301, *Linebreeding*.

ancestor at a nearly constant level instead of letting it be halved with each generation, as would happen if all the matings were outbreeding. Linebreeding provides, so to speak, a ratchet mechanism for holding any gains already made by selection, while attempting to make further gains.

Linebreeding also builds up homozygosity and prepotency within the herd where it is practiced, just as other kinds of inbreeding do. It is no more effective than other forms of inbreeding in this respect except that, on account of the selection of the ancestors toward which the inbreeding is directed, the homozygosis produced by linebreeding is more apt to be for desired traits than is the case with undirected inbreeding. Linebreeding tends to separate the breed into distinct families, each closely related to some admired ancestor, between which effective selection can be practiced.

WHEN LINEBREEDING SHOULD BE PRACTICED

The better the animals in a breeder's herd, the more reason he has for linebreeding to them. The most vulnerable part in the linebreeding program is whether the breeder is right when he decides which of the animals recently used in his herd really were extraordinarily good ones. If he can select the good from among the others with a high degree of accuracy, linebreeding will be a powerful tool in his hands. If his judgment about which animals were good is only fair, then linebreeding has only a little advantage over other forms of inbreeding.

Those who can best afford to linebreed are breeders whose herds or flocks are already distinctly superior to the general average of their breed. If, by wise choice or lucky chance, such a breeder has used on good dams a sire whose offspring turn out to be even better than their dams, such a breeder ought to linebreed at once and strongly to this sire while the animal is yet alive. If it is already dead when he discovers how good it was, then he should hasten to linebreed to it while it still has many sons and daughters by which such linebreeding can be accomplished. While an animal is still living, the possibility of producing offspring more closely related to it than any which yet exist remains open. If a sire is thought good enough to make the risk worth taking, he can be mated to his daughters and granddaughters generation after generation, as seems to have been the intention of those who bred Blackcap Empress (Figure 30). But after an animal dies the limit of relationship to it which can be attained in future animals is only that of its closest relatives then living. Even that is a limit only to be approached. If an animal is dead by the time we realize how good it was and if there are no living animals more closely related to it than 50 per cent, then there is no possible way to produce animals more closely related to it than that. If we have let its sons and daughters and full brothers and sisters die before we wake up to its merit and there are left no living animals more closely related to it than 25 per cent, then we cannot produce any future animals more closely related to it than that—hence the importance of starting the linebreeding while there is time to do so effectively. Figure 39 shows a case where that seems to have been planned definitely.

It is an open question whether breeders with purebred herds of average merit can afford to do much linebreeding. Certainly there are many

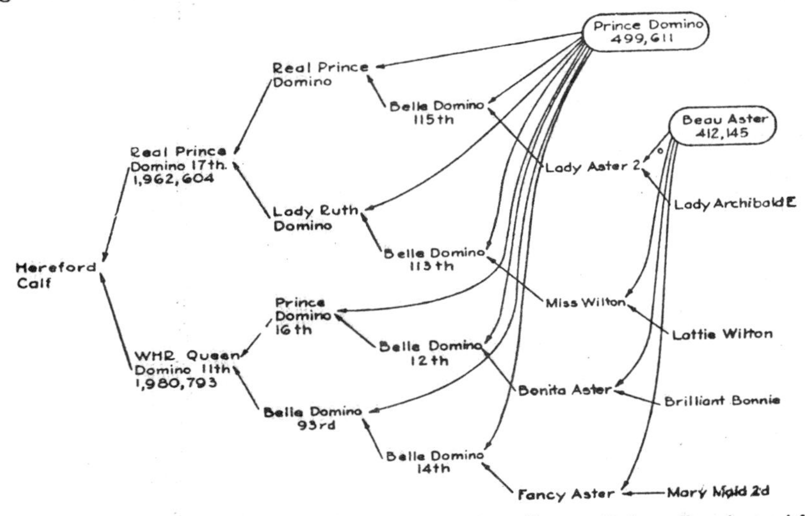

FIG. 39. Long-continued and deliberate linebreeding to Prince Domino with a very little linebreeding also to Beau Aster. Pedigrees like this do not "just happen." It took planning to get four different grandparents with so nearly identical pedigrees and to bring them together in this way without any secondary linebreeding.

good animals in such herds and much good inheritance which stands small chance of being kept together unless linebreeding is practiced. On the other hand, if the initial merit of the herd was only average, one must count on a certain amount of inbreeding degeneration which might bring the average merit of the herd below the level of the breed. The question at issue is whether the increased effectiveness of selection possible under linebreeding will be more than enough to offset the expected amount of inbreeding degeneration.

Breeders of grades cannot often afford to linebreed. The inbreeding risk involved is probably just a little greater for them than for the breeder of purebreds on account of the slightly greater heterozygosis of the grades. Even if a breeder of grades is successful at linebreeding, he cannot sell at a premium the increased prepotency and uniformity which would thus be put into his animals. He does not have the chance to gain as much by successful linebreeding as breeders of purebreds do. However, it sometimes happens that the breeder of grades uses a sire whose offspring turn out to be so much better than their dams that the inbreeding risk of using the sire on his own granddaughters or even on his own daughters seems worth taking. It seems likely that there are more breeders of grades who lose by failing to conserve a good sire than there are who lose by getting too many of the usual bad results of inbreeding while trying to linebreed to a good sire. For the breeders of grades, the certain merit of the animal to which he might linebreed needs to be further above the probable merit of the next sire which he would otherwise use than is the case with the breeder of purebreds.

Linebreeding is especially needed where there is much epistasis. Wherever a desired characteristic depends on a combination of genes which individually have undesired effects, those gene combinations tend to be scattered at each segregation. If inbreeding has made the family homozygous for several of these genes, the whole combination has more chance of being transmitted to enough of the offspring to permit its becoming established in that family. If the form of inbreeding used is linebreeding, with selection constantly directed toward keeping the family closely related to animals which once showed that desirable combination, the chances of recovering the whole combination among the descendants are much better than if the descendants were continually outbred to unrelated animals. Outbreeding would increase the likelihood that this particular combination of genes would be scattered into its constituent and individually useless parts. Linebreeding is the only very promising way of securing desirable gene combinations differing from the most frequent type of the breed by much more than four or five gene substitutions, each of which is harmful if made one at a time but beneficial if all can be made at once. That is, linebreeding is the answer to the situation pictured in Figures 20 and 21, in chapter 12, where it was pointed out that selection could carry a population to the nearest peak of desirability but could not carry it to a peak of higher desirability across an intervening valley which was more than a few gene substitutions in width.

DANGERS OF LINEBREEDING

The breeder may have the wrong ideal and be breeding toward a type which has a lower sale value than some other type. Of course this same danger exists in all other breeding systems. But, since linebreeding is more effective in carrying the breeder toward his goal, it is more important for a breeder practicing linebreeding to be sure of his goal than for one who is breeding by individuality alone.

Linebreeding may be so intense that genes will become homozygous more rapidly than the breeder can discard the undesired homozygotes. The inbreeding may thus result in fixing in his herd some undesired genes in spite of all the selecting he can do against them. Whether this will happen depends not only on the inbreeding intensity but on the merit of the stock with which he starts and on the skill which he exercises in his selection, including such use as he makes of progeny tests, pedigree estimates, etc. Then, too, a part of the success or failure will be due to the chance inherent in Mendelian inheritance whereby one individual from a particular mating may happen to be a better or worse individual than would ever be produced again from that same mating.

There is no magic about the linebreeding process which will automatically produce good results. If selection is not practiced, a breeder will generally do better to avoid linebreeding altogether, since he would thereby avoid the inbreeding effect. But a breeder starting with good stock and directing the linebreeding toward the best of the recent ancestors in his herd can effect more improvement by selection while holding the improvement he already has than would be possible if he were continually outbreeding.

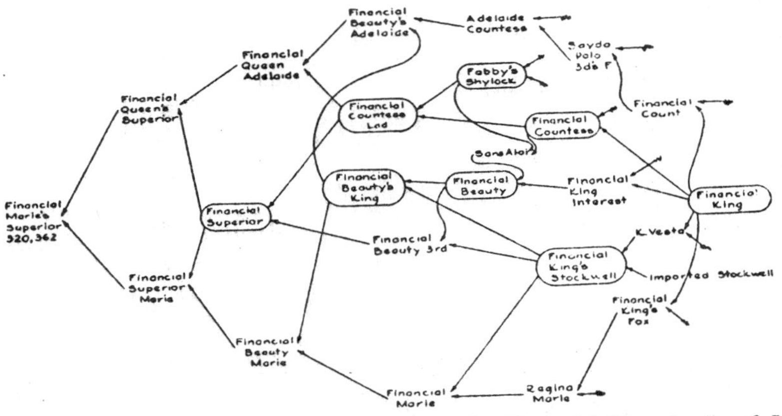

FIG. 40. Long-continued linebreeding within the Financial King family of Jerseys. Much of the linebreeding here is secondary and to recent animals such as Financial Superior, Financial Countess Lad, and Financial Beauty's King.

If one wishes to linebreed purely to one animal, he must see to it early that a large number of sons and daughters of that animal are saved. Otherwise the time quickly comes when further linebreeding to that ancestor also involves considerable linebreeding to some of its descendants. Figures 38, 40, and 41 show cases of that. There is no particular reason why this secondary linebreeding should be avoided if the animal toward which it is directed is an unusually good one. But, if the herd is small and only one man is linebreeding to this line, there will be only a few individuals in each generation. In some generations it will happen that no one of those will be outstanding enough to justify linebreeding to it. If the number of animals in this linebred strain or family

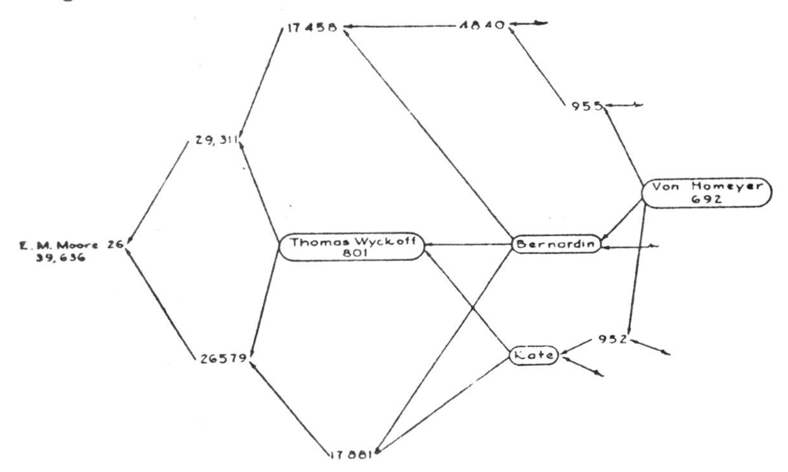

FIG. 41. A Rambouillet pedigree in which one male is the center of the linebreeding in each generation.

is very small, the breeder must either linebreed to some of those which were not good enough to justify it, or else he will have to give up his linebreeding plan and make a distinct outcross. This is the intrinsic danger of a permanent linebreeding policy based on too small a herd. If the herd is large enough, such secondary linebreeding can be avoided or at least can be kept so small in amount in those generations when there is no outstanding individual that it will be practically harmless. Hence, a linebreeding plan which is to last more than two or three generations without much risk requires the equivalent of a herd large enough to justify keeping about three to five sires in use at all times.[2] This might be one large herd; or several breeders with small herds might co-operate in breeding toward the same line, exchanging breeding stock with each other but rarely if ever introducing a breeding animal from herds not in the group.

GENETIC ASPECTS OF LINEBREEDING

Linebreeding, more than any other breeding system, combines selection with inbreeding. In a certain sense, linebreeding is selection among

[2] These figures are based on the $1/8M$ formula for loss of heterozygosis within a closed group. (See Chap. 21.)

the ancestors rather than among living animals. Since many of the ancestors being considered will have had several different offspring, they are to some extent proved sires and proved dams. The linebreeding is, therefore, selecting from among progeny-tested ancestors those whose influence is to be preserved. This advantage is partly offset by the fact that the individuals used to preserve the traits of their ancestors will vary in how much they are really like those ancestors. One cannot depend entirely upon the pedigree in making his selections. If a breeder's linebreeding plans are so rigid that he decides, even before certain animals are born, that he will use those animals extensively for breeding purposes, he is certain to run into trouble sooner or later. The practical procedure is to decide approximately what kind of pedigree the next sire or dam of a sire must have, but to proceed so that one can choose from among several different animals with pedigrees of about that kind the one which seems to be the best individual. If a breeder can use in a tentative way several of the individuals, postponing until he has seen several offspring of each the final decision for carrying on the breeding program, he is on still safer grounds.

Outbreeding systems, as contrasted with linebreeding, risk half the merit of the offspring on the selection of the individual sire next to head the herd. If selection has already made the herd superior to the average of the breed, probably half of that will be lost in the next generation unless selection is again as effective as it was before. Every breeder will occasionally make mistakes in his selections. The breeder who continually practices outbreeding can therefore expect to have the merit of his herd at times go far back toward the average of the breed. One who wants to make and keep his herd far different from the average of the breed to which it belongs must put some kind of a pedigree barrier between it and the rest of the breed, so that the differences continually being produced as successive sires are used will tend to accumulate and not be halved with each successive sire. An analogy may make that point clear. Water tends to seek its level. If there were no barriers in the way, the level of the water in all the lakes of the world would quickly seek the level of the ocean, just as the water in the rivers is continually doing. The breeder who practices outbreeding is placing no barriers, except his own skill at selecting, in the way of his herd's tending toward the average level of the breed. The breeder who practices linebreeding is to a considerable extent isolating his herd from the rest of the breed, and its merit tends toward that of the isolated group rather than toward that of the breed as a whole, just as the level of the water in Lake Erie remains nearly constant but several hundred feet above the level of the water in the ocean, even though water is steadily flowing into it and out of it again.

SUMMARY

Linebreeding is a form of inbreeding directed toward keeping the offspring closely related to a highly admired ancestor. All inbreeding not necessary for holding this relationship high is avoided as far as possible. Hence, the intensity of the inbreeding is usually moderate in linebreeding systems. Relationship to a chosen ancestor is the main feature which distinguishes linebreeding from other forms of inbreeding.

It is practiced to conserve the good traits of an outstanding sire or dam among its descendants, increasing those descendants in numbers without lessening their resemblance to this ancestor.

The more superior a breeder's herd or flock is to the average merit of its breed the more reason he has to practice linebreeding to his very best animals or to the very best of their recent ancestors.

The risk involved in linebreeding depends upon how much undesirable inheritance is in the herd when the linebreeding begins, upon how skillful the breeder can be in his selections, how much use he can make of progeny tests before he has to decide whether to use a sire extensively, how large his herd is, and whether he must work alone. If he can co-operate with several other breeders who are linebreeding to closely related animals, he can get an occasional mild outcross from them without disturbing his whole program.

Linebreeding is choosing which ancestors shall have their influence conserved and spread through the whole herd and which ancestors shall be allowed to diminish in importance with each generation until they no longer have much effect.

CHAPTER 24

The Family Structure of Populations

Even in populations which are breeding entirely at random, an individual does not have the same probability of being like every other individual. Each is more closely related to some than to others. This gives the population some kind of a family structure. Biological populations are not as homogeneous as a population of balls or numbered tickets in an urn, such as are often used to illustrate the elementary laws of probability.

The definition of family always has in it something of the idea that members of the same family are like each other and different from members of other families. Yet usage varies widely as to the degree of relationship which is meant. Sometimes family means a set of full sibs. This is frequent in poultry breeding, but so restricted a definition is uncommon in other animals where the number of full sibs is usually too small for this to be very useful. In plants which can be self-fertilized, family often means all the progeny of a single plant. It may mean a more highly inbred group than that, but usually "line" is then used rather than family. In animals which have long been linebred to a certain individual, family may mean the whole group of individuals which are linebred enough to be closely related to this individual and to each other. An example is the Owl-Interest family of Jerseys.

In animals where little or no linebreeding has been practiced, family is more likely to mean the descendants of a particular individual, usually a purchased one (a "foundation" aimal) or one thought to be unusually good and with offspring well above average. Sometimes this usage is carried to extremes, the family name being traced back only through the female line (Shorthorns and Aberdeen-Angus) or only through the male line (Herefords) to an ancestor so remote that, if there has been no subsequent linebreeding, most of its descendants are little if any more related to each other than they are to other animals of that breed. Such usage is much like the transmission of family names in man. There is little more reason to expect any real average difference between Blackbirds and Ericas in Aberdeen-Angus than there is to expect differences between the Smiths and the Wilsons in the United

States or between the Hansens and the Larsens in Denmark! The idea of relationship between members of the same family becomes very dim here, and family names tend to become artificial designations which may be convenient but do not correspond to any biological reality.

Taxonomists use family in a special and definite sense to denote a group which is intermediate between a genus and an order, as the cat family (*Felidae*), the deer family (*Cervidae*), cattle family (*Bovidae*), etc.

We will consider first some of the less definite usages of family and then the family as a basis for selection.

THE FAMILY—A GROUP OF CLOSE RELATIVES

When we say that an individual is from a good family, we usually mean that the average merit of all its near relatives, regardless of whether they are related through the sire or dam or bear the same family name, is considerably above the breed average. This is the same sense in which the term is often used in man when someone is said to be of "a good family" or from "a shiftless family." Family, in this sense of the word, usually does not extend much farther among the collateral relatives than to first cousins. Not often is anything implied about ancestors farther back than the great grandparents, or about descendants much more distant than grandsons and granddaughters. This use of family is a practical application of relationship in estimating the heredity of an individual from the appearance and performance of a considerable number of its close relatives.

The family in this sense is somewhat indefinite, and one family grades into another. For example, an individual's maternal uncles and its paternal uncles are members of its family but the paternal uncles need not belong to the family of the maternal uncles at all. In fact, no two individuals would belong to exactly the same family unless they were full sibs. The individual is at the center of its family with its relatives clustered around it at various distances according to their relationship. There is no accurate and simple formula for giving proper weight to different relatives when averaging their good and bad qualities to find the merit of the family, although of course the closest relatives are the most important—unless there are strong environmental correlations between them, as may sometimes be the case with maternal sibs. If the family contains only a few members, chance can still play a large part in giving one such family a good rating and another one a poor rating.

THE FAMILY NAME

An Aberdeen-Angus cow is called an Erica if she traces through an unbroken female line to the cow called Erica, regarded as the foundress of the family. Technically she is still an Erica even if she does not trace to Erica through any other line of her pedigree. In the Shorthorn and Aberdeen-Angus breeds the family name is traced through the dams. In the Hereford breed the family name is from the sire. In most dairy breeds the family name comes from the dam, but in some both systems prevail. The Holstein-Friesians have the De Kol family and the Pietertje family which take their names from foundation cows, but the Netherland family takes its name from the bull. In the Jersey breed such families as Tormentor and Golden Lad are named after bulls; but there are also such families as Coomassie, Fontaine, and Oxford named after cows. Breeders of cattle and horses mention family more than do breeders of sheep and swine.

This idea of family is a natural development in one-sire herds. A breeder with several cows but only one bull will, of course, observe many differences between his calves. Since all his calves in any one calf crop are sired by one bull, it would be natural to assume, without even realizing that he had done so, that all the differences between the calves were due to differences in their dams. If when the successive calves from the same cows are compared it is seen that there is a general tendency for one cow to have good calves and another one to have mediocre calves, it is natural for the breeder to group his animals in his own mind in terms of their dams or grandams, as far back as he remembers those. The inference that all the differences between the calves are due to differences in their dams is, of course, unjustified, since the sire is never entirely homozygous and some of the differences between the calves will be due to difference in the inheritance they have received from him.

This tendency for the owners of small herds to think of families in terms of female foundation animals is reversed in large herds where several sires are maintained at all times. It is difficult for the breeder to know all of his individuals closely in such herds and easy for him to compare the calves by one sire with their contemporaries by other sires. This naturally leads to a system of referring to the calves in terms of their sires and grandsires instead of their dams. Perhaps this is responsible for the fact that the Hereford breed, which is prevailingly bred in large herds, tends to trace its family through the male line, whereas breeds more commonly bred in small herds tend to trace the family through the dam.

When the system of tracing the family name through the dam is followed far, it naturally leads to printing the pedigrees in the "abbreviated" form, a sample of which is shown herewith in the pedigree of the noted Shorthorn bull, Rodney. Pedigrees in cattle breeds other than the Shorthorn are now usually printed in the bracketed form which gives information on all lines to the same number of generations. Breeders of horses often use the abbreviated pedigrees. Formerly only that part of the pedigree which appears in columns in this example was shown. In recent years footnotes about the three most recent sires are added, as in this case. Figure 42 shows the pedigree as it would appear in bracketed form. The line drawn across the pedigree separates the information which is contained in the footnotes from that which is given in the columns.

RODNEY—753,273

Red. Calved February 8, 1917. Bred by C. H. Jolliffe, Darlington, Eng.

Dam	*Breeder of dam*	*Sire*	*Breeder of sire*
		Sanquhar Dreadnaught	A. M. & O. J. Law
Rosetta 7th	C. J. Jolliffe	Prince of the Blood	W. Duthie
Ballechin Rosetta	A. Robertson	Victor Chief	J. Durno
Scottish Rosebud	W. Duthie	Scottish Archer	A. Cruickshank
Red Rosebud 1st	S. Campbell	Gravesend	A. Cruickshank
Red Rosebud	S. Campbell	Borough Member	J. Bowman
Rosebud	S. Campbell	Novelist	S. Campbell
Rosebud 1st	S. Campbell	Diphthong	A. Cruickshank
Rosebud	S. Campbell	Scarlet Velvet	A. Cruickshank
Thalia	A. Longmore	Earl of Aberdeen	Mr. Hay
Myrtle	A. Longmore	Balmoral	A. Longmore

SANQUHAR DREADNAUGHT 680399, Sire, Hawthorn Champion 530142 by Bapton Champion (78285), out of Hawthorn Blossom 10th V46-585E. Dam, Zoe 11th V53-916E by Scotch Thistle (73584), out of Zoe 3d V44-502E.

PRINCE OF THE BLOOD 715108, Sire, Pride of Avon 311139 by Primrose Pride 222709, out of Rose Blossom V45-489E. Dam, Scottish Princess V52-677E by Scottish Archer (59893), out of Princess Royal 41st V46-455E.

VICTOR CHIEF 206990, Sire, Lord Lynedoch 206982 by Sittyton Pride 136401, out of Lenora V46-450E. Dam, Violet Blossom V44-406E by Lord Douglas 132003, out of Chief Blossom, V42-374E.

This abbreviated form of pedigree was fairly adequate when all of the breeders were acquainted with the sires which were being used in the prominent herds. There was no need to print the pedigree of the sire, since each potential customer knew that. The customer would not know all the females of the breed, so he did want to see the pedigrees of the cows. When the breeds grew larger the time came when no one knew all the sires; therefore, it became necessary to add these footnotes. The abbreviated pedigrees emphasize remote ancestors beyond all usefulness. For example, Rodney is called a member of the "Rosebud family" after the cow, Rosebud (by Scarlet Velvet), which was the first one bred by S. Campbell, who developed this family. If there has been no linebreeding to Rosebud in Rodney's pedigree, his relationship to her will be $(1/2)^8$, which is about 0.4 per cent of his genes which probably came from her. Rodney must have had literally tens of thousands of contemporary relatives which had other family names but were more closely related to him than was the cow from which his family name comes.

The abbreviated pedigrees emphasize the names of the breeders. The value of any pedigree is affected by the general reputation of the herd in which the animal was bred. Something worth while is lost when the name of the breeder is given a less prominent place than it has in

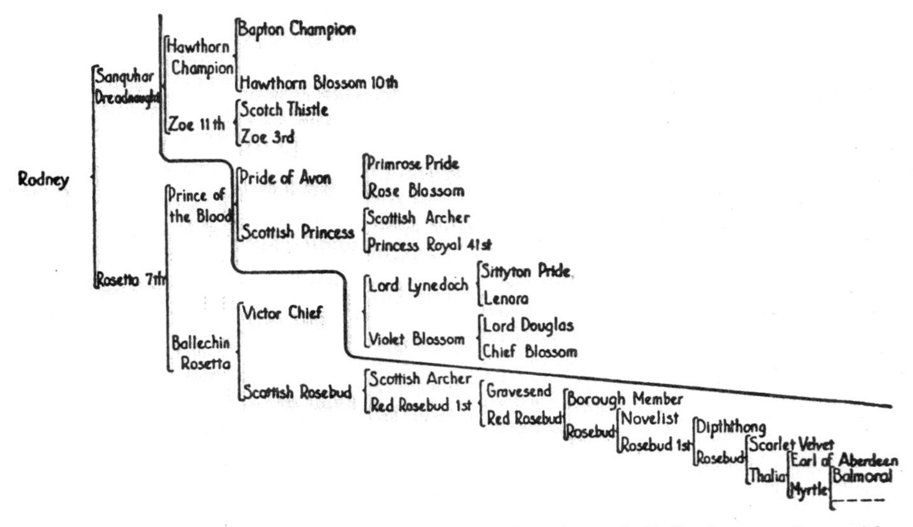

FIG. 42. The abbreviated pedigree of the Shorthorn bull Rodney, as it would appear if the same information, except the breeders' names, were given in the bracketed form.

the abbreviated pedigrees. Then, too, a cow does not get to be regarded as the foundress of a family on her own merits alone, but rather on the high merit of many of her offspring. It is fairly safe to infer that Rosebud 1st and also Rosebud (by Novelist) were distinctly better individuals than average or else Rosebud (by Scarlet Velvet) would not have been regarded as the foundress of a family.

The commercial importance of the family name is usually small, unless perhaps in times of booms or pedigree speculation. It lends itself well to speculation, particularly in breeds where the family name is traced only through the female line. Even the best of cattle are none too prolific, and a family can become famous and remain famous for many years without a large number of females of that family ever existing at any one time. If a strong demand for a family name can be created, extreme speculation can easily result because the supply is limited. Naturally such speculation is rare except when there is general prosperity and prices for breeding stock have been rising for some time. The most noted case of this kind was the speculation that went on in the "pure" Duchess Shorthorns in the 1870's. There have been several periods of less extreme speculation in Aberdeen-Angus family names. Yet in 21 Aberdeen-Angus sales, studied from this point of view from 1929 to 1938 in Iowa (unpublished), the only conclusion possible is that practically no cash was really being paid for family name. However, this was in a period of economic depression and perhaps this finding is not significant. Four more sales from 1940 to 1942 when prices were rising showed distinct price differences between families.

The family name has genetic importance when the animal which gave its name to the family is still within three or four generations of the animals concerned. In such a case the coefficient of relationship between the animal and the foundation animal is still high enough to mean that the two are apt to be alike in a noticeable proportion of their genes.

Paying attention to maternal family names compels a certain amount of added attention to the females in breeding selections. Some breeders might be more careless about the dams if it were not for this extra attention forced on them by the family system. The actual importance of this may be slight.

The family name would have some genetic importance whenever the general condition exists that breeders strive always to mate a cow of one family to a bull of the same family; that is, to breed the family "pure." If S. Campbell had always sought Rosebud bulls to mate to his Rosebud cows, and if this had been continued to Rodney's time, Rodney would have been kept very closely related (linebred) to the original Rosebud cow. If that were a general practice among most breeders, it would lead to steady linebreeding which might keep the foundation animal important for many generations after its death. Where the family system brings this about, it can be a powerful instrument in improving the pure breeds, but this does not happen often. The cows in a small herd may belong to a dozen families, but the same bull is usually mated to all of them.

THE LINEBRED FAMILY

Sometimes "family" is used to designate a group which has been partially separated from the rest of the breed for a long time in their breeding and among which there has been considerable linebreeding. Not often have such cases really been carried far enough to make the family very distinct from the rest of the breed. Even a slight separation of this kind has sometimes been the occasion for a large amount of speculation in pedigrees. The most famous case is that of the "pure" Duchess Shorthorns, for which the pedigree speculation reached its climax in 1876 in the New York Mills sale where one cow sold for $40,600. The "pure Scotch" Shorthorns are another example. In spite of many bitter condemnations of the "straight Scotch" craze, the straight Scotch almost entirely displaced the other beef Shorthorns in the United States during the two decades preceding 1920. The "straightbred" or "airtight" Anxiety 4th Herefords may be a similar case in which the final outcome is still in doubt. These straightbreds are a group whose pedigrees in nearly every line go to daughters of North Pole and to sons and daughters of Anxiety 4th. That is, they carry nearly 50 per cent of the blood of North Pole and Anxiety 4th combined. Somewhat milder cases have happened in the Jersey breed in connection with the Owl-Interests, the St. Lamberts, and Tormentors, and in the Holstein-Friesians with the Homestead family.

The principles involved are just the same as those that have been discussed under linebreeding. If the linebreeding has been carried far enough to make the family really distinct from the rest of the breed, then there is an important genetic basis for the family name. This kind of a family is to some extent a breed within a breed.

THE GENETIC DEFINITION OF FAMILY

The biological basis for treating a group as a family is the average genetic likeness among members of the group. The best estimate of genetic likeness where the actual genotypes are unknown is the coefficient of relationship. This will usually give a reasonably true picture of the average genetic likeness of family members where the base to which relationship was computed is not many generations in the past.

As an example of this way of defining a family quantitatively, we might choose to consider each set of full sibs in a random breeding population as a family. For comparison with other kinds of families we can define this kind as a group which are related to each other 50 per cent. If all the offspring of each male are to be considered as a family, that kind of a family can be defined as a group related 25 per cent to each other.[1] We can compare the importance we should attach to family when making selections in the two cases by using alternately .50 and .25 for r in the formulas which are in the next few pages. If all the grandsons and granddaughters of a male are to be considered as a fam-

[1] Although such a family will not be entirely homogeneous if some of them are full sibs to each other, this will not increase the average relationship much if there are more than three or four different sets of such full sibs or if the number in each such set is small. For example, if the progeny of a boar consist of four litters of five pigs each and we call this a family, it will not of course be a perfectly homogeneous family but will be one large family with four branches or subfamilies. The average relationship of each pig to the other 19 will be an average of 4 full sib and 15 half sib relationships or about 30 per cent. If the progeny of a bull are five pairs of full sisters, the average relationship within the group of 10 will be an average of 1 full sib and 8 half sib relationships or 28 per cent.

ily, we can define this family as a group which are related to each other 6¼ per cent, plus a little more from the fact that some of them will be sibs or cousins through more than one grandparent. When family is thus defined quantitatively, it is easy to see why the practical usefulness of family groupings becomes so small when the group members are related only through ancestors as distant as grandparents.

The observed family resemblance may be expressed either in terms of the correlation between members of the same family or in terms of how much smaller the differences between members of the same family are than the differences between members of different families. The formula which relates the two is simply that the correlation between members of the same family equals $\frac{V - B}{V}$ where V is the variance between individuals which belong to different families and B is the variance between individuals which belong to the same family. $V - B$ is the variance caused by things which are alike for all members of each family but may vary from one family to another. $V - B$ might be wholly genetic in some cases but is likely also to include some differences caused by common environment for family mates, or by epistasis, or dominance.

The formulas showing quantitatively the advantages and disadvantages of selection on a family basis are rather complex but they are given in the following sections because they are important guides for estimating whether a plan for selecting on a family basis is likely to increase progress sufficiently to be worth its costs.

CONDITIONS AFFECTING PROGRESS WHEN CHOOSING BETWEEN FAMILIES

The formulas for comparing individual and family selection, when the same percentage of the population must be culled in either case, are expressed as follows for convenience:

Let G = the additively genetic variance between individuals.

E = all other variance (largely environmental in most cases) which is random with respect to family.

C = the variance caused by whatever fraction of the environmental, epistatic, and dominance deviations are alike for members of the same family, but vary from one family to another.

r = genetic relationship between members of the same family.

t = phenotypic or observed correlation between members of same family $= \frac{rG + C}{G + C + E}$.

n = number of individuals in each family.

Then: Variance between individuals from different families = $E + G + C$

Variance between members of the same family $= E + (1 - r)\ G$.

Variance between actual family averages =

$$rG + C + \frac{(1-r)\ G}{n} + \frac{E}{n}.$$

Having large numbers in the family permits the environmental differences (E) and the genetic differences between members of the same family—the $1-r$ fraction of G— to cancel each other, so that the actually observed differences between family averages tend toward $rG + C$, which will be almost wholly genetic if C is very small. When n is small a considerable part of the differences between the actually observed averages of various families may still be due to the E/n term which is environmental and misleading.

The larger r is, the more of the genetic variance (G) will be between families instead of within families. This will permit the family averages to be farther apart, so that one can reach farther when selecting between families when r is large than when it is small. Also any increases in r will make a larger fraction of the observed differences between family averages genetic, so that one selecting between family averages will actually get a larger fraction of what he reaches for. To have r large is an important prerequisite for selection between families to be very useful.

When C is large the heritability of differences between family averages will be low, since that heritability tends toward $\frac{rG}{rG + C}$ as n becomes indefinitely large. Hence, with C large many of the differences between family averages will not be genetic, many mistakes will be made in selecting between families, and only a small fraction of what is reached for in family selection will actually be gained in the merit of the offspring. The deceiving effects of C do not diminish with increases in n as those of E do. In the mammals, C is especially likely to be important when the family consists of maternal sibs.[2] C is likely to be large also in birds which hatch and brood their own young. Even in birds hatched in incubators and reared apart from their dams, certain initial environmental differences caused by the size of the egg may not be wholly equalized before the birds are adult. In data collected from

[2] Environmental correlations between full sibs are also prominent in data on man, especially in data pertaining to mental and social traits. Man's long infancy and childhood and the wide differences from home to home in cultural environments and parental precepts and examples give an unusual opportunity for such correlations to develop in characteristics which are susceptible to much modification by such influences.

many different farms, C is often troublesomely large because environments vary considerably from farm to farm, and in most cases each family will have been raised wholly on one farm. In carefully planned breeding experiments every effort will be made to reduce C to zero by controlling or randomizing the environment with respect to families. Often that can be fairly well achieved for all of C except the part due to maternal environment and the part due to weather and to changes in general environment from year to year in cases where the families are not contemporary. But breeders when purchasing animals usually must compare families kept in different herds. Then the best they can do to eliminate C is to observe the conditions under which each family is kept and make allowances for the differences which they think these conditions produced between families.

FAMILY SELECTION COMPARED WITH INDIVIDUAL SELECTION

For clarity we will first consider what family selection would accomplish if it were practiced by itself without any attention to individuality. Figure 43 will illustrate the difference between family selection alone and selection on an individual basis. The data are 180-day weights of four pigs in each of four litters. One-fourth are to be saved for breeding purposes, and selection is for the heaviest weights. If selection is wholly on a family (litter) basis, the selected pigs will be all four of those in family 2, since that family has the highest average weight. Pig *E*, which has a low weight, will be selected along with the other three because it is in the family with the high average. If selection is wholly on an individual basis, pigs *D*, *G*, *H*, and *L*, will be selected regardless of the merits of their sibs. If the method of selection is some compromise which gives attention to family averages as well as to individual merit, pigs *F* and *P* might possibly be saved instead of pigs *D* and *L*.

Two things about this situation must be emphasized. First, one cannot pay attention both to family and to individuality without compromising on both. Almost never will it happen that *all* members of the family which has the highest average will be individually superior to *all* members of the other families. One must compromise on one thing or the other when deciding what to do with good individuals (like *D* and *L*) from mediocre or poor families and with mediocre or poor individuals (like *F* and *E*) from a good family. Second, some of the same animals will be saved, no matter which method or compromise is used. The family with the highest average *must* contain more than a fair share of individuals which are above average. *G* and *H* illustrate this. Purely individual selection does some of the same things which purely family selection would do. If either method is *absolutely* ineffective, the other will be also. The contrast between them is between two methods both of which will produce some improvement, if either of them will produce any, but which will not, except by coincidence, produce exactly the same amount of improvement per generation.

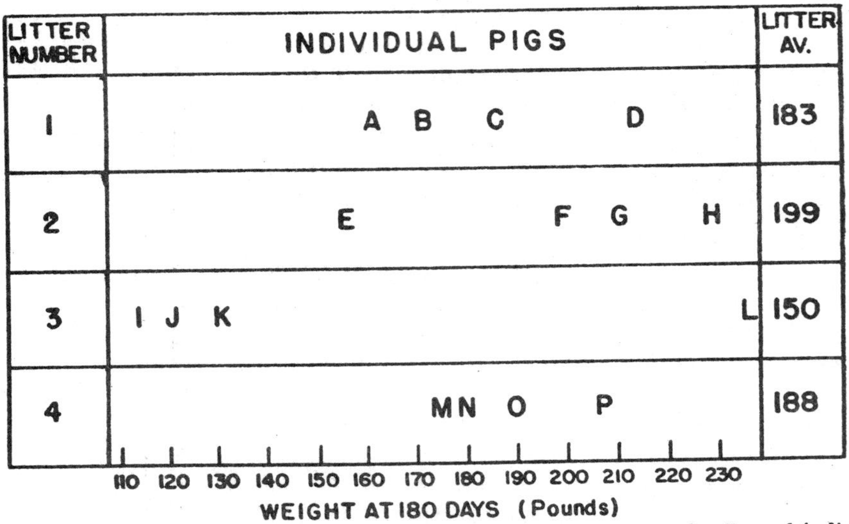

LITTER NUMBER	INDIVIDUAL PIGS	LITTER AV.
1	A B C D	183
2	E F G H	199
3	I J K L	150
4	M N O P	188

110 120 130 140 150 160 170 180 190 200 210 220 230

WEIGHT AT 180 DAYS (Pounds)

FIG. 43. Distribution of some pig weights by litters to illustrate family and individual selection. Each pig is designated by a letter located vertically according to its litter and horizontally according to the pig's weight.

The increase in the population mean each generation under purely family selection is expected to be the following fraction of the increase to be expected under purely individual selection: $\frac{1 + (n - 1)r}{\sqrt{n[1 + (n - 1)t]}}$.

Although this formula is complex, it can be seen that family selection is most likely to be superior when r is large and t is small. Differences in n do not affect the ratio very much unless t is extremely small and r is large. In that case high values of n increase the effectiveness of family selection markedly. If t is nearly equal to r, family selection cannot equal individual selection in effectiveness, even when r and n are large.

If purely family selection is to produce improvement x times as rapid as would be produced by individual selection, then r must equal $\frac{x\sqrt{n[1 + (n - 1)t]} - 1}{n - 1}$. For x to be 1.0 when families consist of 5, r would have to be:

.41 if t is .1
.50 " " " .2
.58 " " " .3
.65 " " " .4
etc.

If n is as large as 25, the corresponding values of r necessary for x to be 1.0 would be only a little lower, namely: .34, .46, .56, .64, etc. In short, if family selection is to be much more effective than individual selection, r must be considerably larger than t. Increases in n do not lower the requirements for r much unless t is very small.

Because t equals $\frac{rG + C}{G + C + E}$, it is necessary for C to be nearly zero and E to be much larger than G if r is to be much larger than t. When heritability is low (G is small, compared with $E + C$) neither individual selection nor family selection will make rapid progress, but family selection can then be considerably more effective than individual selection if C is zero or nearly so. Among important characteristics for which E is likely to be very large and C may be small, are such complex things as fertility, vitality, longevity, disease resistance in general,[8] and probably growth rate so far as that does not depend mainly on dimensions of bones.

SUPPLEMENTING INDIVIDUAL SELECTION WITH FAMILY SELECTION

It is sensible of course to use both the family average and the individual's own characteristics in selecting, compromising somewhat on each in order to make faster progress than could be made by using either alone. The progress per generation which will be achieved under the optimum combination of individual and family selection will be the following fraction of what would be achieved by selection on individuality alone:

$$\sqrt{1 + \frac{(n-1)\ (r-t)^2}{(1-t)\ [1 + (n-1)\,t]}}$$

The most important thing in determining how large this ratio will be is the term, $r - t$, which measures how much more the members of the same family are like each other genetically than they are outwardly. When $t = r$, nothing at all is gained by paying attention to the family average. The larger the difference between r and t the more there is to gain by paying some attention to family. Even when t exceeds r, something is to be gained from considering the family average, but in this

[8] Individual differences in resistance to some specific diseases may be rather highly genetic.

case the attention given to the family average is negative; i.e., the individual is judged partly by its own merit and partly by how much it deviates from its family average, instead of being given some credit if the average merit of its family is high and being penalized if it is from a poor family.

The numerical values in Table 18 for some selected conditions may make it easier to see what circumstances lead to much gain from paying attention to family. The basic formula is that if paying attention both to the individual and also to its family average is to make progress $1+y$ times as rapid as if selection were on individuality alone, $r - t$ must equal

$$\frac{y(2+y)(1-t)[1+(n-1)t]}{n-1}.$$

For progress to be made 20 per cent faster by considering the family average requires the difference between r and t to be 1.45 times as large as is necessary to increase progress by 10 per cent. If progress is to be 30 per cent faster, the difference between r and t will need to be 1.81 times as large; for 40 per cent it will need to be 2.14 times; for 50 per cent 2.23 times is required; etc.

Since r cannot exceed 1.0, extremely large gains from paying attention to the family average are possible only when t is very small. Also n must be large, but this of itself will not help much unless t is so small that $(r - t)^2$ is far larger than $t(1 - t)$. This marks out the domain in which family selection is most useful, for t can be small only when heritability is low and when other causes (C) for family members resem-

TABLE 18

GENETIC RELATIONSHIPS NECESSARY IF PAYING ATTENTION ALSO TO FAMILY IS TO INCREASE THE RATE OF IMPROVEMENT BY 10 PER CENT OR BY 100 PER CENT

t	For Progress to Be 10 per Cent Faster (y = .1)		For Progress to Be Twice as Fast (y = 1.0)	
	When n = 5 r must =	When n = 25 r must =	When n = 5 r must =	When n = 25 r must =
.01	.27	.11	.89	.40
.05	.29	.19	.97	.56
.10	.36	.26	Impossible	.72
.20	.47	.40	"	.96
.30	.58	.52	"	Impossible
.40	.69	.64	"	"
.50	.78	.74	"	"
.60	.87	.83	"	"

bling each other are zero or very nearly so. Then if r and n can both be made large, selection on the family basis can increase progress very much.

Family selection and individual selection are mainly supplementary procedures rather than competitive ones, individual selection doing nearly all that the two together could do when heritability is high but declining in effectiveness in direct proportion to the decline in heritability, while family selection helps little when heritability is high but increases in relative effectiveness as heritability of individual differences declines. Thus, in framing efficient breeding plans, attention should gradually shift from individual selection to emphasis on family selection more and more as one turns from highly hereditary to less and less hereditary characteristics.

OPTIMUM ATTENTION TO PAY TO FAMILY AVERAGE AND TO INDIVIDUAL MERIT

For the maximum rate of improvement, each bit of merit or defect in the family average should receive $\frac{n}{1+(n-1)t} \cdot \frac{r-t}{1-r}$ times as much attention as the same absolute amount of merit or defect in the individual's own characteristic. This ratio is large when r is large, t is small, and n is large, although the latter doesn't make much difference unless t is small. This ratio goes to zero when $r = t$ and takes negative values when t exceeds r. For t to exceed r means that C is large and that each family average is being shoved up or down by circumstances other than the average breeding value of that family. That the fraction then is negative merely indicates that it is then more accurate to judge the individual partly by its deviation from its family average, as an automatic way of correcting partly for the nongenetic circumstances included in C. The deviation of the individual from its family average is composed of variation coming from E and from $(1 - r)\,G$ and does not include the C term. To judge the animal entirely by its deviation from its family average would open the door to large errors from E and would forego opportunity to select for differences caused by rG. Hence the optimum combination of attention to family and to individuality is a compromise aimed at some discounting of C, some use of rG as well as $(1 - r)\,G$, and some reduction of E by n.

The conditions when attention to the family average should turn negative may actually be reached in data where r is low and C is large, as in dairy production records used in proving bulls which have been kept and used in different herds. Also characteristics markedly influenced by prenatal or pre-weaning differences in environment are likely to have a large C between litter mates. For such characteristics in pigs, full sibs which are not litter mates or even paternal half sibs may deserve more attention than litter mates.

INBREEDING AND THE FAMILY STRUCTURE OF POPULATIONS

Inbreeding helps in several ways to make family selection more effective. First it increases G to $1 + F$ times what it was in the foundation population.[4] This also helps mass selection by increasing the standard deviation a little and thus making a larger selection differential possible. A more important effect is that it increases heritability, and thereby a larger fraction of the selection differential is actually gained in the offspring. The gain had by increasing G is rather quickly exhausted when the poorer families are culled. To renew it the remaining families must be intercrossed and distinct families formed again by inbreeding these crosses. It is therefore a gain which cannot be harvested in every generation.

In the second place, inbreeding is the only way to make r much larger than .31 in large families of the less prolific animals, or larger than .50 in families of animals like pigs and chickens. As a numerical example of how rapidly inbreeding will increase r, the full sibs in the first inbred generation of full brother-sister inbreeding are related 60 per cent, in the second generation 74 per cent, and in the third generation 79 per cent— compared with 50 per cent where there is no inbreeding. The r between full sibs equals $50 \left[1 + \frac{F + F'}{1 + F}\right]$ per cent, F being the inbreeding of sibs and F' the inbreeding of their parents. This shows vividly for full sibs how closely the increase in r beyond 50 per cent depends on the intensity of inbreeding. In continuous half sib inbreeding—one sire in a large herd closed to outside blood—half sibs in the first inbred generation are related 39 per cent, in the second generation 50 per cent, and in the third generation 58 per cent. How rapidly inbreeding will increase the genetic relationship between half sibs may be seen from the fact that this relationship equals $25 \left[1 + \frac{5F + F'}{1 + F}\right]$ per cent, provided the three parents are equally inbred and equally related to each other.

Even one or two generations of rather mild inbreeding can raise r enough to increase greatly the proper amount of attention to pay to the family average for the most effective selection, especially if the charac-

[4] The increase will generally be somewhat more than this if there is much dominance or epistasis. But G may actually decline if enough of the poorest families are culled while the inbreeding is being done.

teristic is only slightly hereditary, since then the accompanying increase in t would be far less than the increase in r. This may be a very practical procedure under many circumstances, since the risk of inbreeding degeneration would not be large, it would take only a generation or two to produce this much inbreeding, and therefore a selection between families could be made every second or third generation. To carry inbreeding to higher levels before making the selections between families would make both r and G larger and would make selection between families more effective in the generation in which it was practiced, but would involve more inbreeding risk and would require more generations for each cycle of inbreeding, selection, and re-crossing the selected families. Therefore, it might make less net progress per generation than the shorter cycles with the milder inbreeding. Not all of these relations have been explored yet, but it appears[5] that not much is to be gained by increasing the inbreeding much farther than 30 per cent before selecting between the lines, unless epistatic interactions are highly important.

A third way in which inbreeding can make family selection more effective is that it permits high values of r without necessarily having high values of C in those characteristics where maternal environmental influences are strong or where contemporaneity carries with it some strong environmental correlations. Without inbreeding it is difficult to get families which have r much larger than .31 and yet are not maternal sibs, and impossible to get such families with r as large as .38. Yet in the second inbred generation of continuous half-sib inbreeding, the members of the same family will already be related to each other 50 per cent and in the third generation 58 per cent, although they are not from the same dams.

The close relation between intensity of inbreeding and distinctness of families is shown by the speed with which, in a population inbred steadily without selection, the genetic variance tends to be shifted from variance within families, $(1 - r)\,G$, to variance between families, rG. In regular full-sib inbreeding, rG equals $(1 - r)\,G$ before the inbreeding begins, is 1.5 times as large in the first inbred generation, 2.7 times as large in the second, 3.8 times as large in the third, etc. In regular half-sib inbreeding the corresponding ratios are .33 before the inbreeding begins, .64 in the first inbred generation, 1.00 in the second, 1.40 in the third, 1.86 in the fourth, etc. The general formulas are that rG equals $2fG_0$ and $(1 - r)\,G = (1 + F - 2f)\,G_0$ where F is the inbreeding of the animals concerned, f is the average inbreeding of the offspring which

[5] Dickerson, G. E., 1942, Experimental design for testing inbred lines of swine, *Jour. of An. Sci.*, Vol. 1.

would be produced by mating members of the same family together, and G_0 is what G was in the foundation generation to which the inbreeding and relationship coefficients are traced. The ratio of the genetic variance between unrelated lines to that within lines thus becomes $\frac{2f}{1 + F - 2f}$.

This forming of families between which selection could then be more highly effective may have been the major part which inbreeding played in the development of hybrid corn. The inbreeding was carried so far that r was nearly 1.0 and G was nearly doubled. In that condition the differences between lines were almost wholly genetic (except for whatever there was in C), and selection between them for their combining power could be more effective than ever before. However, this is not the whole story, for the inbreeding also aided greatly in purging the the lines of rare and undesirable recessives, and was almost the only means for isolating and comparing epistatic combinations. These effects may have been very important, too.

FAMILY DISTINCTNESS AS AFFECTED BY DOMINANCE AND EPISTASIS

In random breeding populations most of the effects of dominance are included in the E term, but when members of a family are related through both parents of each, there is some correlation between their dominance deviations. This contributes a little to the C term in such families. If a population is being inbred, the heterozygotes become scarcer and the variance caused by dominance deviations tends to disappear, part of it going to join the increases in the G term.[6] Thus the general effect of dominance is to make the E term distinctly larger and the C term a little larger than if there were no dominance. It makes the effectiveness of family selection increase with advancing inbreeding a little more than the preceding formulas indicate.

Variance due to epistatic gene interactions likewise goes mostly into the E term when inbreeding is zero, but the epistatic deviations of family members are correlated, and this contributes something to the C term. As inbreeding gets more intense, the correlations between epistatic deviations of members of the same family rise with r at an ever-increasing rate. This increases C, but also some of what were epistatic deviations in a random-breeding population can be gathered into the additive scheme in partially inbred populations. This increases G at the expense of E and perhaps of C. The net result is that family differences

[6] This is curvilinear, and there are certain special but uncommon combinations of circumstances under which the variance due to dominance deviations would increase a bit in the early stages of inbreeding, before beginning to decline.

and distinctness become even more pronounced with increasing inbreeding than was indicated in the preceding formulas. Presumably this makes family selection more advantageous than the preceding formulas indicate.

RELATED FAMILIES

In actual populations the families are not wholly unrelated to each other. Instead each family is related to others—to a few closely, to some less closely, and to most scarcely at all unless the relationship is relative to a basis farther back than the time when this population first diverged from other populations. The family structure of a population is somewhat like a fishing net in which each knot is rather close to a few others but distant from most.

Since each individual has two parents but the number of its offspring can vary from zero to many, the network of descent as traced backward will necessarily be more regular than when traced forward. If there has been some degree of inbreeding and separation of the population into small and partially isolated subgroups between which there is little inbreeding, the irregularity of the family structure of the population may become extreme. It is somewhat like an irregularly torn and tangled net, some clumps of strands being heavily intertangled with each other but swinging almost free from adjacent strands and only remotely connected with the rest of the net. Some of the subgroups themselves at a later date may be subdivided still further into partially non-interbreeding groups. Thus arise families within families.

When selection is to be practiced between such related families, the effective r between members of the same family is approximately $\frac{r_2 - r_1}{1 - r_1}$ where r_2 is the relationship within the family, and r_1 is the relationship of the two families to each other. For example, suppose two families are related 20 per cent to each other, but the relationship within families is 50 per cent. Then, for selecting between two individuals each belonging to one of these families, the r in the preceding formulas for determining how much attention to pay to the family average is approximately $\frac{.5 - .2}{1 - .2}$, or about 38 per cent. The practical consequence is that when separate families are built from a common and closely related foundation stock, the inbreeding needs to be pushed further for selection between families to reach a given effectiveness than if each had been started from an unrelated stock. When selecting between individuals from related families, the family averages should receive less attention than when selecting between individuals from unrelated families. The common sense of this is obvious when one considers the extreme case of selecting between full sibs. The family average is useless for helping discriminate between them, since it is exactly the same for both. In selecting between two half sibs the family averages of each are partly determined by the common parent. To that extent the family averages are less helpful in indicating which of the two has the higher breeding value. The discussion here and formula for effective r merely generalize the principle of this and extend it to less obvious cases.

SUMMARY

The term "family" in animal breeding implies a group of animals related to each other, but usage varies widely as to how close that relationship must be before two individuals are considered as members of the same family. Often the definition of family is vague and variable, even in a single discussion.

Family often signifies a name which is handed down, perhaps for many generations, usually through the female line but occasionally through the male line. As such it has no more real significance than human family names do. However, it lends itself to speculation in boom times and has sometimes played a part in making some purebred individuals sell for higher prices than others of the same breed. It may help emphasize the breeder's name. It may make selection of females a bit more strict than would otherwise be the case.

If females are consistently mated to males of the same family as their own, such linebreeding tends to make and keep families distinct from each other. If the linebreeding is continued long or becomes intense, such a linebred family tends to become a breed within a breed. Such distinct families within pure breeds have been formed only occasionally.

The family has a measurable genetic basis and practical usefulness when it is defined as a group with an average genetic relationship, r, to each other. Attention to family is most helpful when r is large and the observed resemblance of family members, t, is low. Each bit of merit or defect in the family average should receive $\frac{r - t}{1 - r} \cdot \frac{n}{1 + (n - 1)\,t}$ times as much attention as the same amount of merit or defect in the individual.

The family should be given negative attention when the observed resemblance between family members is higher than their genetic resemblance. This means that the individual should be judged partly on its own characteristics and partly on its deviation from its family average.

The family averages should receive less attention when comparing two individuals from related families than when comparing individuals from unrelated families.

Inbreeding is necessary for making families with r much above .31 in the less prolific animals or above .50 in any animals. If individual variations in the characteristic are only slightly hereditary, the increases which inbreeding makes in r will be much larger than the increases in phenotypic resemblance. Then the gain from paying attention to the family average in selection may become large.

If epistatic effects and dominance are important, increases in inbreeding will increase family distinctness and family differences even more than the formulas in this chapter indicate.

Characteristics which can take only a twofold classification in the individual animals and which are subject to considerable chance variation need very much the help of family selection, because the error in individual selection is high. Vitality, or disease resistance in general, are examples if the animal's success or failure can be measured only by whether it lived or died.

Family selection is most helpful for characteristics for which heritability is low and family members do not resemble each other for any other reason than their genetic relationship.

REFERENCES

Malin, D. F. 1923. The evolution of breeds. Des Moines: Wallace Publishing Company. (All chapters dealing with the Shorthorn and the Aberdeen-Angus breeds are full of references to the family system. pp. 47–50 and pp. 154–60 are the most concentrated.)

Marshall, R. R. 1911. Breeding farm animals. Chicago: Sanders Publishing Company. (Especially pp. 105–10.)

Watson, James A. S. 1926. "Family" breeding and linebreeding. In Proc. Scottish Cattle Breeding Conference for 1925, pp. 176–82.

Wentworth, E. N. 1920. (An account of how the Juana Erica family was founded, setting forth the reasons for the family system.) The Aberdeen-Angus Jour., 1 (18) :15 et seq. April 25.

CHAPTER 25

Blood-Lines

The word "blood-line" is often used by breeders and is found in many advertisements and current animal breeding writings. It is rare, however, in textbooks on animal breeding and still rarer in textbooks on genetics.

In general, blood-line is synonymous with pedigree but is not so definite. Sometimes it is used more nearly in one of the senses of family, as, for example, when a man suggests performance testing of many animals in a breed to find out "which blood-lines are the most productive and valuable," or wants to learn "which are the most prominent blood-lines of the breed."

Sometimes blood-line is used to convey the idea of relationship, as when a man says that two animals "have nearly the same blood-lines," or that some animal "has valuable blood-lines." In the first case he implies that the two animals are closely related, and in the second case he implies that this animal is closely related to some ancestors whose descendants are highly valued. As a measure of relationship blood-line is an indefinite and sometimes misleading substitute for the probability of likeness which is expressed accurately in the coefficient of relationship. Usually it makes the relationship seem much higher than it really is. Blood-line is a convenient term, however, because almost everyone understands it in a general way.

Sometimes blood-line is used to describe a linebreeding or an inbreeding program, as when a breeder says he "believes in mating together animals of similar but not identical blood-lines." He thus conveys a vague idea of what would be more precise but probably not so readily understood if he used the inbreeding coefficient and the relationship coefficient to state how intensely he planned to inbreed and how closely he was trying to keep his herd related to some noted ancestor.

Sometimes blood-line is used to infer that a whole complex of inheritance is transmitted as a unit unchanged from parent to offspring, generation after generation. This idea comes from studying pedigrees backward. The present famous animal is traced through his sire to a grandsire and through it to a great grandsire, all of which were outstanding individuals of their breed. Looking back to what happened, we sometimes see an unbroken succession of outstanding merit. If we could turn the pedigree around and look forward from the first famous animal in the line, we might see what really happened. The outstanding individual which was the first in this blood-line was used in one of the leading herds of the breed. He had many sons and daughters; and, as far as the breeder could pick them out, only the best of his sons were saved for tentative use in leading herds, where they were mated to better-than-average females. That son whose offspring proved him to be the best became the leading sire of his generation and his supposedly best sons were eagerly sought and in turn were tried out in the leading herds of their time. This may have lasted several generations, or at least as long as even one outstanding son of the outstanding sire in each generation could be found. In a breed where one herd or a small group of herds which get their sires from each other maintain a leading position over many years, it sometimes happens that from grandsire to sire and to son there was an unbroken succession of outstanding breed leaders. This will become familiar to everyone who studies pedigrees of that breed, and people will soon be referring to this as a "very valuable blood-line." Really what happened in such cases was nothing more fundamental than an intense selection among the sons in each generation.

Because of its vagueness, blood-line is in bad repute as a scientific word. Its claim to retention in the animal breeder's vocabulary is that it is widely used now and that everyone understands—at least in a general way—what is meant by it. The relationship coefficient and inbreeding coefficient are not yet widely used and understood. They would often require a long translation or explanation. There is no way to make blood-line quantitative, but it is often useful where only a qualitative meaning is necessary.

It is more nearly correct biologically to think of the individual as one knot in an enormous network of descent, rather than as belonging to some blood-line. The network is irregular in practically all respects except that each individual has two and only two parents. Figure 44 shows a network which corresponds in a small way[1] to the irregular and interlocking lines of descent which constitute the pedigree structure of a breed. Each small circle represents an individual, and the short straight lines connect parent and offspring. Each individual's ancestry widens out rapidly until, not many generations into the past, its pedi-

[1] Except that the figure shows more inbreeding and hence more separation into distinct families than is usual.

gree includes nearly the same animals as the pedigrees of its contemporaries do, but with some ancestors repeated rarely and others repeated many times. Some individuals leave many sons and daughters, others few, and still others leave none. The breed is continuous in time and

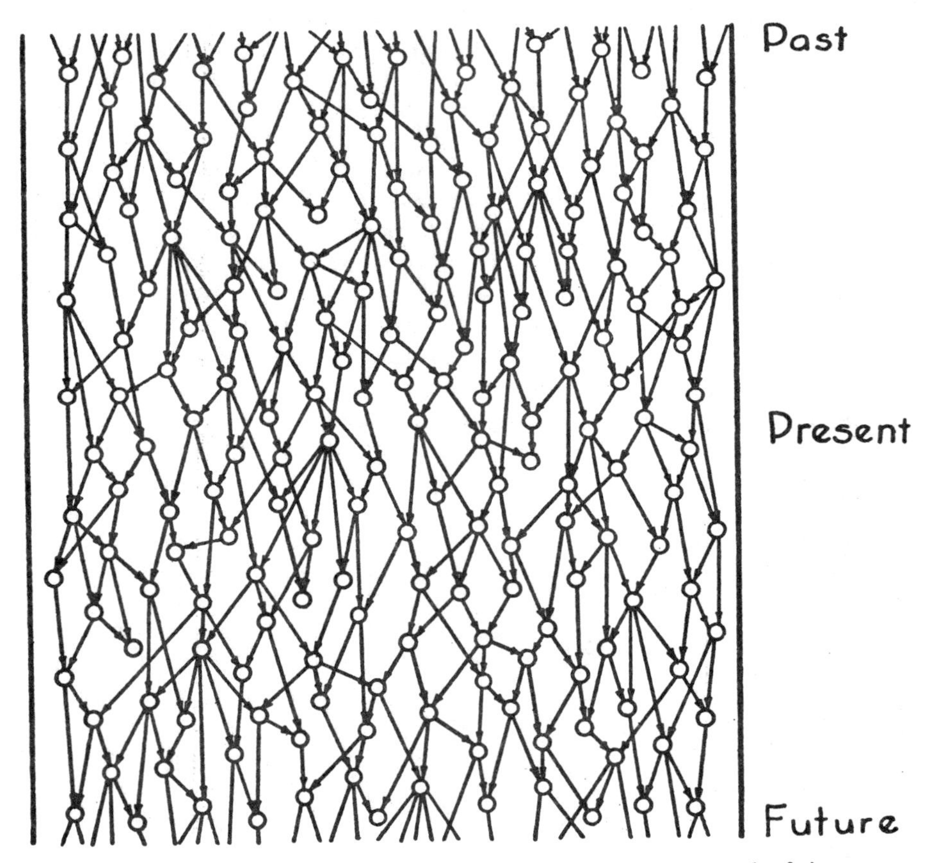

FIG. 44. The pedigree of a population, showing that it is a network of descent and is not composed of "blood lines" which are separate.

space and changes but slowly. The individuals are discontinuous, and each is different from all the others. Each individual is related to all the others but in widely varying degrees. One blood-line can no more be lifted out by itself than one strand of a fishing net could be picked up without picking up all the others. Those nearest would be affected soonest and most strongly. The fishing net, however, is much more regular than the pedigree structure of a breed.

SUMMARY

"Blood-line" is an elastic term used sometimes as synonymous with family, sometimes as a substitute for relationship, and sometimes to describe vaguely a breeding system.

Because of its vagueness, blood-line is in bad repute as a scientific term. But, because it is so widely understood by breeders, blood-line will sometimes be found useful in conveying a general qualitative idea about breeding topics where the speaker does not wish to call attention to the quantitative aspect of that idea.

REFERENCES

Malin, D. F. 1923. The evolution of breeds. Des Moines: Wallace Publishing Company. This book contains abundant references to blood and the word "strain," or family. These show how one can speak more definitely on the subject and yet avoid the use of "blood-line."

Whitney, Leon F. 1933. The basis of breeding. New Haven: E. C. Fowler. (Presents many arguments against any use at all of "blood" to mean inheritance.)

CHAPTER 26

Outbreeding Systems

Outbreeding is the general scientific term for mating animals distinctly less closely related to each other than the average of the population concerned. Its general effects are the opposite of those of inbreeding. Outbreeding increases the heterozygosity of the individual and increases the uniformity of the breed when it is first practiced, although in a generation or two it comes to a limit in these respects. Continued outbreeding merely serves to hold this individual heterozygosity and breed uniformity. Any families which may have started to separate from the rest of the breed are blended again toward the breed average by crossing them with each other.

The practical usefulness of outbreeding rests on the general fact that favorable effects of genes are apt to be dominant over the unfavorable ones. Therefore, outbreeding increases the average individual merit of the animals but lowers the breeding values of the best among them. It increases at first the uniformity of the breed, but hampers further progress in breed improvement. This superiority of the outbred animals over the average of their parents in individual merit is so general a phenomenon in many kinds of plants and animals that it has been called "hybrid vigor" or "heterosis." It is not often extreme unless the parents are from different inbred lines or have in some other way been made distinctly different from each other in the genes they carry. The maximum practical usefulness of outbreeding systems is in the production of market animals or purebred animals which are to be shown to advertise the herd but which are not intended for breeding use.

CROSSBREEDING

Crossbreeding is the mating of two animals which are both purebred but belong to different breeds. The mating of a purebred sire of one breed to high grade females of another is often included under the term crossbreeding.

Crossbreeding is often practiced in producing swine, sometimes in producing poultry, and in some regions is extensively practiced by sheepmen. Thus, in the northwestern range states many sheepmen plan to keep one-quarter to one-half Merino or Rambouillet blood in the ewes but use mutton rams on these ewes to produce market lambs. Crossbreeding is rarer with cattle and horses; but there are certain well-established practices of it, such as the production of blue-gray cattle for feeding by crossing Angus and white Shorthorns, or the practice of certain ranches—e.g., the SMS ranch near Stamford, Texas—in maintaining an undercurrent of Shorthorn blood but of using bulls in the ratio of 90 Herefords to 10 Shorthorns. Crossbreeding among cattle is also practiced on a commercial basis along the Gulf coast, where many cattlemen try to keep a quarter to a half Brahman blood in the cow herd, but for siring the market steers and heifers use bulls from the beef breeds which originated in Europe.

There have been many crossbreeding experiments with sheep, but most of those have been planned to find what kind of ram is most profitable for use upon range-bred ewes carrying a considerable amount of Merino or Rambouillet blood. There have been several crossbreeding experiments with swine to find how much general advantage there might be in such crossing. The few crossbreeding experiments which have been conducted with cattle have been directed mainly toward a genetic analysis of the difference between breeds rather than toward finding whether crossbreeding is a commercially successful practice.

Crossbreeding, like any other form of outbreeding, tends to lower the breeding value of the individual by making it more heterozygous and by making selection among the crossbred individuals less effective. Like other forms of outbreeding it promotes individual merit because of general dominance of genes favorable to size, vigor, fertility, etc.

When the crossbreds are used for breeding purposes, their offspring are more variable than the crossbreds were and generally average somewhat lower in individual merit. If both parents are crossbreds, the offspring usually average below their purebred grandparents in individual merit. Often the distribution of the offspring of crossbreds is distinctly skewed, there being few which exceed the average of the crossbreds and many which fall below it—some of them far below.[1] But if

[1] Besides the general dominance of favorable effects, it is probable that much of this asymmetry in the distribution of the offspring of crossbreds is caused by gene interactions such as those studied by Rasmusson (1933, *Hereditas,* 18:245–61). In more general terms it can be pictured as shown in Figures 20 and 21. The pure breeds crossed will usually have been in different peaks, each more desirable than the adjacent genotypes. A general tendency to dominance of the favorable effects of genes may keep the crossbreds high in individual merit. But when the crossbreds reproduce, many of their offspring will fail to get some of the genes vital for the successful functioning of the complete sets of genes which came to the crossbreds from each of their purebred parents. Many of the offspring of the crossbreds will, therefore, fall into some of the intervening valleys of low merit.

one is to make full use of the heterosis of the crossbred females, it may be necessary to use them for breeding. For example, in swine the number of pigs farrowed and weaned and their weight at birth and probably also at weaning are perhaps more dependent on the dam's characteristics as a mother and nurse than on the genes which the pigs themselves have, although the latter certainly play a part. It will be necessary to use the crossbred females for breeding if this part of their heterosis is to be used. Similar consideration would apply to egg production in poultry and milk production in dairy cows. Some swine producers attempt to solve this by keeping the crossbred gilts for breeding purposes and breeding them to a purebred boar of a third breed. If all three breeds nick well with each other in crosses, the pigs from such a "triple cross" should show many of the advantages of the original crossbreds besides permitting their dams to show the effects of heterosis on their fecundity and nursing ability. The triple-cross pigs are usually more variable, since their crossbred dams transmit various combinations of genes to them. The breeds can be chosen so that the triple-cross pigs will be as uniform in color as the first-cross pigs. This is done by choosing for the third cross a boar of a breed which has a conspicuous and dominant color, such as solid white. This practice might theoretically be continued to a fourth or even a fifth cross with only a little loss of heterosis in the dams, but it becomes increasingly difficult to find more breeds which are distinct from those already used and yet which nick well with all of them. Actually this practice is rarely carried past the triple-cross.

"Criss-crossing" is another method proposed[2] for utilizing heterosis in the dams but not incurring the full decline in average individual merit which usually occurs when crossbreds are mated together. The plan is to use purebred sires all the time but to alternate breeds. Thus, sows produced by crossing breeds *A* and *B* would be mated back to *A*. Their daughters (carrying 75 per cent of *A* blood) would be mated to a *B* boar. The gilts thus produced (carrying 37½ per cent of *A* blood) would then be mated to an *A* boar. If this were practiced regularly, it would approach the condition in which each crop of pigs had 1/3 of its inheritance from one breed and 2/3 from the other, but all the sires used would be purebreds. The Minnesota Station reports good results from this system, as far as it was carried in six years of experimenting. Some practical difficulties with overlapping of generations are to be expected in herds where only one boar is kept and is used both on gilts and on older sows. This would not occur where all the sows are the same age. Figure 45 shows examples of how pedigrees would appear after several generations of regular criss-crossing or three-breed crossing.

---

[2] 1935, Minnesota Agr. Exp. Sta., Bul. 320.

Whether crossbreeding is a sound commercial policy depends on the balance between the extra size, vigor, fertility, etc., which is usually gained by crossbreeding and the extra cost of replacement which is incurred when the crossbred parents are replaced. Heterosis does not occur uniformly in all crosses. That is, not all breeds nor all animals within the same breed "nick" equally well. The heterosis from crossing breeds of farm animals is not apt to be larger than around 2 to 8 per

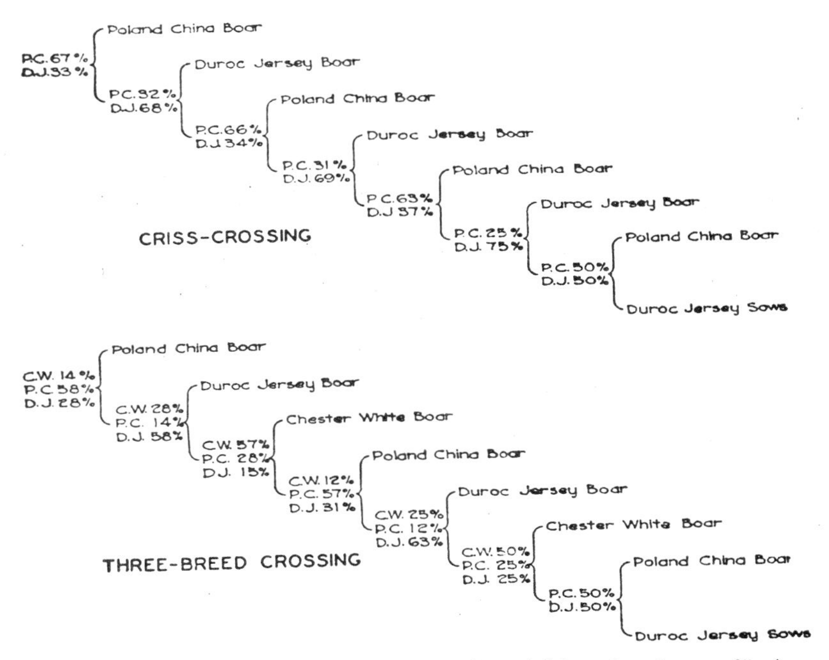

FIG. 45. Illustrative pedigrees of "criss-crossed" and "three-breed crossed" pigs.

cent increase over the average of the parental breeds for such things as size, growth rates, fertility, or other complex physiological traits. It is generally largest for vitality as measured by percentage raised of those born. There is nothing in animal breeding to correspond to the very large amount of heterosis which the corn breeders often find when they cross two inbred lines. Presumably the underlying principles are the same, but nothing corresponding closely to the inbred lines of corn exists in the breeds of farm animals.

Crossbreeding is most likely to be profitable where fertility is highest and the percentage of replacements necessary to keep up the female herd is lowest. For example, with cattle under most range conditions 70 calves weaned per 100 cows per year is considered a good calf crop. With half the calves being females, a herd would need to be kept three years in order to produce enough females to replace the original cows even if all heifer calves which lived to weaning time were used without selection. If the average cow only stays in the herd about six to eight years, nearly half of all her daughters will be needed to maintain the number. If crossbreeding were practiced and all calves were sold for beef, this would necessitate an annual replacement of about one-fourth as many cows as there were calves dropped. On the other hand, sows can be managed to produce two litters a year and under fair management should wean six pigs per litter. Thus, instead of requiring the female offspring of three years of her life to replace her, as is the case with a range cow, the sow needs only one-sixth of one year's female offspring to replace her. Many swine breeders who practice crossbreeding try to raise two litters per year and keep their old sows as long as these are able to wean large litters. Under such a plan it might not be necessary to save more than one female for every six or eight litters produced. This would spread the cost of each replacement over a large number of crossbred pigs produced, so that even a small benefit from crossbreeding on each individual pig might more than offset the replacement costs of procuring high grade or purebred sows to take the place of the dam when her usefulness is over.

OUTCROSSING

Outcrossing is a term frequently used by breeders of purebreds to express the same idea as outbreeding, except that it usually implies a return to an original plan of linebreeding immediately after the outcross is made. If a breeder says: "This pedigree has only one outcross in it," he means that all but one of the branches of the pedigree are from one family. If a breeder mentions a "mild outcross," he refers to a mating with an animal which is not quite of the family he is breeding but which is related to it. A man after having practiced linebreeding for a time may say that he needs an outcross. In this case he means that he needs to mate his stock to animals from some other line; but the usual implication is that, after one generation of outbreeding, he will return to using animals of his original family, attempting by selection to hold the good traits introduced by the outcross, while by linebreeding to his chosen family again he tries to recapture and hold all the good traits he already had in that family. The corn breeders call this kind of a process "convergent improvement" when used on their more intensely inbred material.

Outcrossing is a minor part but eventually a necessary part of most linebreeding programs. Any linebreeding which is carried far is apt to fix some undesired traits so that mild outcrossing may be necessary to remedy them. If the outcross is a success, the breeder is sometimes so carried away by enthusiasm for it that he gives up his plan of returning to his original family and decides to mate the outcrossed animals together. To do this is the same in principle, although less extreme in degree, as attempting to fix desired crossbred traits by breeding crossbred females to crossbred males.

BACKCROSSING

Backcrossing is the mating of a crossbred animal back to one of the pure parent races which were crossed to produce it. It is a term commonly used in genetic studies but not widely used by breeders. In genetic analyses, particularly where one of the parents possesses all or most of the recessive traits, the backcross permits a surer analysis of the genetic situation than an F_2 generation does. General experience with backcrosses in practical animal breeding has not been quite as satisfactory as experience with crossbreds. The backcrosses retain some of the heterosis in many cases, but rarely as much as the first crosses or the triple crosses.

TOPCROSSING

Topcross usually refers to the last sire in a pedigree. When a breeder mentions a "Scotch-topped" Shorthorn, he means a purebred Shorthorn whose dam belongs to a family not originating in Scotland but whose sire and perhaps maternal grandsire were "straight Scotch." When a breeder says that "this animal has four topcrosses of Scotch blood," he means that it is by a Scotch sire, that its dam is by a Scotch sire, that its maternal grandam is by a Scotch sire and that the dam of the maternal grandam is by a Scotch sire. Presumably the pedigree farther back in the maternal line is not Scotch.

Top crossing is the same in principle as grading, except that topcrossing is usually applied to different families within a pure breed, whereas grading is applied to the continued use of sires of one pure breed starting with foundation females which were of another breed or of no particular breed at all.

In plant breeding topcrossing is sometimes used to mean the production of seed by putting pollen from an inbred sire on plants from a good commercial variety.

GRADING

When the pure breeds were new and relatively scarce in this country, grading common or mongrel stock up to the purebred level by the continued use of sires of a pure breed was the quickest way available for improving commercial herds. Many of the experiment stations conducted experiments or demonstrations in the results of such grading.

Generally the first cross showed a marked improvement over the original stock. The further improvement made by each successive cross was progressively less. Grading can rapidly bring the stock near the level of the pure breed which is being used for the grading. Grading will remain the most important form of breeding for the commercial market as long as the merit of the pure breeds is distinctly above that of the commercial herds, and unless heterosis itself is so important that wider outbreeding plans, such as criss-crossing, are more profitable than continued grading to one pure breed. The fact that in so many grading experiments the major improvement has come in the first cross seems to indicate that some of the improvement in the first cross was from heterosis. No doubt the original mongrels in such experiments had at least a few desirable genes which should have been kept if there had been any way to select them and keep them while letting the rest of the genes from the mongrels be bred out by the continued grading.

SPECIES HYBRIDS

The mule is the only commercially important species hybrid in North American animal husbandry. Male mules are always sterile as far as is yet known. A few well-authenticated cases of fertile mare mules have been reported,[8] but these have been so rare that they have had no commercial importance. These fertile mare mules might possibly be the means of transferring some characteristics from the ass species to the horse species or the reverse. Because mules are sterile the problems of mule breeding are only those of choosing the most suitable kinds or breeds of mares and of jacks for crossing. The reciprocal cross, called a "hinny," has been made many times, but is generally regarded as inferior to the mule as a work animal.

Crosses between zebu cattle and cattle breeds of European origin are of considerable economic importance in the Gulf coast region of the United States and in nearly all the tropical regions of the world. Some would regard these as species crosses, but the majority opinion is that zebus and the cattle of European origin are not distinct species. Intermediate types exist in the regions between their native lands, as in southeastern Europe, Asia Minor, southern Siberia, and the northeastern regions of Africa.

Crosses between European cattle and the American bison have been made. Some would regard this as a generic cross. The males are sterile, but many of the females are fertile. By backcrossing these females to cattle and to bison, attempts to form a new breed, the "cattalo," have been made on a fairly large scale, but commercial success was not achieved.

[8] For two examples, see *Jour. of Heredity* 19:412–16, 1928.

Other species crosses which involve farm animals but have hardly passed the stage of zoological curiosities or menagerie specimens include: horse and zebra, European cattle and yak, American bison and yak, American bison and European bison or wisent, yak and zebu, mouflon and domesticated sheep, bactrian and dromedary camels, chicken and guinea hen, pheasant and hen, and peacock and hen. Hybrids between yaks and other cattle are economically important in some parts of China. Crosses between sheep and goats may start to develop, but the embryos die and are resorbed or aborted long before the normal gestation period is completed. A similar fate happens to the embryos from crosses of chickens and turkeys.

Species hybrids do not seem to offer as much opportunity for economic improvement in animal breeding as they do in plant breeding.

SUMMARY

Outbreeding generally leads to individual excellence but low breeding worth.

Outbreeding systems hamper progress in further improvement of a breed because they destroy families by constantly crossing together any which start to develop. They thus make the breed temporarily more uniform than if outbreeding were not practiced.

Crossbreeding is a special form of outbreeding where the parents belong to different breeds. It generally results in increased size, vitality and fertility; but the amount of this increase is variable in different crosses. The economy of crossbreeding depends upon whether the increase in these things is more than enough to balance the possible confusion and increase in cost of replacements under a crossbreeding system.

Crossbreeding is more apt to be profitable where fertility is highest and females can be kept for the longest period of time and where the cost of their replacement is lowest. Mainly for these reasons crossbreeding is practiced most widely with swine and poultry and next with sheep.

Outcrossing usually applies only to matings within a pure breed. It may mean the same thing as outbreeding but usually implies also an intention to return to the original family or strain after making the one outbreeding mating.

Backcrossing is mating a crossbred animal back to the same kind of animal as one or the other of its parents.

Topcross refers to the sire, maternal grandsire, and sires of the other females in the purely maternal line. Generally it is used only within pure breeds.

Grading is the continued use generation after generation of males of one pure breed on an original foundation of another breed or of no particular breed. Grading is the most economical way of lifting the commercial stock rapidly toward the level of the purebreds.

REFERENCES

For recent reports on experiments with crossbreeding, see the following bulletins from agricultural experiment stations: Arkansas 411, California 598, Iowa 380, Minnesota 320, Mississippi 347, Nevada 153, Pennsylvania 279, and Wyoming 210. See also articles in *Journal of Animal Science* 1:213–20, and *Scientific Agriculture* 16:322–36 and 19:177–98. Also see mimeographed leaflet BDIM-Inf. 30, May 1946, from the USDA.

CHAPTER 27

Mating Like to Like

Although many writings on animal breeding stress the importance of mating like to like, it is usually *selection* which is being discussed. The familiar recommendation to "breed the best to the best" usually implies that the worst (and the mediocre as far as numbers will permit) are to be discarded. That would be selection, whereas mating like to like would require also the mating of the worst to the worst and the mediocre to the mediocre—at least among those selected to be parents.

Actually some selection is always practiced; that is, the different types are not permitted to reproduce at equal rates. The nearest actual approach to mating like to like without selection occurs in breeds where there is a marked disagreement about the ideal type, some breeders working toward one goal and some toward another. So far as concerns those traits on which there is disagreement, these cases show some approach to the mating of like to like in the breed as a whole, although in each individual herd the practice is merely selection. There is a little of this at all times in all breeds because some breeders emphasize certain characteristics more and other characteristics less than other breeders do. Also a breeder who uses more than one sire at a time might, if he chooses, mate the best sire to the best females, the second best sire to the second best group of females, etc., until finally the poorest sire among those he uses will be left for mating to the poorest bunch of females which he keeps. This would be mating like to like within the selected group. By contrast he might try to balance the groups of females so that the mates of each sire would be about equal to the mates of every other sire in average merit. This he might well do if his primary object was an accurate progeny test of the sires. Or he might mate the best male to the poorest group of females he kept, the second best male to the next to the poorest group of females, etc. That would be mating unlikes within the group selected to be parents. That is the subject of the next chapter. "Best" and "worst" might describe net merit or, for one characteristic at a time, we could be more specific by using such terms as: largest or smallest, coarsest or most refined, most active or most sluggish, darkest or lightest, etc.

These illustrations will show how any given intensity of selection may be accompanied by any intensity of mating like to like, ranging from almost perfect positive through random mating to almost perfect negative. To see clearly what additional would be accomplished by superposing a system of mating like to like on a certain intensity of selection, which would be practiced anyhow, it is simplest to consider how mating like individuals together, regardless of pedigree, would change a population not under selection.

The fundamental difference in principle between inbreeding and mating like individuals together regardless of pedigree is that inbreeding is the mating of individuals which are apt to have *the same genes,* while assortive mating is the mating of individuals which tend to have *similar characteristics,* irrespective of their relationship. Characteristics are only partly caused by the genes, and it often happens that characteristics which appear to be the same are caused by very different combinations of genes. To the extent that variations in characteristics are caused by environment or by epistatic deviations or by dominance deviations, the mating of like individuals may cause only a slight tendency for mates to be alike in the genes they have. That can be expressed quantitatively as follows for purely assortive mating. If m is the correlation between the net hereditary values of mates, t the correlation between the visible or measurable characteristics of mates, and h the correlation between the characteristic and the net hereditary value of the same individual, then under purely assortive mating $m = h^2t$ and m cannot exceed h^2, even if the breeder has succeeded in getting the mates to be perfectly alike (except for sex) in all characteristics he can see or measure. On the other hand, under purely inbreeding systems $t = h^2m$ and t cannot exceed h^2. As a numerical example consider a moderately heritable characteristic for which $h^2 = .2$. Under purely phenotypic assortive mating m could not exceed .2 even if t were made perfect. In actual practice it would be less. By contrast in the first generation of full brother-sister inbreeding m would be .5 and t would be only .1. The actual consequences of phenotypic assortive mating depend largely on what size of m really is achieved. Hence they are very slight for characteristics moderate or low in heritability.

The example shown in Table 19 deals with variation in only one characteristic. The practical breeder must nearly always consider many traits. He will be using only one or at most a few sires but will have several females, no two of which are alike, to mate to each sire. Many of the animals he might choose to mate together are alike in some characteristics, moderately unlike in others, and perhaps extreme opposites in still others. If he considers many characteristics, it will be impossible for him to achieve in all respects a high degree of resemblance between mates. This is in addition to the general situation, discussed in the preceding paragraph, that under assortive mating, likeness in net hereditary values will be less than outward likeness. In actual practice assortive mating can rarely cause m to have high values for any characteristic other than net merit.

ONE PAIR OF GENES

If only one pair of genes is involved and if there is no dominance or other reason for mistaking hereditary values—i.e., if h and t of the next to the last paragraph each equal 1.0—the results are the same as in self-fertilization. If dominance is complete but there is no other complication, the change is in the same direction and the final result is the same but progress is slower. If the attempt to mate like to like is not quite perfectly successful—i.e., if t is less than 1.0 or if anything other than dominance makes h less than 1.0—the population will come to equilibrium while a few heterozygotes still remain.

TWO PAIRS OF GENES—SIMPLEST CASE

If this mating of like to like is for a characteristic influenced by two pairs of genes, lacking dominance and with equal effects, then we have the situation shown in Table 19 which starts with a random breeding

TABLE 19

PROPORTIONS OF EACH PHENOTYPE UNDER A PERFECTLY ACCURATE SYSTEM OF MATING LIKE TO LIKE, q_A AND q_B REMAINING AT .5

Genotypes	aabb	aaBb Aabb	aaBB AAbb AaBb	AaBB AABb	AABB
Phenotypes	No Plus Genes	1 Plus Gene	2 Plus Genes	3 Plus Genes	4 Plus Genes
Generation:					
Start	6.2%	25.0%	37.5%	25.0%	6.2%
1	13.6%	20.8%	31.2%	20.8%	13.6%
2	19.3%	16.5%	28.4%	16.5%	19.3%
3	23.9%	13.8%	24.6%	13.8%	23.9%
...	...	...	...	...	...
...	...	...	...	...	...
∞	50.0%	0.0%	0.0%	0.0%	50.0%

population in which the two pairs of genes are independent and the two alleles of each pair are equally abundant, that is, in which $q_A = q_B = .5$.

The result is a decrease in the proportion of the intermediate phenotypes and a corresponding increase in the two extreme phenotypes. If the mating of like to like is perfect and is carried on forever, it approaches as a limit the condition in which all of the genotypes have disappeared except the two homozygous extreme ones. If the mates are not exactly alike phenotypically, as, for example, if an occasional mistake is made in classifying an individual, progress will be slower, and the ultimate goal will be an equilibrium which falls short of complete fixation of the two extreme phenotypes.

This system of breeding tends to fix the extreme types, provided those are both outwardly and genetically extreme; but, in contrast to inbreeding, it cannot fix intermediate types. The likeness between parent and offspring and the likeness between full brothers increases very rapidly, although that may not be clear from this example. The variability of the population is greatly increased, since the population tends to become concentrated at the two extremes.

MANY PAIRS OF GENES

Many more than two pairs of genes may affect the characteristic, and the effects of the different pairs of genes will rarely be equal. The more genes there are, the slower is the rate of increase in homozygosis. The likeness of parent and offspring or of full brothers also increases at a slower rate when n is large, although these likenesses are not nearly as much affected by changes in gene number as is the rate of increase in homozygosis.

GENERAL RESULTS OF MATING LIKE TO LIKE

Very little genuine fixation of type is ever accomplished by this system of breeding because increase in homozygosis is dependent upon the number of genes being very small and upon the breeder's not being deceived by dominance, environmental effects, or epistasis. Figure 46 shows what happens to homozygosis as a result of mating like to like. Here n is the number of gene pairs involved, while m is the correlation between the net hereditary values of mates. Both m and n set limits on the amount of homozygosis which may finally be attained, and n has a tremendous influence on the rate at which it is attained. Only with very high values of m and very low values of n can such a breeding system alter homozygosis much. The values of m cannot often be high in actual practice, and n will probably be large for all characteristics of much economic importance.[1]

[1] The fraction of the heterozygosis of a random breeding population which will still remain when assortive mating has done all it can is $\frac{2n(1-m)}{2n(1-m)+m}$. Since m cannot exceed 1.0 (it must usually be much smaller) and n can be large, this fraction will rarely be much less than unity. *Genetics*, 6:153.

Mating like to like increases the resemblance of parent and offspring very much, each individual resembling its sire not only because it received half its inheritance from him but also because it received half its inheritance from its dam who resembled the sire more than if mating had been random. That is, the dam was chosen to have many genes

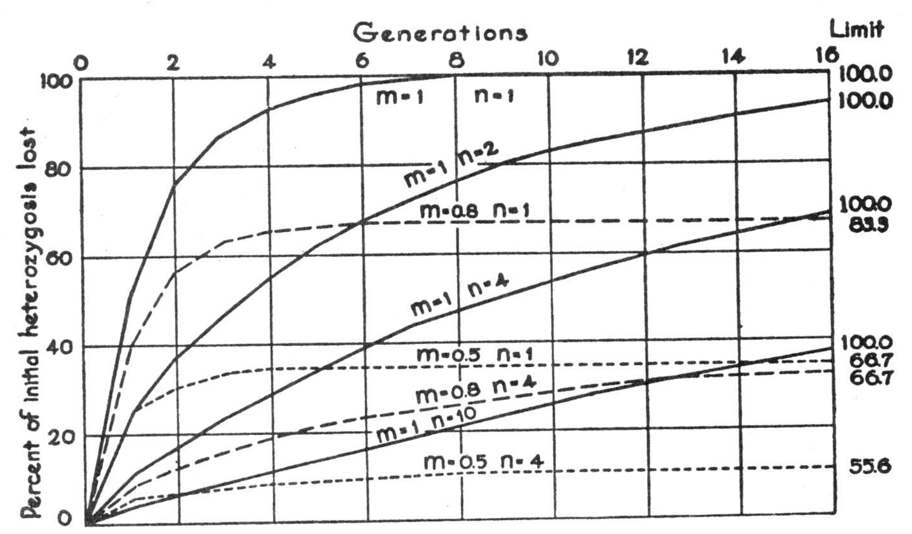

FIG. 46. The percentage of initial heterozygosis which is lost by continued assortive mating of various intensities, *m*, and with *n* pairs of equal genes involved. (After Wright in *Genetics*, 6:175.)

which produce the same kinds of effects as the sire's genes do, although they may not be genes from the same allelic series. The limit which the parent-offspring correlation approaches is determined by *m*, large *n* merely making the approach to that limit a little slower. The parent-offspring correlation goes far toward its limit in the first two generations in which mating like to like is practiced.

Figure 47 shows what happens to the correlation between full brothers under purely assortive mating in the extremely simple case of no dominance, no epistasis, and no environmental variations which are incorrectly discounted. The limits are determined by *m* and the only effect of large *n* is to make progress a little slower. This system of breeding has considerable effect, even when *m* is small and *n* is large. The existence of dominance and epistasis and environmental effects has the effect of making *m* lower than it need be otherwise. Environmental effects might increase the correlations if these effects tended to be the same for brothers.

A high degree of resemblance between parent and offspring and a high degree of resemblance between full brothers seem to indicate that the breeder is gaining control over his material, but that is partly contradicted by the fact that there is little increase in real homozygosis. This has long been recognized in a general way by breeders in the confidence they have in the mating of like to like as a means of getting the

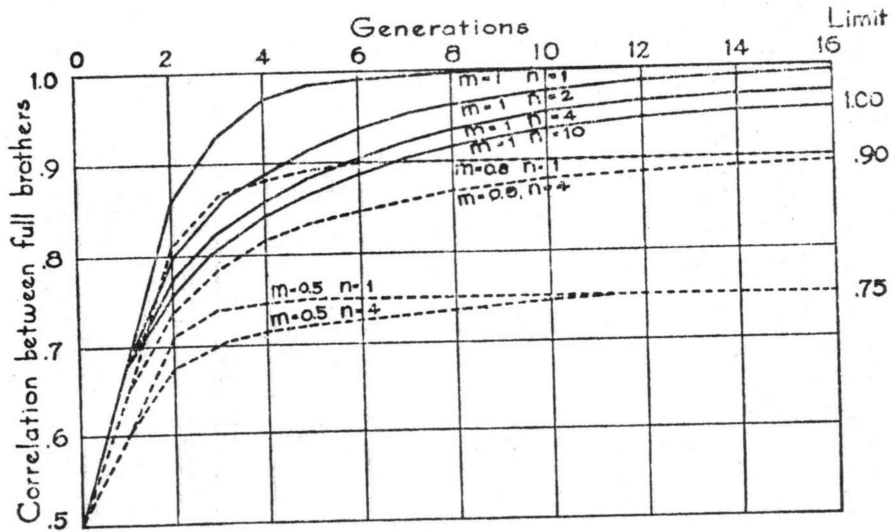

Fig. 47. The correlations between brothers in successive generations under different degrees, *m*, of assortive mating and with *n* pairs of genes involved. (After Wright in *Genetics,* 6:170.)

kind of herd they want, but again and again the more experienced among them express the idea that inbreeding is really necessary if type is to be "fixed."

Mating like to like is one of the most powerful tools which breeders have for creating extreme diversity in a population. Inbreeding tends to fix intermediate families as well as extreme ones and thereby tends to double the additive genetic variance of the population. But mating like to like tends to scatter the population toward the two opposite extremes of each characteristic for which it is practiced. For example, assortive mating for stature within most local races of man is enough to make the standard deviation about 20 or 25 per cent larger than it would be if mating were entirely random with respect to stature. If the characteristic is highly enough hereditary that *m* in assortive mating can rise above .5, mating like to like can make the standard deviation in an unselected population larger than the most extreme inbreeding can.[2]

[2] *Genetics,* 6:154.

In each individual herd the mating of like to like will usually be accompanied by selection which will discard one or the other extreme. If all breeders select toward the same ideal, this will not change the variability of the whole breed any more than selection changes the variability within a single herd. But if the breeders disagree markedly about the ideal and some of them discard animals which other breeders think are very desirable, this process can easily produce a lack of uniformity, in the breed as a whole, so pronounced that everyone familiar with the breed will be aware of it. Examples which have shown some tendency in that direction are: "Island type" and "American type" in Jerseys; "hot bloods" and "big types" in Poland-Chinas. Even when all breeders work toward the same ideals, a little of this herd heterogeneity within a breed will arise because some breeders will try harder than others or will be financially able to outbid others for the animals thought best. Such inequality of striving toward the same ideal produces to a very mild degree something of the same results as a divergence of ideals.

The changes brought about by purely assortive mating are temporary. If there has been no accompanying selection, the population returns far toward its initial condition in the very first generation after random mating is resumed. In actual practice there will, of course, have been some selection; and that is apt to have changed gene frequency enough that the population will never return exactly to its initial condition.

SUMMARY

Mating like to like has almost no effect on homozygosis except in very simple and rare genetic circumstances.

Mating like to like immediately increases the resemblance between parent and offspring and between full brothers.

Mating like to like comes near to the full limit of its effects within a very few generations after it is begun.

Mating like to like tends to scatter a population toward the two extremes with respect to each character for which such mating is practiced. It, therefore, greatly increases the variability of the population if both extremes are kept and heritability is high.

In actual practice mating like to like is always accompanied by selection.

The effects of mating like to like disappear almost at once when random mating is resumed, except as the accompanying selection may have made some permanent changes in gene frequency.

REFERENCES

Wright, Sewall. 1921. Systems of mating. III. Assortive mating, based on somatic resemblance. Genetics, 6:144–61. (See also pp. 167–78.)

CHAPTER 28

Mating Unlike Individuals

The mating of unlike individuals (negative assortive mating on the basis of somatic resemblance) is most commonly practiced either to mark time while the breeder is deciding what his goal is to be or, where the ideal is an intermediate, to correct defects by mating each animal to one which is equally extreme but in the opposite direction. This is sometimes called *compensatory* mating.

Everyone does some of this from time to time, at least for minor traits. There are no absolutely perfect animals. A breeder usually realizes that his females are good in some respects but below his standard in others. Under such circumstances he is almost certain to seek for his next sire one which is particularly strong where his females are weak. Since he cannot find a sire which is absolutely perfect in all respects, he will accept one which is a little below his standard in characteristics for which the females are unusually good. In the breeding of Rambouillet or Merino sheep, it is common practice for breeders to seek "light C" rams to mate to their "heavy B" ewes and "heavy B" rams to use on their "light C" ewes. Many are confident that this type of mating is more apt to produce a high percentage of lambs which are on the borderline of being the desired heavy C's or light B's than would be produced from parents both of which were the desired type. Another prominent example of mating unlikes is in the breeding of dual-purpose cattle, where there is considerable mating of those which vary most toward the extreme dairy type with others which vary most toward the extreme beef type.

CONSEQUENCES

The consequences of mating unlikes are the reverse of those of mating like to like. Heterozygosis is increased only a very little. The maximum effect of mating unlikes, even if continued indefinitely, would be to make the heterozygosis $\frac{2n(1-m)}{2n(1-m)+m}$ of what it would be under random mating, m being negative. In the impossibly extreme case when

$m = -1.0$, that would increase the heterozygosis by only $\frac{1}{4n-1}$ of what it would be under random mating. That would increase heterozygosis by one-third if only one pair of genes were involved, by one-seventh if there were two pairs, only by one-eleventh if there were three, etc. If the value of m is nearer zero, the power of this breeding system to affect heterozygosis will be still further reduced. When m is as near zero as $-.2$, the increase in heterozygosis cannot exceed one-eleventh of the original amount even when n is 1 and cannot exceed one-forty-seventh if n is 4. Since m must usually be low and n may well be large, it is obvious that average homozygosis is scarcely affected at all.

The mating of unlikes together makes the correlation between parent and offspring distinctly lower, since the two parents are quite different from each other and the effects of the genes which an offspring inherits from one tend to be canceled by the effects of the genes it inherits from the other. Likewise, this system of mating tends to lower the correlation between brothers, although not as much as the correlation between parent and offspring is lowered.

Mating unlikes together tends to make the whole population uniform since an extreme individual in one direction tends to be mated with one which is extreme in the other. The offspring of each mating thereby usually average nearer the population average than they would if mating were random. If mating were random, there would be some matings where both parents happened to be extreme in the same direction. This reduction in variability nearly reaches its limit within the first and second generations after the mating of unlikes begins. In the very first generation produced by mating unlike individuals, the variance becomes $\frac{2+m}{2}$ times what it was under random mating. Of course m is negative, but it has to be small unless heritability is high. The maximum effect on the variability of a whole population occurs when n is large and m is strongly negative. That maximum effect is to halve the variance when $m = -1.0$ and n is very large. If $m = -.4$, the maximum effect is to reduce the variance two-sevenths when n is very large and one-sixth when n is only 1. The original variability will reappear almost at once when the mating of unlikes is abandoned.

A system of mating unlikes is most useful when the desired type is an intermediate. Under such conditions the maximum proportion of desired individuals among the offspring can be obtained by mating males which are of the desired type to females which are also of the desired type and mating any breeding animals which deviate from the ideal in one direction to those which deviate equally far in the other direction.

Mating unlikes together is extensively practiced commercially to correct defects. This is a sound practice wherever there are enough females undesirably extreme in one direction to justify the keeping of a male equally extreme in the other direction.

SUMMARY

Mating unlikes together in the absence of selection leads to:

1. A more uniform population than under random mating, with a larger percentage of intermediate offspring and fewer extremes in either direction. This increased uniformity reaches nearly its full extent in the first generation after the mating of unlikes is begun. If the mating of unlikes ceases, the population returns almost at once to its original variability under random mating.

2. Only a slight increase in heterozygosity under the very simplest genetic situations and practically no change in heterozygosity under the situations apt to be encountered.

3. A distinctly lower resemblance between parent and offspring and a somewhat lower resemblance between other relatives than under random mating.

Mating unlikes is useful in holding the population together as a more uniform group until the average type or some other intermediate type can be fixed by close inbreeding.

Mating unlikes is useful in correcting defects wherever the ideal is intermediate and some animals are too extreme in one direction while others are too extreme in the other direction.

REFERENCES

Wright, Sewall. 1921. Systems of mating. III. Assortive mating based on somatic resemblance. Genetics, 6:144–61. V. General considerations. Genetics, 6:167–78.

CHAPTER 29

The Relative Importance of Sire and Dam

In general, sire and dam are equally important in inheritance. There are three exceptions which are sometimes of practical importance. The first is that a sire can have many more offspring in a single season.[1] Therefore, the sire is much more important than any one dam in determining the inheritance of the next generation in the herd, although not more important in determining the inheritance of any one animal. This is the basis for the common statement that "The sire is half the herd."[2] The second exception is sex-linked inheritance. The general rule in sex-linked inheritance is that sons are more like their dams, and daughters are more like their sires than in ordinary inheritance. The third exception is the one already discussed in Chapter 14 that where the merits of the two parents are not equally well known, less attention should be paid to the less well known one in estimating the merits or breeding value of their offspring.

HIGHER FERTILITY OF THE SIRE

When a breeder with a one-sire herd buys a sire, he is buying half the inheritance of many offspring. When he buys a dam he is buying half the inheritance of her sons and daughters only. Therefore, he can afford to spend more money to procure a desirable sire than to procure an equally desirable female.

Every individual has as many female as male ancestors. Because of the larger number of offspring per male used for breeding than per female, it is the usual rule that a breed is influenced more by some of its males than by any equal number of females. However, about half of what the male transmits came to him from his dam. It occasionally

[1] Monogamous species, such as pigeons, doves, and some foxes, are exceptions to this. Among them the sire generally leaves about the same number of offspring as the dam.

[2] This statement is an exaggeration in the many cases where a sire is kept in service for less than the average length of a generation. For example, in small dairy herds, one sire is rarely used much more than two years; but the average productive life of the cows is more nearly four years. Rarely are more than half the cows in a herd daughters of one bull. Only half of their genes come from him; hence one sire rarely furnishes more than a fourth of the genes of the whole herd, although he does furnish half of the genes of his own offspring.

happens that a female will exert more influence on a breed than any contemporary male. A notable case is that of the Holstein-Friesian cow, De Kol 2nd, whose relationship to the whole breed (about 10 per cent) is more than that of any other cow or bull. She is almost a great grandam of the whole breed today.[3] Naturally she could not exert this much influence by having an enormous number of calves, although she did live at least 16 years and produced 14 calves. She exerted her remarkable influence on the breed through the fact that she had five different sons which were quite prominent sires of the breed in their time and saw extended service in leading herds. At least four other sons and one daughter also left registered descendants. This is an exceptional case, yet it illustrates how a female may exert a tremendous influence on a breed if several of her sons are saved for extensive use. There is some reason for thinking that more improvement in the dairy breeds can be accomplished by the careful selection of cows which are to be the dams of sires and therefore the grandams of the next generation than can be done directly by selecting among the bulls. However, the exact balance of the quantitative relations involved here is not yet clear.

SEX-LINKED INHERITANCE

In all farm animals except poultry, so far as is yet known, the male has an *X* and a *Y* chromosome and the female has two *X* chromosomes. In poultry this situation is reversed. The discussion which follows may be applied to poultry by substituting the other sex.

Genes carried by the *Y* chromosomes would, of course, be transmitted from sire to son in unbroken lines if there were no crossing over between *X* and *Y*. It would not be easy to distinguish the effects of such genes from secondary sexual characteristics. It is unlikely that the *Y* chromosomes can often, if ever, be entirely empty, or they would have been lost long ago without harm to the species. Yet genetic research has found only a few genes on *Y* chromosomes.

In some species of fish, in man, and perhaps in most mammals, some parts of the *X* chromosomes are homologous with parts of the *Y* chromosomes and these can cross over. Genes carried in these regions show "partial sex-linkage." Their behavior is intermediate between that of autosomal genes and that of the completely sex-linked genes which are carried on the nonhomologous parts of the *X* chromosome. The latter are meant when sex-linkage is discussed in the following pages. Examples of partially sex-linked genes in man are those for retinitis pigmentosa and those for *total* color blindness. The genes for the ordinary red-green color blindness are completely sex-linked.

The genes carried on the *X* chromosomes are those responsible for

[3] *Jour. of Heredity*, 27:61–72.

sex-linked traits. The male receives all of his sex-linked genes from his dam; and, so far as those genes are concerned, he is no relative at all of his sire. The female receives half of her sex-linked genes from her dam and half from her sire just as with her other genes. But her sire can only transmit one kind of *X* chromosome to her, whereas the dam can transmit either one of the pair she has. So far as daughters are concerned, this amounts to the same thing as if the sires were completely homozygous for their sex-linked genes, whereas the dams are as heterozygous for sex-linked genes as they are for other genes. Consequently, paternal half sisters will receive from their sire identical sex-linked genes; but maternal half sisters need not be any more like each other in sex-linked genes than they are in other genes. The following pedigree diagrams of the *X* and *Y* chromosome situation for males and for females show what happens when this is extended to the grandparental generation. The notes show the part which each grandparent plays in the sex-linked inheritance of the grandson or granddaughter.

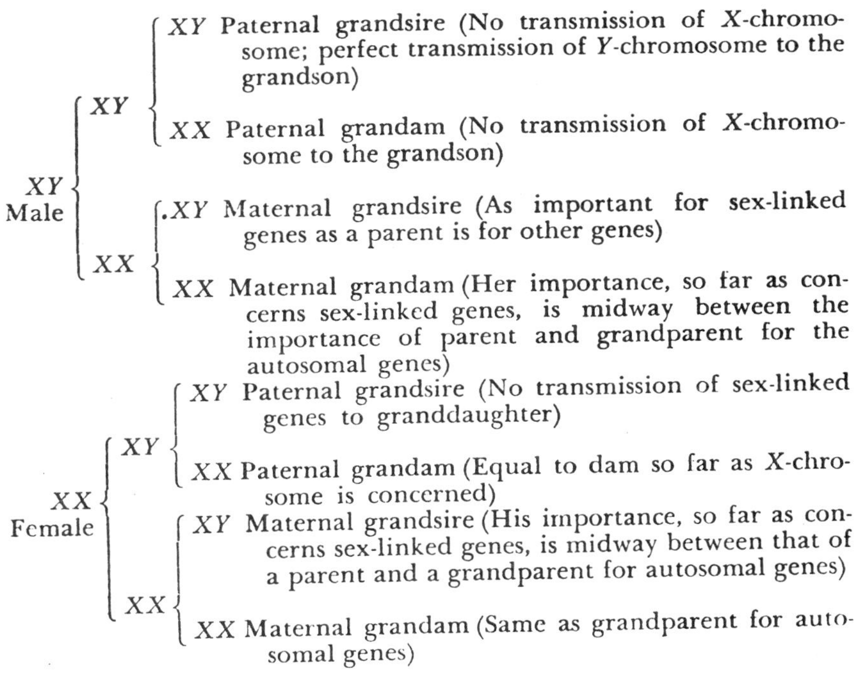

If it is assumed, on the basis of chromosome number, that about 5 per cent of the genes are sex-linked, then the expected statistical effects of sex-linkage are those shown in the last two columns of Table 20. The existence of sex-linkage will have so small an effect that for most characteristics it would be difficult to prove that there is any sex-linkage. Differences in the correlations observed between various kinds of relatives have sometimes been interpreted as indicating sex-linkage, but it is rarely possible to be sure that such differences were not caused by (1) sampling errors, (2) greater similarity of environment for some kinds of relatives or (3) differences in the selection which had been practiced for various kinds of relatives. It is to be expected that there will be some sex-linkage in the inheritance of most characteristics which are affected by many genes but that only rarely will a large fraction of the genes be sex-linked. In such rare cases the corrections will more nearly approach those in columns 1 and 2 of Table 20. The main point which Table 20 demonstrates is that only rarely will sex-linkage alter resemblances noticeably.

In breed lore there are many cases where a sire was noted for his daughters but not for his sons, or vice versa. In Hereford history it is said that the daughters of North Pole were exceptionally good individuals but that his sons were not outstanding. All his sons were sold to the range trade and passed out of the pure breed. The sons of Anxiety 4th in the same herd were regarded more highly, but among their offspring those regarded most highly were out of daughters of North Pole. It is possible that North Pole carried in his *X*-chromosome desirable genes which were largely missing from the other Hereford cattle of that date. If that were the case, the use of his daughters' sons would be an effective agency for spreading those genes through the breed. This is probably the most conspicuous instance of the kind in animal breeding history; yet it is more likely that the breeders, on observing the calves of the two bulls, merely decided that the calves of Anxiety 4th were the better. They could try out only a few bulls; so, of course, they kept the sons of the better bull. Naturally they kept all of the best cows sired by both bulls. At this date we will probably never know whether North Pole really did carry sex-linked genes which were very valuable to the Hereford breed but were not possessed by Anxiety 4th nor by many of the other animals in the Gudgell and Simpson herd. This may have been true, or it may have been largely an accident that the sons of North Pole were all sold while many sons of Anxiety 4th were kept.

Alternative explanations of the same sort are usually available for cases where it is reported that a certain sire produced remarkable sons but ordinary daughters, or remarkable daughters but ordinary sons. Sex-linkage is a possible explanation in such cases, but it is probable that the major factor has usually been the selection which the breeder practiced and whether he was at that time selling many males to head other herds or was selling only females.

TABLE 20

EXPECTED EFFECTS OF SEX-LINKAGE ON CORRELATIONS* BETWEEN MALES OR FEMALES AND VARIOUS OF THEIR RELATIVES

	All Genes Sex-linked		No Sex-linkage	5% of Genes Sex-linked	
	Male	Female	Male or Female	Male	Female
Ancestors					
Sire	.00	.71	.50	.487	.512
Dam	.71	.50	.50	.512	.500
Paternal grandsire	.00	.00	.25	.244	.244
Paternal grandam	.00	.50	.25	.244	.268
Maternal grandsire	.50	.35	.25	.268	.256
Maternal grandam	.35	.25	.25	.256	.250
Collateral relatives					
Full brother	.50	.35	.50	.500	.494
Full sister	.35	.75	.50	.494	.515
Paternal half brother	.00	.00	.25	.244	.244
Paternal half sister	.00	.50	.25	.244	.268
Maternal half brother	.50	.35	.25	.268	.256
Maternal half sister	.35	.25	.25	.256	.250
Half first cousin (female)					
Through paternal grandsire	.00	.00	.062	.061	.061
Through paternal grandam	.00	.25	.062	.061	.083
Through maternal grandsire	.18	.125	.062	.073	.067
Through maternal grandam	.09	.062	.062	.064	.062

* For traits entirely determined by heredity, without dominance or epistasis and in a population breeding at random.

SUMMARY

Because the sire can have so many more offspring per year than the dam, he is a more important individual than any one female so far as the whole herd is concerned, although not more important so far as concerns any one offspring.

This makes it possible to cull prospective sires more closely than prospective dams and profitable to pay more for an unusually good sire than for an equally good dam.

Every individual has the same number of female and male ancestors. A female who has more than two sons which are widely used may exert more influence on a breed than any one of her sons. This has actually happened at times, although most animals which have influenced a breed much have been males.

Sex-linked inheritance has the effect of making daughters resemble their sires, and sons resemble their dams, more closely than if there were no sex-linkage. This is not often important.

Most cases reported from animal breeding history where a certain sire produced good daughters but ordinary sons, or the reverse, are probably to be explained as incidental results of the sales or culling policy in that herd. Sex-linkage may have played a part in some of these cases.

Sometimes one side of the pedigree will seem to be more important than the other merely because more is known about it, and therefore more use can be made of it for prediction.

REFERENCES

Gowen, John W. 1924. Milk secretion. Baltimore: Williams and Wilkins. (This book includes correlations between the performance of various female relatives in the Holstein-Friesian Advanced Registry.)

Madsen, Karl. 1932. Inheritance of milking capacity. Nature, January 30.

Pearl, Raymond. 1922. The biology of death. Philadelphia: J. B. Lippincott Company. (See pp. 174 and 175 for correlations between various kinds of relatives in man. Six different characteristics are included.)

Smith, A. D. B., and Robison, O. J. 1933. The genetics of cattle. Bibliographia Genetica X. (See pp. 46–51 for a review of the evidence on sex-linkage in the inheritance of milk and fat production.)

Snyder, Laurence H. 1942. The mutant gene in man. Amer. Nat. 76:131–33.

Winge, Ö. 1934. The experimental alteration of sex chromosomes into autosomes and vice versa, as illustrated by Lebistes. Comp. Rend. Laboratoire Carlsberg, 21:1–49. (An account of the sex-chromosome situation in this genus of fish with much about sex-linked genes and about crossing over between the *X* and *Y* chromosomes.)

CHAPTER 30

Registering High Grades

For years after the registry associations were first organized, many of them admitted to registry grade animals with a certain number of top-crosses of registered sires. Nearly all of the American registry associations have ceased doing this, and most of those which register a breed native to other lands never did admit grades to registry in the United States. Most of the European breeds still register females having three or four top crosses of registered sires.[1] In most cases those breeds also practice selective registration (see chapter 16). The proposal is occasionally made that American breeds should also admit to registry high grades which are outstanding individuals.

The pure breeds of today are comparatively modern developments. Few have herdboks much more than 80 years old. Usually there was no official herdbook until years after the breed had really been formed. Naturally, at the time the herdbooks were established, the line between registered and other animals was somewhat arbitrary. Often there was disagreement as to whether certain animals really should have been included in the herdbook. With this background for the history of registration, the question automatically arises: What is wrong now with registering grades, when that was done in the founding of all breeds and still continues among many of them?

BREED HOMOZYGOSITY

The generations of strictly pure breeding that have elapsed since registration began have enabled the breeders to make the breed somewhat more homozygous than it was when the herdbooks were first closed. The amount of heterozygosity which was in the original foundation stock but which has been lost solely through the pure breeding since pedigree registration began can be measured by studying sample pedigrees of the breed at any desired date. Those figures for the breeds so far studied are as follows:

[1] For example, the Shire Horse Society in Britain admits mares with three top-crosses by registered Shire stallions. Volume 4 of their "Grading-Up Register" in 1944 contained the pedigrees of 49 mares with one topcross and 41 with two top-crosses, while 760 mares are entered in the contemporary Volume 64 of the regular Stud Book.

Shorthorn cattle in Great Britain	26.0%	by 1920
Jersey cattle in Great Britain	3.9%	by 1925
Ayrshire cattle in Great Britain	5.3%	by 1927
Holstein-Friesian cattle in the United States	4.0%	by 1931
Hereford cattle in the United States	8.1%	by 1930
Brown Swiss cattle in the United States	3.8%	by 1929
Aberdeen-Angus cattle in the United States	11.3%	by 1939
Clydesdale horses in Great Britain	6.2%	by 1925
Rambouillet sheep in the United States	5.5%	by 1926
Hampshire sheep in the United States	2.9%	by 1935
Poland-China hogs in the United States	9.8%	by 1929
Brown Swiss cattle in Switzerland	1.0%	by 1927
Landrace swine in Denmark	6.9%	by 1930
Thoroughbred horse in the United States	8.4%	by 1941
Standardbred horse in the United States	4.4%	by 1940
American Saddle horse	3.2%	by 1935
Telemark cattle in Norway	2.3%	in 23 years

These figures are probably no more than .5 per cent above or below what would have been found if it had been possible to study the pedigrees of all the animals of the breed. Evidently pure breeding by itself causes only a slow drift toward homozygosity. The high figure for the Shorthorns was mostly incurred in the first 30 years, most of it while the breed was largely confined to the herds of the Colling brothers. For the other breeds it appears that about one-half of 1 per cent of the remaining heterozygosity is being lost per animal generation.

The figures given do not include any changes in heterozygosity which may have been caused by selecting animals which were more or less homozygous than their pedigrees indicate. Selection changes homozygosity only incidentally as a result of the changes it makes in gene frequency. Selection requires a long time to change q enough to make much change in $2q\,(1 - q)$ except where the genetic situation is simple and selection is directed toward an extreme and the favored gene already has a frequency above .5. Selection may have increased materially the homozygosity of some genes which affect color, distinct anatomical peculiarities, and other details of breed type for which the genetic situation may be rather simple. But it is unlikely that selection has changed very much the average homozygosity of the breeds for genes affecting complicated characteristics. Selection for the effects of heterosis may even have operated in the other direction to hold the heterozygosity at a little higher figure than the pedigree studies indicate. It seems unlikely that selection can have had much net effect on the general homozygosity of the breed.

These considerations make it reasonably certain that the purebreds are more homozygous than the commercial stock, but doubtful that their difference in this respect is extreme, even allowing liberally for the fact that only a restricted group entered the registry in the first place. Grades having at least four topcrosses of registered blood would already be homozygous for about seven-eighths of the genes which were homozygous in that breed but rare in other animals. The admission of a few such grades would lower the breed's homozygosity very little.

INTRODUCING DESIRABLE GENES

Some grade individuals are much superior to the average of the purebreds in type or production or both. The best of these grades may possess desirable genes which are either unknown in the pure breed or are rare. The admission of these animals to registry might improve the pure breed through introducing or making more frequent such desirable genes. If females were required to have at least four topcrosses of registered sires in order to be registered, only about one-sixteenth of their genes would be other than those of the breed itself. This fraction would be further halved in their offspring. The frequency of genes already existing in the breed would not be changed much through the admission of such grades. If there are desirable genes which do not now occur at all within the pure breed, their introduction in this way might be important.

GRADES SHOULD MEET HIGH STANDARDS AS INDIVIDUALS

If the admission of grades is to improve the merit of the breed, the advantage from introducing desirable genes should be greater than the damage which would be done by upsetting the extra homozygosity which the breed has already obtained. Such safeguarding could be obtained by requiring the grade animal to meet distinctly higher standards of individuality and production than the average individual merit of the animals already registered. Just how much higher than the breed average these standards of individual merit should be for grades for which registry is being asked would depend in principle upon how different the pure breed really was from the foundation stock from which this grade was produced. Any standards adopted would necessarily be somewhat arbitrary.[2]

DISRUPTING EPISTATIC COMBINATIONS

It is possible that the present merit of the various pure breeds depends in part upon certain combinations of genes which produce good results as a combination but not separately. One such combination

[2] For an example of standards of this kind in actual operation in Sweden, see *Hoard's Dairyman*, 74:62.

might be typical of one breed (although of course not entirely homozygous in all animals of that breed), while very different combinations which have the same general kind of effect may be typical of other breeds. If so, the admission of even a little outside blood to the breed might scatter those epistatic combinations enough to lower merit more than the small percentage of outside blood admitted would indicate.

While this is theoretically a possibility, yet there seems no way to estimate whether such situations are frequent enough to be important. Even if such situations are frequent, grades carrying as many as four topcrosses of the pure breed would already have most of the genes which are necessary for such combinations. That makes the theoretical danger of harming the breed in this way seem rather remote. The existence of even a slight possibility of such damage is an additional reason for requiring that any grades to be admitted should meet standards of individual merit distinctly higher than the average of the pure breed into which they come.

ADMISSION OF PUREBREDS WHICH ARE NOT NOW ELIGIBLE TO REGISTRY

It often happens that a purebred animal cannot be registered because the breeder is not certain of its sire. The breeder may be certain that the sire of the animal was one of two or three males, all of which were purebred, but he may not have any record of the breeding date, or the actual birth date may be far from the expected one. Economical use of range resources often requires several males for each group of females. If the females are all purebred and the males are all purebred, all the offspring are purebred, too; but their individual pedigrees are not known and they are not at present eligible to registry in any association. Sometimes outstanding individuals are produced from such flocks or herds, and the breeder would like to use them for stud purposes if they could be registered. Several sheep associations in the United States have discussed proposals for such "flock registration," but none have been adopted. Such "flock registration" would not lower the homozygosity so far attained in the breed. A breeder using such animals could still use mass selection in improving his livestock but could not make much use of ancestors and collateral relatives to estimate breeding worth. Such a proposal might have unexpected consequences on the finances of the registry association, but that might be controlled by adjusting the fees for flock registration. Flock registration is a well established practice with sheep in many other lands, notably Australia.

ECONOMIC AND PSYCHOLOGICAL CONSEQUENCES

To admit high grades to registration might lower the breed reputation with those who set the highest value on absolute purity of breeding. This might be offset, at least in part, by the fact that some would construe such action to mean that individual merit was so highly respected in this breed that something in pedigree desirability would be sacrificed to attain unusual individual merit.

The registration of high grades would increase the supply of registered animals and, therefore, might tend to lower the prices which could otherwise be obtained by those who already have the purebred individuals. This could be controlled by having the requirements of individuality and productive ability so high that only a few grades would be admitted. Apparently this actually happens abroad. In most breeds where grades having at least four topcrosses of registered ancestry may be admitted to registry, only a small number are thus admitted each year. This slight increase in supply might be more than offset by an increased demand from those commercial producers who might be favorably impressed by the evident interest of that breed in individual merit. Just how those two factors would balance is not clear, but it seems unlikely that enough high grades would ever be admitted to registry to be an important economic factor.

Breeders would disagree about the wisdom of admitting any high grades at all to registration. If any breed does undertake such registration, it is likely that the official pedigrees will indicate which animals are absolutely purebred in all lines and which trace to some of those admitted high grades. Thus, the advocates of absolute purity could avoid pedigrees tracing in any line to the grades. Something of that kind actually happens unofficially in the case of pedigrees where there is suspicion but not absolute proof of fraud. Those who know the situation let such animals strictly alone, treating them as grades. There is thus a tendency to eliminate them from the breed, although a beginner unfamiliar with the situation will sometimes pay purebred prices for them. After many generations an official system of indicating those which trace at all to grades might break down with its own complexity; but long before that time had arrived, the breeders would either have revoked the registration of grades or would have come to substantial agreement that it was a sound policy.

SUMMARY

The registration of high grades unusually superior in individual merit is a common practice with many European breed associations but is practiced by very few associations in the United States.

The genetic consequences of such a practice are: (a) Some loss in homozygosity, and (b) the possibility of introducing some desirable genes which are rare or unknown in the pure breed.

If the grades were required to meet distinctly higher standards of individual merit and productivity than the average of the purebreds, the gain to the breed through the increase of desirable genes would be likely to be greater than the harm through loss of homozygosity.

Some animals which are absolutely purebred cannot now be registered because of uncertainty about which of two or more purebred animals was the sire. The admission of such animals to registry through a plan of "herd registration" or "flock registration" would not result in any loss of the breed's homozygosity. It might lead to a little breed improvement if only superior individuals were registered, although the selection of such superior individuals could be little more accurate than simple mass selection in general.

Economic reasons will predominate in decisions about the registration of grades. There might be some loss through the breed's appearing less strict in its standards of purity than its competitors. There might be some gain through advertising that individual merit was receiving special attention in that breed.

REFERENCES

For accounts of the increase in homozygosis in various breeds, see the references at the end of chapter 21. For other articles dealing more directly with the possibilities of registering high grades or with differences between grades and purebreeds, see the following:

Anonymous. 1927. A challenge to pure breeds. *Hoard's Dairyman,* 72:126–27. Feb. 10.
Anonymous. 1929. Admit grades to registry. *Hoard's Dairyman,* 74;637. July 10.
McDowell, J. C. 1928. Comparison of purebred and grade dairy cows. USDA, Cir. 26.
Peterson, Guy A. 1929. Swedish herdbook registration. *Hoard's Dairyman,* 74:62.
Savage. E. S. 1936. Registration of grade cattle. *Hoard's Dairyman,* 81 (9) :242. May 10.

CHAPTER 31

Sire Indexes

A sire index is a way of expressing what the sire's progeny indicate about his heredity. It is most needed for characteristics which the sire cannot show himself: milk and fat production in dairy cattle, egg production in poultry, prolificacy in swine and sheep, nursing ability in all mammals, and certain traits of disposition which are expressed differently in males and females,[1] etc. For characteristics which the sire can manifest in himself, a sire index is useful for revising estimates of his hereditary value when such estimates have been based only on his own characteristics and on his pedigree.

A complete sire index would logically be based on all available information about the sire's own characteristics and about his ancestors and collateral relatives, as well as on the information about his progeny; but most of the current discussion about sire indexes deals only with ways to use the information about the progeny and their other parents. Sire indexes of that kind are only special applications of the principle of the progeny test discussed in chapter 15. Sire indexes are mentioned most often in writings on dairy and poultry breeding.[2] The wording used in the rest of this chapter applies primarily to dairy cattle. The principles are the same in other cases, but the importance of some of the considerations may change.

THEORETICAL BASIS

The reasons for computing a sire's index as $2D - M$, when D is the average of his daughters and M is the average of the dams of those daughters, may be seen from the equations on page 198. The reasons for not trusting the index completely arise from what are called "Mendelian errors" and "errors of appraisal" in those equations.

The Mendelian errors come from chance at segregation which permits gametes coming from the same parent to contain different genes, so that some offspring are genetically better than the average of their

[1] Hammond, John, 1932, *Report on Cattle-breeding in Jamaica and Trinidad*, Publication No. 58 of the Empire Marketing Board, London.

[2] See *Jour. of Dy. Sci.*, 16:501–22, for a discussion of the general principles involved.

parents while others are worse. These variations are truly random, provided the daughters are an unselected sample. Therefore, they tend to cancel each other and their importance in an average diminishes as the number of daughters increases.

The errors of appraisal come from the fact that even after we have corrected the record of a daughter or a dam for age, for times milked per day, and for every other non-standard environmental circumstance about which we know, the record will be higher in some cases and lower in others than corresponds to the real breeding value of the cow. Such of these errors as are random tend to diminish in importance as the number of daughters in the average increases, just as the Mendelian errors do. Some of the errors of appraisal are not random but tend to be in the same direction for all the daughters or for all the mates of the same bull but may be in a different direction or be different in size for the daughters or mates of other bulls. They may even be different for the daughters and for the mates of the same bull. Frequent causes of these biased errors are: (1) The general level of environment to which each group of daughters or of dams was exposed may have varied much from one group to another and we may not know what the environment was in each case, or may know it but not be able to allow perfectly for its effects. (2) The daughters or the records which were used to represent them may have been selected ones, rather than a fair sample of all. (3) The dams of these daughters may have been a selected group and the intensity of that selection, too, may have varied from one sire to another. Such biased errors do not tend to cancel each other as the number of daughters increases.

The actual numerical value of each sire index is partly determined by the real breeding value of the sire but partly also by whatever biased errors that index contains and by the uncanceled remainder of the random errors. Accordingly sire indexes will vary more than the real breeding values of the sires do. The sires with the very highest indexes are generally good sires but they are not likely to be as good as their indexes. Similarly the sires with low indexes are generally poor sires but they are not usually as poor as their indexes. We shall make the smallest mistakes if we estimate the breeding value of each sire at part way between his index and the average of his breed.

The principles governing how much confidence the index deserves can best be understood by considering first the dependability of an index (I) based on the record of only one daughter (D) and her dam (M). The variance of D and of M will be nearly the same. For clarity in the argument this variance can be divided into the additively genetic portion (G) plus a portion (E) due to random discrepancies between the genetic value and and the record of the cow, plus a portion (C) due to those discrepancies between record and breeding value which are alike for all the daughters of a bull or for all the mates of a bull. Then, according to the usual formula for the variance of a difference, the variance of the index is:

$$5G + 5C + 5E - 2G - 4xC = 3G + (5 - 4x)\,C + 5E$$

where x is the correlation between C for a daughter and C for her dam. The sire's breeding value is responsible for only G of the variance in this index, as may be seen by referring to diagram B in Figure 32. That diagram shows how the genetic value of an individual is ¼ determined by the genetic value of its sire, ¼ by the genetic value of its dam and ½ by chance at Mendelian segregation.[3] The genetic variance which the dam determines is removed in subtracting M from $2D$. That leaves in the variance of I only $3G$ of the $4G$ which is the genetic variance in $2D$. The sire is responsible for ⅓ of that $3G$, while chance at segregation is responsible for $2G$.

When the index is based on n daughter-dam pairs, instead of one, the random parts of the variance in I become only $1/n$ as large. Thus the variance of I based on n pairs is: $G + (5 - 4x)\,C + (2G + 5E)\,/n$. The equation which will predict with the least error the breeding value of the sire becomes:

$$\text{Sire} - A = \frac{n}{n[1 + (5 - 4x)\,C/G] + 2 + 5E/G}\,(I - A)$$

where A is the breed average. The fraction in this equation can be considered as showing the extent to which the index should be believed. The square root of this fraction is the correlation between the index and the real breeding value of a bull.

Wright has discussed this equation for the case of complete heritability; i.e., when C and E are both zero. Then it becomes simply: $\text{Sire} - A = \frac{n}{n + 2}(I - A)$ and confidence in the index becomes almost complete when the number of daughter-dam pairs is much more than five. This, however, is not very realistic because in practice E will always be real and for many characteristics it will be much larger than G. C also is likely to have a real value in most sets of data which are collected from many herds, since it is rarely if ever possible to discount accurately for all of the differences in management and environment from one herd to another.

---

[3] If sire or dam were inbred, the fraction determined by the genetic value of that parent would be $\frac{1 + F}{4}$, while correspondingly less part would be determined by chance at segregation. In most populations inbreeding is unusual and is so mild that we can ignore it here with little error.

The effects of E may be seen most clearly by considering the case in which C is zero but heritability $\left(\text{i.e.,}\ \frac{G}{G+E}\right)$ varies. Then the fraction in the prediction equation becomes: $\frac{n}{n+2+5E/G}$ which takes values such as the following:

Heritability	*Fraction*
.10	$\frac{n}{n+47}$
.20	$\frac{n}{n+22}$
.25	$\frac{n}{n+17}$
.33	$\frac{n}{n+12}$
.50	$\frac{n}{n+7}$
.71	$\frac{n}{n+4}$
1.00	$\frac{n}{n+2}$

If E is large the justifiable confidence in the index is obviously very low when n is small. Nevertheless perfect trust in the index is approached, although slowly, as n becomes extremely large.

The existence of C changes the situation so that confidence in the index increases more slowly with n and no longer approaches unity as a limit. Instead it approaches $1/y$ where $y = 1 + (5 - 4x)\ C/G$. This y can be a large number if C is large relative to G and if x is small. That C/G may often be as large as 1.0 in dairy data is indicated by correlation usually of the order of .2 to .3 between the records of unrelated or slightly related cows kept in the same herd. The general size of x is less certain. It would be 1.0 if all of C came from general differences in environment from herd to herd and if the peculiar environment of each herd were unchanging, year after year. However, part of C can come from herd environment which changes between the time when the dams made their records and the time when the daughters made theirs. A part of C can come from selection of the dams or of the daughters having been more intense for some bulls than for others. Hence x will not be 1.0 although it may usually be above .5. The case $C \doteq G$ and $x = .5$, which is not unreasonable for dairy data, will illustrate the power of C to limit confidence in the index. In that case confidence in the index will not exceed $1/4$, even when the number of daughter-dam pairs becomes exceedingly large. If C is only half as large as G, the corresponding limit is $2/5$. If C is as large as G but $3/4$ of C is alike for the daughters and for the mates of the same bull, the limit is $1/3$.

Obviously anything which can be done to diminish C and E by keeping the environmental conditions standard and alike for all daughters and dams, or by correcting the actual records for the effects of varying environment or of unequal selection of the mates of various bulls will increase the confidence which the index deserves.

PRACTICAL CONSIDERATIONS

Relative rating. For comparing bulls with one another when all of them have the same number of daughter-dam pairs, the index is all that is needed since it will then rank all the bulls in the same order and proportionately the same distance apart, no matter what is the value of C or E or x. The need for knowing how much to regress the indexes toward the breed average arises only when we wish to compare bulls proven on different numbers of pairs, or when we wish to compare a bull's index with a cow's record. For the first of these purposes ignorance of C, E, and x makes only a little difference unless n varies extremely from one sire to another. Doubling n will double the numerator but only the first part of the denominator. If that first part of the denominator is already more than half of it, the doubling of n will increase the fraction by less than one third of its initial value. Hence further increases in n when it is already large add only slightly to the confidence which the index deserves.

Comparing bulls' indexes with cows' records. The necessity for comparing a bull's index with the records of a cow is met almost every time we evaluate a pedigree in which the sire or one or both grandsires are proven. For example, in choosing between young bulls A and B we may find that the sire of A has an index which is 40 pounds higher than the index of B's sire but that the records of B's dam average 60 pounds higher than the records of A's dam. To estimate whether A or B probably has the higher breeding value, we have to decide whether a difference of 40 pounds in sire indexes is as important as a difference of 60 pounds in the records of cows. If the cow's record is left in its actual form, sire indexes can be compared directly and fairly with it if the indexes have been regressed only enough that they would have about the same variability as the cows' records—a little more variability if an index is generally a bit more accurate as an indicator of a sire's breeding value than a cow's record is of her breeding value, but a little less variability if the indexes are generally less accurate. If the indexes are regressed as far toward the breed average as they should be to make each numerically equal to the estimated breeding values of the bull to whom it belongs, then the cow's record also should be regressed far enough toward the breed average to make it likewise an unbiased estimate of her breeding value. Present practice in this respect is not uniform. The Holstein-Friesian Association publishes indexes based on six or more pairs and does not regress them at all. This makes these indexes just a bit more variable than the records of cows with one record each. The same is true of the indexes computed by the American Dairy Cattle Club from the records published by the Dairy Bureau of the USDA, except that in this case the minimum number of pairs is five. The Ayrshire Association since late 1944 has been regressing sire indexes half way toward the breed average and basing them on at least ten pairs. Sire indexes regressed this much are only about half as variable as records of cows who have one lactation each.

NUMBER OF DAUGHTER-DAM PAIRS NEEDED. From the principles governing the accuracy of sire indexes it is clear that accuracy is low when the number of pairs is very small but it is also clear that accuracy does not suddenly become perfect when a certain number is reached. Instead the accuracy increases at an ever-decreasing rate as the number rises. Since the gain from increasing n comes solely from the decrease which that makes in the term $(2G + 5E)/n$ in the variance of I, the advantages in having n any larger have mostly been reaped by the time $(2G+5E)/n$ has already become distinctly smaller than $G+ (5-4x)C$.

Largely by trial and error but partly based on considerations like these, all official plans for computing indexes specify some minimum number for n before an index will be computed at all. One important consideration which has kept the minimum numbers small is that if many daughters were required few bulls could be proved. Those who made the Dairy Bureau policy thought that the increased accuracy which the proof would have, if six were required instead of five, would be more than offset by the fact that the large number of bulls who have only five comparisons would thereby not be brought to the attention of breeders who might otherwise hear of them and make some use of that information. There is good reason for believing this still to be sound policy, although the original decision was made years ago. One suggestion for reaping both advantages is that bulls be given a preliminary index when five pairs are available and another index when the number has risen to another level, for example eight or ten. The Ayrshire Association does something of this kind. This, however, adds to the bookkeeping and some customers will not distinguish between the bulls with only preliminary proof and the bulls with more complete proof.

Simplicity of the index is important for explaining it to the potential users and giving them the proper amount of confidence in it, for ease and speed of computing it, and to reduce the chance of clerical errors. Yet the underlying principles are such that at least a little accuracy must be sacrificed if simplicity is to be achieved in data in which n varies or if the same index is to be used in different populations, such as Dairy Herd Improvement Associations or Herd Improvement Registry, in which the relative proportions of G, C and E may vary. When breed associations or other agencies compute and publish indexes as an impartial service to buyers, they have to compromise a little on accuracy in order to achieve the uniformity and simplicity which is necessary for getting the index used. Whether it is better merely to publish $2D$-M and let the user do all the discounting of this for C, E and small n; or to regress it automatically half way toward the breed average as the Ayrshire Association is now doing, or to regress it some other fraction (perhaps one fraction for test, a different one for fat, another for type, etc.), is still open to argument. The issues involved are largely psychological ones centering on what procedure actually will get most breeders to use the indexes with most nearly the proper degree of confidence and with the least confusion and disappointment.

Selection of daughters will bias the index to an extent for which good correction can be made only in the unlikely event that one knows how intense the selection was and on what it was based. If the number tested but omitted from the average is known, it can be assumed that the missing ones had lower records than any of those given. For example, if we read that a bull had 10 daughters which averaged 600 pounds in a Dairy Herd Improvement Association but we happen to know that ten other daughters of his were tested, we can see by Table 12 that the high half of a normal distribution averages .8 of a standard deviation above the average of the whole group. With the intra-herd standard deviation in such data being around 80 pounds of fat, we can estimate that the average of all 20 daughters was about 600 minus 64 or 536 pounds, but that could easily be 30 or more pounds in error with numbers this small and with the distribution perhaps not being exactly normal. As another example, suppose we are told that a certain bull has 30 Advanced Registry daughters, five of which have records over 800 pounds. It is safe to assume that these five are the very highest. By consulting appropriate tables for the normal curve we learn that the point above which the highest one-sixth of the population lies is about one standard deviation above the average of the whole group. Since the intra-herd standard deviation in Advanced Registry data is about 100 pounds, we would estimate that the average of all 30 daughters was about 700 pounds, but this could be considerably in error, especially since the numbers are small and the distribution of Advanced Registry data is not quite symmetrical.

If the missing daughters were not even tested, corrections for their absence are even more in doubt since one cannot then assume that their average would have been low. Some may have died while yet young, some may have been culled intentionally on type or pedigree, some may have been sold into herds in which no testing was done, some may have been started on test and then removed (in the testing plans which permit that) when it was seen that they would do poorly, etc.

Considerations such as these have led breed associations to compute indexes only with data from those systems of testing in which it is required that all animals in the herd be tested if any are. Also it is only reasonable for such agencies to inquire about the missing daughters. Some associations are not able to do more about this than to publish, along with the average of the tested daughters, the total number of daughters old enough to have had a record. This can give the reader some notion of how much possibility there was for selection among the daughters of each sire.

There is no way wholly to avoid natural selection among the daughters. More of the constitutionally weak or sickly daughters than of the normally healthy ones will be barren or will have died before they could be tested. Also it is to the immediate economic interest of the breeder to cull the extremely unpromising heifers as soon as he knows they are such, without incurring the expense of keeping them through their first lactation. However, the correlation of outward appearance or pedigree promise with actual performance is low enough that the effects of such culling prior to testing are much smaller than the effects of omitting daughters after their records are already partly made. It has been proposed but never officially adopted that in proving a sire each daughter old enough to have a record, but without one and without a satisfactory excuse, should be arbitrarily assigned a record well below the breed average. This would increase the pressure on the breeders to get all the daughters tested, but most responsible breeders have thought it too drastic for adoption.

Selected Dams. When the dams are a selected sample of their generation, their breeding values will generally average below their records, especially if the selection was primarily on their own records. Most voluntary culling of dairy cows is done during their first or second lactations. The more calves a cow has, the more chance she has of appearing as a dam because a daughter was reared and tested. The dams will therefore consist largely of cows which have survived several cullings. M will generally be a bit farther above the average breeding value of the dams, than D is above the real average breeding value of the daughters.

If this bias were of the same size for all bulls it would not matter when comparing one bull with another. It would merely make all of them appear a little lower in breeding value than they actually were. But when one bull is mated to a highly selected set of dams while another is mated to a group of dams scarcely selected at all, then subtracting M from $2D$ makes the bull mated to the more highly selected group of cows appear poorer than he actually is. Reasonably good correction for the effects of selecting the dams can be made by using only records which the dams made subsequent to their selection, or by regressing their records toward their contemporary herd average as was indicated in chapter 13, but often the data are not assembled conveniently for doing that.

Missing Dams. Since testing is not universal and is not always continuous within a herd, some of the daughters of a bull may have records, while their dams do not. In such cases the best procedure is generally to use in place of M for each missing dam the average record of the herd contemporary with the daughter or, if that is not available, to assume that the missing record was equal to the average record of the other dams. There is, of course, some possibility of introducing errors by this, but generally the errors thus introduced would be smaller than those which can be removed by utilizing the record of this daughter corrected thus for the average effects of the herd environment.

Differences in general environment from one herd to another are of considerable importance in dairy data, as is evidenced by a correlation usually around .2 to .3 between the records of unrelated cows in the same herd. When daughters and dams are in the same herd, as is usually the case, whatever is peculiar to that herd will tend to push the records of both in the same direction. Doubling the daughter's record and subtracting the dam's record automatically takes from the doubled record of the daughter one half of the effect which the herd environment produced on it, to the extent that environment was alike for daughter and dam.

Two devices for removing from the index the remaining error from herd-to-herd differences in environment deserve mention. First, the prospective purchaser should study the conditions of that herd and discount the effects of conditions which were not standard. This he can never do perfectly but he can often do enough to make it worth while to visit the farm, to inspect the feeding practices, examine the record books, etc.

The other device is the statistical one of considering each cow's record partly on its own absolute value and partly on the difference between it and the average of the herd at the time it was made. To go to the extreme in this direction and judge a cow entirely on her deviation from the average of her herd is equivalent to assuming that all the differences between herd averages are environmental and that all herds are equal in average real producing ability of the cows. To go to the other extreme and not to consider the herd average at all is to assume that there are no differences in environment from one herd to another. The truth is somewhere between these extremes. In principle it is clear that the cow should be judged partly on the absolute size of her record and partly on its deviations on the herd average. The practical problem is to know how much emphasis to place on each. The principles governing this were outlined in chapter 24 on the family structure of populations. Consider each herd as a family. The phenotypic correlation between herd mates (t in the formulas in chapter 24) is usually something like .2 to .3 but if desired, may be determined more exactly for the population in question by examining the data pertaining to that. To determine the genetic resemblance between members of a herd (r in chapter 24) will require some study of pedigrees. Also there may be some doubt about how much genetic resemblance has been introduced by assortive mating or by differences in intensity of selection which will not show in the pedigrees. Probably the intra-breed genetic resemblance between herd mates is usually something like .10 to .15 in small dairy herds in which nearly all of the females were born in the herd. With $r = .10$ and $t = .20$ an average rough correction for the general effects of herd environment could be made by subtracting from the index half of the difference between the average of the herd in which it was made and the general average of all herds in that population. More should be subtracted if t is higher and less if r is higher. This cannot be highly accurate in individual cases, since some herd averages are high mostly for environmental reasons while others are high mostly for genetic reasons. Some such use of the herd average (perhaps with a fraction larger or smaller than one-half) appears likely to remove more errors than it introduces, but it has not yet been tested extensively enough to learn all its actual advantages and disadvantages.

To correct for differences in environment from region to region, especially for the effects of altitude, so that sires in different regions might be fairly compared with each other, Engeler in Switzerland proposed to use the differences between the average of the bull's daughters and the average of the association in which those daughters were tested. This proposal has several practical advantages where environmental conditions are rather uniform within each locality but vary widely from one locality to another. Daughters out of untested dams can be used. The association average is based on so many records that there is little random error left in it. This method would hardly be suitable for data in which environmental conditions varied widely from one herd to another within the same locality. It neglects differences between the average breeding value of the mates of one bull and the average breeding value of the cows in the whole association but perhaps those are not often large. As an example of his method Engeler gives as proof for the bull Zamboli:

	Milk	Test	Fat
Six daughters with 17 records average	3,725 kg.	3.82%	142 kg.
The association average (150 records)	3,556 kg.	3.86%	137 kg.
Difference	+169	–.04	+5

Dams and daughters treated differently. It is always possible that the dams were treated differently from the daughters. This is especially likely to have happened when the dams were tested in one herd and the daughters in another, or when the dams were tested at a much earlier date than the daughters. Such a difference in general treatment lowers x in the preceding formulas and thereby decreases the confidence which the index deserves. The available remedies for this are only the same two mentioned previously; namely, to examine the conditions under which the dams and the daughters made their records and to correct as best one can for the effects which such conditions had on the records or to make more use of the average of their contemporary herd mates.

Indexes for dams could be constructed also. The general principle would be similar to that of diallel crossing. The breeding value of the dam would be estimated to be twice as far above the average of the other mates of the sire as her own offspring were above his other offspring. Of course one would trust such an index only a little, since one dam can have only a few offspring. Such a dam's index should be regressed far toward the average of the breed in order to get an estimate of her breeding value which is as likely to be too high as too low. If heritability is high such an index would be trusted considerably but in that case there would be little need for an index, since the dam's own phenotype would be a rather good guide to her breeding value. If heritability of the characteristic is low, the evidence from the index would be needed more but the low heritability and the small number of offspring would require that it be discounted considerably.

Indexes for other characteristics. Sire indexes can be constructed for other characteristics, just as well as for milk and fat production, if those characteristics are measured or scored definitely enough that the daughters can be averaged into a single figure and also that their dams can be averaged into a single figure. Indexes need not be confined to characteristics which are manifested in only one sex, although they are most needed and useful for those. For example an index could be applied to transmitting ability for type as well as for production. It would only be necessary that the type of the offspring and of the mates be scored, classified or otherwise graded in numerical terms so that these ratings could be averaged. Naturally, the proper amount to trust the index is not likely to be the same for all characteristics, since the relative sizes of G, C and E and the size of x will not be the same for all.

COMPARISON WITH OTHER WAYS OF PROVING SIRES. The average of all daughters without any attention to their dams is sometimes used as the proof of a sire. For example, this is the present practice of the American Jersey Cattle Club. Sometimes the increase of the daughters over their dams is used as the proof. The index is simply the sum of the daughter average and the daughter-dam difference. Therefore it partakes of the errors of each and tends to be midway between them with respect to vulnerability to different kinds of errors except where those errors bias the daughter average and the daughter-dam difference in opposite directions. In such respects the index is more accurate than either. An example is the errors introduced by unequal selection of mates. Such selection makes the daughter average too favorable and the daughter-dam difference too unfavorable toward the bull mated to the highly selected groups.

The relative accuracy of the three measures of a bull is determined mostly by the size of r_{DM}, the correlation between the average of the mates and the average of the daughters of the same sire. When r_{DM} is less than .25 the daughter average is the most accurate of the three. It remains more accurate than the daughter-dam difference as long as r_{DM} is less than .5. The index is more accurate than the daughter average when r_{DM} exceeds .25 and is more accurate than the daughter-dam difference until r_{DM} exceeds .75. In most dairy data r_{DM} is around .55 to .65. When it is exactly .6 the relative accuracy of daughter average, daughter-dam difference, and index is 1.00 : 1.12 : 1.24.[4] The daughter average is simpler to compute, since there is no need even to know

[4] This assumes that M is as variable as D and that the genetic value of the sire is not correlated with the records of his mates. The first assumption is very nearly true in dairy data. Moderate deviations from it will not alter this ratio much anyhow. The second assumption will be very nearly true in all populations unless there is enough inbreeding to make considerable separation into unrelated lines or unless there is a strong degree of assortive mating because breeders differ widely in their ideals and in the intensity of their striving.

the records of the dams. Whether this saving in computation costs is enough to offset the lessened accuracy depends, of course, on the amount of that saving and on what use would be made of the greater accuracy if it were available.

COMBINING INDEXES WITH OTHER INFORMATION. The sole purpose of a sire index is to estimate the breeding value of the sire. An ideal sire index would pay some attention to the records of the bull's dam, sisters, and more remote relatives, instead of being based solely on his daughter and their dams. To some extent these various relatives duplicate each other in the information they offer. They vary in their relationship to the bull and in the number of records each has. No simple and general formula has been devised to fit all the possible combinations of this. Records of the bull's dam and of his full sisters can be combined into a single estimate according to the following formula from Wright: Sire $= \frac{A}{m+1} + \frac{m}{m+1} R$ where A is the breed average and R is the average production of his dam and m - 1 full sisters. This however is based on the supposition of complete heritability. When heritability is less than complete, R would be given less attention and A would receive more attention than this formula shows. Few dairy bulls have as many as two tested full sisters.

If a bull has many more than three daughters his index will naturally receive more attention than his pedigree but his pedigree still contains some information which ought not to be neglected entirely.

Not often will a sire index indicate a bull's inheritance more accurately than the available information will indicate the inheritance of a cow tested in two or more lactations. That will depend mainly on how little the records of the bull's daughters contain of the effects of nonstandard herd environment; that is, of what is called C in the preceding formulas.

The sire index can be useful in pedigrees by indicating which young bulls are most likely to be worth trying as sires. For example, consider the pedigree of the bull X, itself too young to be proved but sired by a bull with an index of 700 pounds of fat and out of a cow whose records averaged 500 pounds but whose sire had an index of 800 pounds and whose dam's records average 400 pounds.

$$X \begin{cases} A \text{ (700 lb. index)} \\ B \text{ (500 lb. record)} \begin{cases} C \text{ (800 lb. index)} \\ D \text{ (400 lb. record)} \end{cases} \end{cases}$$

We would estimate X at $\frac{700}{2} + \frac{500}{2}$, or 600 pounds, modified, of course, by whatever allowance toward the breed average we think is necessary on account of our not being sure that the information about *A* and *B* is exactly equal to their breeding values. If *B*'s records are few or made under uncertain circumstances, we will have less faith in them and will examine her pedigree. That indicates that she was expected to produce $\frac{800}{2} + \frac{400}{2}$, or 600, instead of the 500 she actually produced. Hence, we will suspect that she really has a little better heredity than her own record indicates. We will not be certain of that because we are not entirely certain of the breeding values of *C* and of *D* and because all animals are so heterozygous that such a mating as that of *C* and *D* might produce an individual much poorer or better than is expected. The amount of attention we give to *C* and *D* will depend mainly on how uncertain we are that the 500-pound figure correctly represents *B*'s inheritance. If *B* were never tested—for example, if *X* were her first calf and she had died soon after calving—we would estimate *X* at $\frac{700}{2} + \frac{800}{4} + \frac{400}{4}$, or 650 pounds, modified by some allowance toward the breed average; but we would be more uncertain about our estimate than if *B* had been tested.

SUMMARY

A sire index is a means of expressing in a single figure a sire's progeny test, usually for characteristics he cannot express himself. It is most frequently used for dairy bulls and for roosters.

The index which seems most useful and accurate under many conditions is the average of his daughters plus the average increase of the daughters over their dams.

The daughter average and the difference between daughters and dams are the parts of which the index is the sum. The daughter average is most vulnerable to error from differences in environment from herd to herd. The difference between daughters and dams is most vulnerable to error from environment's not having been the same for daughters and dams or from the dams' having been selected more highly than is fully discounted. The difference between daughters and dams is least subject to error from variations in environment from one herd to another. The index is a better guide to the sire's breeding value than the daughter average or the daughter-dam difference when the correlation between daughter average and average of dams is more than .25 but less than .75.

Some room should still be left for considering the production of ancestors and collateral relatives, even when a bull has many daughters.

An index cannot be guaranteed correct, since indexes will often contain considerable error from Mendelian sampling and from incomplete corrections for other circumstances. Hence a sire should be estimated nearer the average of the breed than his index is, especially if his index is extremely high or extremely low.

REFERENCES

Copeland, Lynn. 1934. Pedigree analysis as a basis of selecting bull calves. Jour. Dy. Sci., 17:93–102.

Davidson, F. A. 1925. Measuring the breeding value of dairy sires by the records of their first few advanced registry daughters. Illinois Agr. Exp. Sta., Bul. 270.

Engeler, W. 1934. Die Leistungsverbessernden Stiere in der schweizerischen Braunviehzucht. Luzern.

Gifford, W. 1930. The mode of inheritance of yearly butterfat production. Missouri Agr. Exp. Sta., Res. Bul. 144.

Goodale, H. D. 1927. A sire's breeding index with special reference to milk production. American Naturalist, 61:539–44.

Gowen, John W. 1930. On criteria for breeding capacity in dairy cattle. Proc. of Amer. Soc. An. Prod. for 1929, pp. 47–49.

Hansson, N. 1913. Kan man med fördel höja meddelfetthalten i den av vora nötkreatursstammar och raser lämnade mjölken? Centralanst. för försöksväsendet på jordbruksomradet, Meddelande, 78:1–85.

Jull, M. A. 1934. Progeny testing in breeding for egg production. Poultry Sci. 13:44–51.

Lörtscher, Hans. 1937. Variationsstatistische Untersuchungen an Leistungserhebungen in einer British-Friesian Herde. Zeit. f. Zücht. Reihe B. 39:257–362.

Lush, Jay L. 1931. The number of daughters necessary to prove a sire. Jour. of Dy Sci., 14:209–20.

———. 1933. The bull index problem in the light of modern genetics. Jour. of Dy. Sci., 15:501–22.

———. 1935. Progeny test and individual performance as indicators of an animal's breeding value. Jour. of Dy. Sci., 18:1–19.

———. 1944. The optimum emphasis on dams' records when proving dairy sires. Jour. of Dy. Sci., 27:937–951.

———, Norton, H. W., III, and Arnold, Floyd. 1941. Effects which selection of dams may have on sire indexes. Jour. of Dy. Sci., 24:695–721.

Mount Hope Farm. 1928. Selecting a herd sire. The Mount Hope Bull Index. Mount Hope Farm, Williamstown, Mass.

Rice, V. A. 1944. A new method for indexing dairy bulls. Jour. of Dy. Sci., 27:921–36.

Turner, C. W. 1925. A comparison of Guernsey sires. Missouri Agr. Exp. Sta., Res. Bul. 79 (see especially p. 27).

Ward, A. H. 1941. Sire survey and investigational work. Seventeenth annual report of the New Zealand Dairy Board.

Wright, S. 1932. On the evaluation of dairy sires. Proc. Amer. Soc. An. Prod. for 1931, pp. 71–78.

Yapp, W. W. 1925. Transmitting ability of dairy sires. Proc. Amer. Soc. An. Prod. for 1924, pp. 90–92.

CHAPTER 32

Bull Associations or Bull Circles

The bull circle is a co-operative plan by which dairymen exchange sires at regular intervals.[1] In the association there are at least three, and usually not more than five, blocks or stations. One bull is purchased for each block or station. A block may consist of a single large herd, or it may be a group of several small herds so located that one bull can be used for all. After the bulls have been in service two years, they are moved to another block. At the end of the second two years they are moved to a third block. If there are more than three blocks in the circle, the bulls may be moved at the end of six years to a fourth block, where they have not been in service. If there are only three blocks in the circle and a bull is still alive and capable of service at the end of six years, he may be brought back to the block where he was first used. By that time he will have been thoroughly proved. If he proved to be a very good bull, it will be wise to use him even on those of his daughters which are still in the first herd. If he turned out to be only an ordinary or an inferior bull, he will be sold to the butcher. In most cases he will have died or become sterile before he has completed six years of service. Only rarely will the question arise as to the desirability of breeding one of those bulls to his own daughters.

OBJECTS

The primary object of a bull circle is to get bull service at lower cost, or better bull service at the same cost, without running any risks from inbreeding. If three men co-operate in a three-block circle, each will at all times own one-third of three bulls instead of each owning one bull. Except for death and sterility, each man will get six years of

[1] The intervals are usually two years in length, which is about as long as they can well be without occasionally breeding some sire to his own daughters. Sometimes the intervals are as short as one year in order that the different owners may get more nearly the same amount of service from each bull. The shorter interval tends to equalize among the circle members the differences between the bulls, but it incurs the bother of the exchange more frequently and increases the opportunities for spreading breeding diseases. This chapter assumes that the exchanges will take place at intervals of two years; but, if a few words are altered, the discussion will apply as well to cases where the exchanges are more frequent.

bull service instead of two for the purchase price of one bull. A breeder can spend less money per year in purchasing bulls even though he may spend more money for each bull he does buy in order to get a better pedigree or individual.

Another object is to keep dairy bulls alive until they are proved in order that the unusually good ones among them can be used extensively after the evidence proves them. Two years after a bull first begins his service, his oldest daughters will be approaching 15 months of age and will be ready to breed. At the end of four years of service his very oldest daughters may have completed one lactation; but unless he had many daughters from his very first services, he will be only partially proved when it is time to move him to the third herd. Before it is time to move him to the fourth herd he should be thoroughly proved. If he is proved inferior, he will, of course, be sent to the butcher and his least productive daughters will follow him. If he is proved only mediocre, he is apt to go to the block also, since his age increases the probability that he will soon become useless for breeding. If he proves to be very valuable, he can be returned to the first herd for use on his own daughters and granddaughters, thus making possible some intense linebreeding to him.

The bull-circle plan makes it easy to pursue a consistent linebreeding policy. For example, the first bulls in a four-block circle may all be half brothers by some famous bull. The continued use of these bulls on each other's daughters would tend toward producing herds which would be almost as closely related to this outstanding sire as if they were daughters, although it might have been quite impossible financially to buy actual daughters of that noted bull. The amount of inbreeding in such a plan would never get very high, tending toward but never reaching 12½ per cent; and all of it would be directed toward the famous bull. If the cows in these herds are purebred, and one of the sires used proves to be an unusually good one, it would be practical to choose the next bulls out of the best cows in those herds where the best sire had been used. This would lead to still further linebreeding, but with four or five blocks in the bull circle it is unlikely that this linebreeding could rise high enough to be dangerous, even in 30 or 40 years of steady co-operation, provided care was always taken to select the sires from the best cows in the herd, sired by the best of the preceding bulls. In short, the bull circle offers almost an ideal plan for linebreeding which is fast enough to make progress but not fast enough to be dangerous.

The bull-circle plan may also assist in a less tangible way through the development of community spirit and co-operation. Naturally the members of the bull circle will need to be members of a cow-testing association if they are to take advantage of any of the objects of the bull circle other than that of the cheaper bull service.

A "bull club" is merely the joining of several men together in the co-operative purchase of a single bull. This lowers the bull cost to each of them but does not lead to the proving of the sire since, after he has been used two years, they will need to exchange him if he is not to be bred to his own daughters.

INTENSITY OF INBREEDING BROUGHT ABOUT BY BULL CIRCLES

The upper part of Figure 48 shows a pedigree with the most extreme inbreeding which could be produced in a five-block bull circle, where the five bulls first bought were all half brothers and each saw service in all five blocks. A bull would rarely remain in service that long. Variations in the sex ratio would make exceedingly rare a succession of daughters which would result in a pedigree like this one, where *X* is a descendant of all five bulls.

The lower part of Figure 48 shows some examples of the most extreme inbreeding which would be apt to result in a three-block circle where all three bulls were half brothers at the start, were used for six years, and then it was discovered that one of them was so outstanding that he would be used again. He would go back for service again in the herd where he had first been used and would be mated to some of his daughters, granddaughters and great granddaughters. There would not be many of each. Sons of his would be placed in service in the other two herds, where they would be used on some cows which were their paternal half sisters and on others which were their cousins through the paternal grandsire. The total amount of inbreeding in 10 or 12 years of such a plan is not likely to go above 25 per cent in any individual and probably would average only about 8 or 10 per cent. Moreover, this inbreeding would nearly all be toward the famous sire, *Y*, or his best son, *E*.

BUSINESS PRECAUTIONS ADVISABLE

The title to the bull should rest in the bull association rather than in the individual members. If each man owns one bull it is almost certain to happen that, by the time the bulls are mature, some of them will appear to be better individuals than others, and the owners of those will be reluctant to exchange. If each man owns his share of all bulls, there will not be as much of this difficulty.

A reserve fund should be provided for the replacement of bulls which die or become sterile or which eventually prove themselves to have been only average or less. An assessment of each block might be

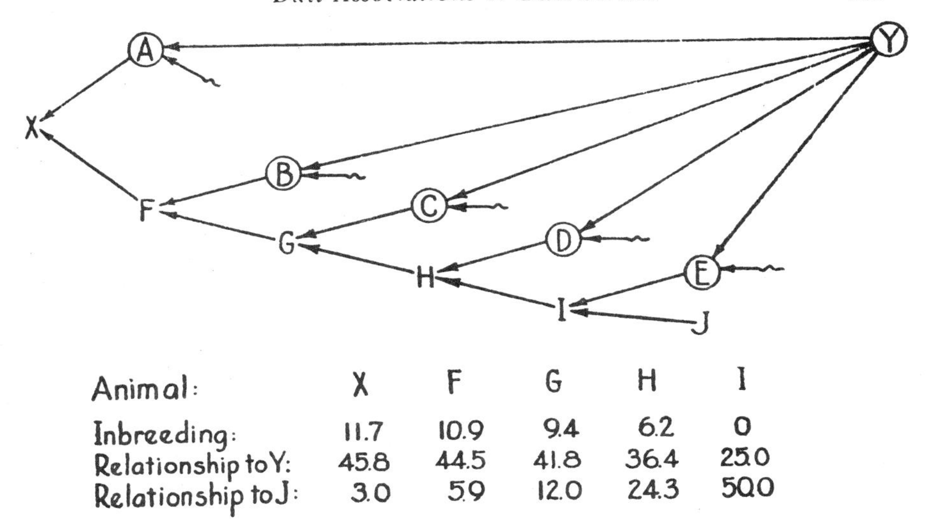

Animal:	X	F	G	H	I
Inbreeding:	11.7	10.9	9.4	6.2	0
Relationship to Y:	45.8	44.5	41.8	36.4	25.0
Relationship to J:	3.0	5.9	12.0	24.3	50.0

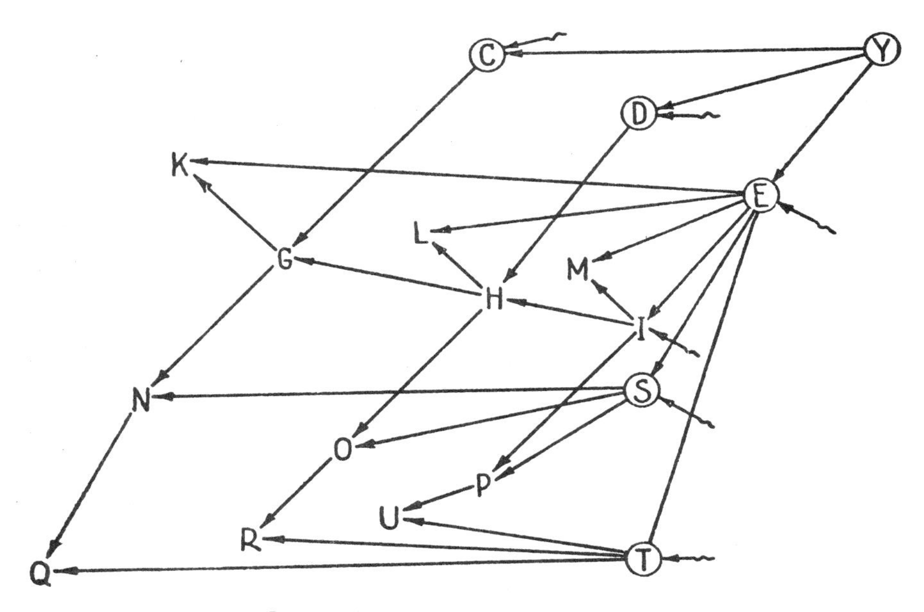

Inbreeding Percentages:

M - 25.0	P - 12.5	U - 12.5
L - 18.8	O - 9.4	R - 10.9
K - 15.6	N - 7.8	Q - 10.2

FIG. 48. Some examples of extreme inbreeding which might happen in a bull circle where the bulls were paternal half brothers. Upper: In a five-block circle where no bull ever returned to the same block and the linebreeding is purely to Y. Lower: In a three-block circle where the linebreeding is first to Y and then to his son, E, who is returned to be used a second time in the circle in rotation with two of his sons, S and T. Encircled letters indicate sires.

made after the need arises, but it is a sounder policy to have the money available immediately. Otherwise, the men who are using the bulls which are still healthy will be inclined to blame the sterility or death of the other bull on the carelessness of his caretaker and to ask that the caretaker pay for him. Each member will be more willing to pay an assessment made in advance because he may be the one who will benefit first from it.

There should be a definite plan for rotating the bulls, so that there will be less chance for controversy over getting the bull which at the time of exchange appears to be the best individual. The plan should also state when and how bulls should be culled as their daughters begin to prove them. Such a plan need not be elaborate; but, if there is no definite plan, the man using the bull whose daughters begin to prove him undesirable may have difficulty in convincing the other men that the circle should buy a new bull to replace that one.

SUMMARY

The bull association or bull circle is a plan for co-operative exchange of dairy sires at regular intervals.

Its objects are: (1) to provide better bull service at lower cost, (2) to prove bulls and keep them alive so that use can be made of that proof in further breeding, and (3) to make it easy to follow a consistent but reasonably safe linebreeding policy.

The intensity of the close breeding involved, even if the bulls in the bull circle are all fairly close relatives of each other, is not high. If the initial bulls are half brothers of each other, the total amount of close breeding tends toward that produced by a single mating of half brother and sister but is distributed over several generations with much opportunity for culling any undesired individuals.

Certain business precautions, such as having a definite plan for rotation and culling and for the purchase of a replacement sire in case one dies, becomes sterile, or is culled, help the association run more smoothly. It is almost essential that the title to the bulls rest in the association rather than in the individuals who comprise it.

REFERENCES

Fourt, D. L., and Loughary, Ivan H. 1938. Idaho bull associations. Idaho Agr. Exp. Sta., Bul. 223.

Lush, Jay L., and Lacy, M. D. 1932. The ages of breeding cattle and the possibilities of using proven sires. Iowa Agr. Exp. Sta., Bul. 290.

Simons, Rodger L. 1933. Dairy development in Sweden. *Hoard's Dairyman,* 78:324.

Winkjer, Joel G. 1936. Co-operative dairy bull associations. *Hoard's Dairyman,* 81:111 et seq.

————. 1939. Co-operative dairy bull associations. USDA, Farmers Bulletin No. 1830.

CHAPTER 33

Community Breeding

Most breeds of livestock arose from community breeding in a small region where a few herds located conveniently to each other exchanged breeding stock during the formative period of the breed and really established the breed by linebreeding to the best individuals within those herds. Community breeding became less typical when herdbooks were established and men unfamiliar with each other's stock could still work with the same breed.

Community breeding has not been general in the United States, although at one time the Poland-China breed was a community breed in Butler and Warren Counties in Ohio. Likewise, the Chester-White was long a community breed in Pennsylvania, although the details of that history were not kept. A few other prominent breeds for at least short periods have been community breeds in America before they expanded to nationwide importance. Spread over a vast territory, with breed organizations endeavoring to expand their spheres of influence, and with high-pressure salesmanship often working most effectively on prospects who are at some distance from the herds where the animals are bred, the prominent breeds in America have generally been far from any condition which could be called community breeding. Among scattered examples which approached the condition of community breeding in America should be mentioned the Vermont Merinos, light horses in the Bluegrass region of Kentucky, the New Salem (North Dakota) Breeding Circuit for Holstein-Friesians, and county associations such as the Delaware County (Ohio) Percheron Breeders' Association and some of the county associations of dairy breeders in Wisconsin.

Many writers on animal breeding subjects have emphasized the advantages of community breeding, but this seems to have had little effect on general practice. In many communities even yet a man beginning to breed purebred livestock will deliberately select a breed which is not present or at least is not abundant in his community, thinking that he will thereby have less competition and a better chance to make his herd well known than if he started with a breed already well established in that community.

MORE ACCURATE CHOICE OF BREEDING STOCK

One who sees his neighbor's herd frequently and knows many of the animals in it, has a better chance to make correct choices in selecting animals out of that herd than if he selects from a distant herd the first time he sees it or if he buys in an auction sale an animal which may be the only one there from the herd in which it was bred. In the neighboring herd he has a chance to see at several different times or ages any animal he is thinking of buying. Also, he can usually see many of its close relatives. This is helpful in keeping to a minimum the amount he is deceived by environment, by dominance, and by epistasis.

Moreover, he is dealing with a man whose business reputation he knows and one who has a neighborly as well as a business interest in keeping him satisfied with his bargain. If the purchase is unsatisfactory and some adjustment is necessary to satisfy guarantees, there need be little delay and no correspondence or traveling expense. The transportation of the animals to be exchanged is a negligible item. Also, he can more easily satisfy himself about the health of the herd from which he is buying and thereby minimize the risk of introducing disease along with his purchases.

PROMOTING THE CONTINUED USE OF GOOD SIRES

One of the important advantages of community breeding is the ease with which it leads to the exchange of sires thought to be unusually good but which the present owner cannot use longer without close inbreeding. If the general sentiment is in favor of using homebred stock, then a sire proved unusually good by his offspring can be continued in use in neighboring herds without extreme inbreeding. This reaches an extreme form in the dairy bull circles discussed in chapter 32. Another fairly common form is the stallion club for the joint ownership of a good stallion by several farmers; although, if only one stallion is owned, this will not be much help in keeping him in service more than three years. Cow-testing associations are primarily organized to aid in the intelligent culling of cows and in improving feeding practices but can lead to some community breeding.

The exchange and continued use of the best proved sires in each neighborhood, if carried far enough, will ultimately develop linebred families which are distinct from one community to another. When that has happened, comparisons of those families can be made, and weak points of one can be corrected by mild outcrosses to others which are strong in those points. The Homestead family of Holstein-Friesians was a notable example of such community breeding.

Community breeding, if carried far, will tend to make the breeds less uniform than they are today. Probably this would be a real advantage to the utility of the breed, although interchange of breeding stock from great distances would diminish and this would affect some of the commercial aspects of the purebred business.

MUTUAL EDUCATION OF THE BREEDERS

Where many people in one community are breeding the same kind of livestock, there are frequent occasions for them to discuss their problems of breeding, animal health, sales, shows, etc. If this continues long, most of the people in that district soon know much about the lore of that particular breed. Almost without realizing it, they come to possess knowledge and skill ordinarily acquired by isolated breeders only after years of experience. Something of this kind is seen in long-established dairy regions and in the regions of Kentucky and Missouri where saddle horses or Thoroughbreds are raised extensively. Where there is much community interest in the breed, it is not difficult to arrange local shows where the exhibits will be creditable and where lively interest will exist, since nearly all of the animals shown will come from herds which the spectators know personally.

BUSINESS ASPECTS OF COMMUNITY BREEDING

At present commercial necessities must govern the operations of most breeders. Sometimes this operates against community breeding. A prominent Jersey breeder says that many American-bred bulls are as good as the average imported bull but at the same time advises young breeders to use imported bulls to head their herds, since they will find a readier sale for the young bull calves by an imported sire than if they were by an American-bred bull. There is no sound genetic reason for this. It is only that the word "imported" may carry with it a certain glamour which helps break down sales resistance. Things of this kind must be considered by the breeder of purebred livestock. He must find a sale for his young stock; and it is to his advantage, other things being equal, to produce the kind of stock which sells most readily. Interchanging breeding stock from great distances constitutes an economic load on the purebred industry. In many cases there is no commensurate gain. The breeding worth of an animal depends upon its genes; and those are not changed by advertising, although the animal's chance to affect the whole breed by its genes may be much changed by that.

One of the main business advantages of community breeding is that lower selling costs can thus be achieved. If a community contains many herds of one breed of livestock, it may acquire a district reputation in addition to the individual reputation of the breeders residing in it.

This will attract buyers from a distance because they know they can find animals which will suit them without heavy traveling expenses in going from herd to herd. Sometimes a buyer from far away will scarcely bother to stop in districts unless he feels reasonably sure that within driving distance of one loading station he can buy a whole carload of animals which will suit him. This happened frequently in the expansion of the dairy business in the southwestern states during the decade beginning about 1920. Buyers coming from that region with orders for an entire carload of dairy stock would often pass by well-known but isolated herds in Kansas, Missouri, Iowa, and Illinois to go into counties in Wisconsin where they thought they could buy a whole carload in two or three days without traveling far from one shipping station. Community breeding also makes possible the organizing of co-operative consignment sales at a low cost for each animal. Often it is scarcely economical for the ordinary breeder to arrange such a sale since his herd is not large enough that the costs of advertising and holding such a sale could be distributed over enough animals to keep the sales cost per head reasonably low.

Community breeding also makes more effective advertising possible. Several breeders located near the same place may run a single advertisement with the names of all signed to it. There need be no business connection between them except in this advertising. By this means the public is told that each of them has breeding stock for sale and is also informed that there are several different flocks or herds from which to choose, all of them close enough that the buyer may perhaps see them in a single day with a minimum of time and expense.

As a general rule, the formal organization of a community breeding enterprise should be kept as simple as possible. Sometimes it is necessary to have a secretary and a board of directors or executive committee. It is usually possible to avoid the employment of any salaried officer. Such expense might increase the overhead expenses enough to offset the business advantage otherwise inherent in community breeding.

SUMMARY

Most of our breeds were formed originally by more or less definite community breeding. Occasional examples of such community breeding have occurred in America, but this has not yet become the general practice in any nationally important breed.

Because breeders see their neighbors' animals often and know them so much better than they do herds at a distance, fewer mistakes are made in selecting breeding stock from neighboring herds.

Community breeding makes possible the exchange of sires at low cost and thus preserves the services of the best sires without the necessity of close inbreeding.

Community breeding can lead naturally to linebreeding which can be quite effective without the intensity of the inbreeding necessarily becoming high.

Community breeding gives opportunity for exchanging of experiences and discussion of problems, thus helping a breeder acquire knowledge and skill which would take many years if he were operating in a community where he was the only man with his chosen breed.

Community breeding has many business advantages, among the most important of which are lower selling costs, more buyers because of the reputation of the district and the larger number of herds from which to select, co-operative sales, co-operation in advertising, and the effective and economical operation of fairs at which local interest may be keen.

CHAPTER 34

Masculinity and Femininity

In many writings on stock judging and animal breeding, it is urged that sires which are masculine in appearance should be chosen and those which appear somewhat feminine should be avoided. It is likewise stressed that a feminine appearance is desirable in females. One of the reasons occasionally advanced for this is that such individuals will be more prepotent than others. As explained earlier, this is without experimental support. The belief may have arisen incidentally from other and better-founded reasons for desiring full development of the secondary sexual characteristics in breeding animals.

ACTIVITY OF PRIMARY SEX GLANDS

The development of the secondary sexual characteristics is controlled to a large extent by hormones secreted by the primary sex glands (ovaries or testes) in a normally healthy individual. Variations in the expression of secondary sexual characteristics may indicate variations in the activity or state of health of the primary sex glands. Secretion of the sex hormones is not identical with the activity of the sex glands in producing sex cells (ova or spermatozoa). Thus, in ridglings (cryptorchids) the testicle which is retained high in the body cavity rarely produces functional spermatozoa. It seems, however, to secrete the sex hormone in almost if not quite normal amounts. Males possessing a cryptorchid testis but having had the normal testis removed show the secondary sexual characteristics and behavior of normal males, although they are rarely able to beget offspring. Among birds, cases have been reported where individuals which have been functional as females later became apparent males. Nearly all of these birds when dissected show that the ovary (the adult female bird normally has only the left ovary instead of two as mammals have) which had originally been functional when the individual was a normal female had become diseased by tuberculosis, cysts, or some other condition, until it had wasted away and was no longer able to produce the sex hormones. In short, the hen had been physiologically castrated,[1] and the effects were practically the same as would have occurred had she been castrated with a knife.

Incidents similar in principle sometimes occur among the mammals. A familiar example is the free-martin (see chapter 35), which is a heifer born twin with a bull. Such heifers are usually barren and often are quite masculine in appearance. Examination usually shows that the sex organs are in a rudimentary or abnormal condition. No doubt some other cases, besides free-martins, where females are quite masculine in appearance are really cases of poor functioning of the ovaries. Perhaps some males do not appear normally masculine because their testes are in some way functioning subnormally.

Thus, one of the reasons which the breeder has for seeking masculinity in his males and femininity in his females is that such evidence is some indication of the normal health and functioning of the primary sex glands. Certainly there are many exceptions to this rule, and probably it is not worth much attention if the seller will guarantee that the animal in question is a sure breeder. H. H. Wing tells of a bull which sired three very good daughters in the Cornell University herd but, on account of his feminine appearance, was sold before his daughters' merits became known. It should be added that the physiology of hormone action is not simple. There are many reactions and complicated interactions of the hormones from the sex glands with the hormones from other sources.

ABNORMAL DIVISION OF CHROMOSOMES

A second mechanism which may cause deviations from normal sex characteristics is abnormal behavior of the chromosomes. Definite evidence of this exists for Drosophila and some other laboratory organisms. It is reasonable to suppose that such behavior would occur occasionally among the mammals. Sex is not determined simply by whether the number of *X*-chromosomes present is one or two, but depends upon a balance between the effects of genes trending toward femaleness and genes trending toward maleness, most of the former being present on the *X*-chromosomes and most of the genes trending toward maleness being scattered on the autosomes. Normally, the presence of one or of two *X*-chromosomes throws the balance completely in one direction or completely in the other. If the chromosomes do not divide regularly, as happens in rare cases, an individual may have a few more or less than the normal number of chromosomes. This may keep the balance from turning definitely toward maleness or toward femaleness and may result

[1] "Spayed" is the word more commonly used in connection with females in animal breeding, but "castrated" may be used for either sex in scientific writings.

in intersexes of various kinds. Some of these intersexes are so extreme as to be sterile, but less extreme ones are sometimes fertile. In Drosophila this process is known to have produced occasional individuals which are more feminine than normal females or more male than normal males—the so-called "super females" and "super males." These are sterile. Winge has reported a case in which abnormal division of the chromosomes in a species of fish resulted in one race becoming homozygous for the *X*-chromosomes. Another pair of chromosomes, which evidently was not homozygous for all the genes affecting the expression of sex, took over the function of normally throwing the balance toward maleness or toward femaleness in each individual. In fishes, amphibians, and birds it appears that the balance normally thrown toward maleness or femaleness by the chromosome mechanism can more easily be reversed by genes on other chromosomes or by environmental conditions than is the case among the mammals.

Such chromosomal intersexes as were not sterile would transmit to some of their progeny the abnormal chromosome balance which caused them to deviate from the normal expression of sex characteristics. If a similar condition exists among mammals, some forms of intersexual conditions may be inherited. The use of breeding animals showing intersexuality might result in increasing the amount of intersexuality in the next generation. It seems improbable that this occurs often enough among the mammals to deserve much attention, but it is a possibility.

GENES AFFECTING SECONDARY SEXUAL CHARACTERISTICS

A third cause of variations in masculinity and femininity is the action of definite genes affecting the expression of the secondary sexual characteristics. Several of those are known in Drosophila and some, such as the genes for "hen feathering" in poultry, are known in other animals. No doubt some of these exist in all kinds of animals and also in such plants as exist in separate sexes (are "dioecious"). In fact, the results of abnormal chromosome behavior just discussed are difficult to explain on any other basis than that there are various genes affecting the expression of sex differences, a high proportion of the genes operating toward femaleness being located on the *X*-chromosome while most of those which operate toward maleness are located on the autosomes.

Further evidence about the existence of such genes comes from race crosses. These have been most extensively studied by Goldschmidt in the gypsy moth, Lymantria. Crosses between certain races of these produce intersexes of various degrees of intersexuality, while others produce normal individuals. A given cross always behaves in the same manner. That is, the results are orderly and definite. It seems likely that the normal balance between maleness and femaleness is caused by somewhat different combinations of genes in different races. In any comparatively pure race the balance falls definitely in one direction or the other. When two races which differ in the genes controlling this balance are crossed, the delicate balance required to throw the mechanism of sex determination in one direction or the other is likely to be upset. We do not know how general this is in the animal world, especially among the mammals. It may be so rare that it scarcely deserves mentioning. Perhaps the best general evidence on this subject is that from the sex-ratio in species crosses. Sometimes this is not disturbed at all, but often it is. Wherever there is a disturbance the heterogametic sex is usually the more violently affected. Among the common farm animals the mule is nearly always sterile, although a few cases of fertile mare mules have been reported. No cases are on record of a fertile male mule, but in fairness it should be added that the circumstances are such that fertility in the male mule would be more rarely detected than in the female. In the crosses between domestic cattle and the American bison, the few males produced have all been sterile, but most of the females have been fertile.

The nature and degree of secondary sexual differences vary from race to race within the same species. There is a somewhat greater difference in temperament between bulls and cows of the dairy breeds than in beef breeds, although a part of this may be a result of differences in the way they are managed. In some breeds of sheep, horns are a masculine trait; in others they appear in both sexes; in still others, neither sex has horns. No breed of sheep normally has horned ewes but hornless rams. In man, many racial differences occur in the expression of sex-differences. In the races from northwestern Europe and around the Mediterranean region, the men are generally rather heavily bearded. Some of the European peoples and most of the peoples of eastern Asia as well as the original natives of the two Americas are scantily bearded. The various negro races of Africa differ much among themselves in this respect. The beard is regarded as a secondary sexual characteristic in man, but its degree of expression varies from race to race, and that variation is not accompanied by corresponding variations in masculinity. In practically all races the height of the men is greater than that of the women, but the proportion of this difference varies from race to race.

Specific genes affect the expression of the secondary sexual characteristics, and many seem to have no direct bearing on reproduction itself. No doubt many of the differences in masculinity and femininity which we see or discuss in our breeding selections are the effects of such genes. While such genes may have no direct physiological importance, yet so long as the customers desire masculinity in the males and femininity in the females it will be to the breeder's interest to produce the kind of animals they want to buy. Probably users of dairy cattle would be better off if gentleness and meekness had always been sought by breeders of dairy bulls. Many of the dangers of handling aged bulls would have been diminished.

SUMMARY

Masculinity in males and femininity in females to some extent are expressions of the functioning of the testes or ovaries in secreting hormones. Absence of masculinity in males or of femininity in females may indicate lowered functioning of these glands, which might in some cases be extreme enough to cause these individuals to be irregular in breeding or even sterile. The importance of evidences of masculinity and femininity has often been exaggerated. The most valid reason for desiring manifestations of normal secondary sex differences is that those may be a partial guarantee that the animal will be a regular breeder.

Abnormal chromosome distribution may on rare occasions disturb the balance of gene action which normally determines complete maleness or complete femaleness. The resulting intersexes, if fertile, may transmit an abnormal number of chromosomes to some of their offspring, thereby leading to some inheritance of this intersexuality.

Some genes definitely affect the expression of sex differences without any detectable effect on the real efficiency of the animal in reproducing itself. These lead to inherited differences in the expression of secondary sexual characteristics. The breeder will need to pay some attention to these differences if his customers do.

REFERENCES

Craft, W. A. 1938. The sex ratio in mules and other hybrid mammals. Quart. Rev. of Biology, 13:19–40.

Deakin, Alan; Muir, G. W.; and Smith, A. G. 1935. Hybridization of domestic cattle, bison, and yak. Tech. Bul. 2, Dept. of Agriculture, Dominion of Canada.

Goldschmidt, Richard. 1934. Lymantria. Bibliographia Genetica, 11:1–186.

Lush, Jay L.; Jones, J. M.; and Dameron, W. H. 1930. The inheritance of cryptorchidism in goats. Texas Agr. Exp. Sta., Bul. 407.

Warwick, B. L. 1935. Inheritance of the ridgling characteristic in goats. Texas Agr. Exp. Sta., 48th Annual Report, pp. 35–36.

Wing, Henry H. 1933. The Cornell University dairy herd, 1889 to 1928. Cornell University Agr. Exp. Sta., Bul. 576.

Winge, Ö. 1934. The experimental alteration of sex chromosomes into autosomes and vice versa, as illustrated by Lebistes. Compt. Rend. Laboratoire Carlsberg, 21:1–50.

CHAPTER 35

Hermaphroditism and Other Abnormalities Pertaining to Sex

The simplest form of reproduction is that in which one organism merely divides into two and it is impossible to say which is mother and which is daughter. This is common among protozoa and bacteria. In the plant kingdom asexual reproduction has been maintained in various specialized forms (such as budding, sprouting from roots, etc.) even among the highest plants. None of the higher animals has maintained asexual reproduction except in the form of parthenogenesis, although truly remarkable regenerative powers are still possessed by animals as complicated as many of the worms. Many species of insects, such as the aphids and the bees, can reproduce parthenogenetically. Sex exists among them, and sexual reproduction occurs at times, but at other times the females lay unfertilized eggs which can develop without fertilization into mature individuals.

SEXUAL REPRODUCTION

The union of two individuals to form many others occurs occasionally among even the simplest protozoa and bacteria. In the simplest form it is impossible to distinguish which of the uniting individuals is male and which is female. In such cases the union is usually called by some such term as "conjugation." In some cases of conjugation it is possible by physiological means to show that the conjugants are different in kind and therefore might be said to belong to different sexes. In some of these cases there are more than two such forms. If they were called sexes, there would be more than two sexes. Other terms more precise but less common than male and female are used to describe such forms.

Sexual reproduction possesses a tremendous evolutionary advantage over asexual reproduction. In a species reproducing asexually, 100 new mutations would only mean the existence of 101 pure breeding genotypes from among which natural selection could choose. With sexual reproduction there is the opportunity for trying out each new mutation in combination with all the others. Therefore, 100 new mutations in a sexually reproducing species make possible 2^{100} new true-breeding genotypes. This is an enormously greater number. The possibility of finding some combination among that number which would be superior to the previous combinations is tremendously increased. This is the fundamental biological importance of sexual reproduction. It is such a big advantage that all but the very simplest plants and animals have evolved ways of reproducing sexually, at least occasionally. Many species, especially among the plants, have retained asexual reproduction for part of their life history but at intervals reproduce sexually. This combines certain advantages of both methods.

HERMAPHRODITISM

Sexual reproduction does not require that the sexes must be in separate individuals. In many of the simpler animals and in most plants the same individual has both male and female reproductive organs; that is, is truly hermaphroditic. Among the vertebrates, only a few fishes are functionally hermaphroditic, but many animals as highly organized as the mollusks and round worms are normally hermaphroditic. The date palm and several of the temperate zone trees, such as the mulberry, are typical examples of the few higher plants which have the sexes in separate individuals.[1] Many other plants—hemp for example—normally exist as separate male and female individuals; but it is easily possible to reverse the sex or to make them hermaphroditic by controlling environmental conditions, such as hours of illumination. Geneticists have even succeeded in producing races of corn which are dioecious, although corn is typically monoecious, the tassel bearing the male organs while the ear and its parts are the female organs.

That the animal kingdom has prevailingly adopted sexual reproduction in a form where the sexes are in separate individuals while the plant kingdom has prevailingly stayed with hermaphroditism naturally calls for an explanation. For many species, separate sexes make possible such a division of labor that a male and female individual together can leave more descendants than two hermaphroditic individuals could. Wherever the anatomy and habits of life of the species were such that division of labor between the sexes conferred this advantage, it was but natural that the species should ultimately give up hermaphroditism. Since plants do not move about, they have little to gain by a division of labor among the two parents. They had much to gain by securing cross-

[1] The botanists call such species "dioecious" and restrict the words "male" and "female" to the gametophyte generation. This usage does not correspond to the common and zoological usages of "male" and "female" but is customary in most botanical writings.

fertilization, at least occasionally, because permanent and complete self-fertilization would destroy the genetic advantage of sexual reproduction in making possible new combinations of genes. The plant world is full of remarkable mechanisms for promoting cross-fertilization. As long as plants have the advantage of occasional cross-fertilization, not much more is to be gained by being dioecious. With the higher animals the situation is quite different. Many of them take remarkable care of their young. In most cases much is gained by having one parent specialized to look after the young directly, the other specialized for obtaining food, for combat, for protection, and perhaps for other duties. The animals which have carried this specialization and division of labor farthest are the social insects, such as ants and bees, with their workers, soldiers, drones, queens, and other classes. The mode of reproduction among the mammals is especially extreme in involving considerable disability of the mother during the bearing and the early rearing of the young. In most mammals the male has become specialized for greater efficiency in combat, in the procuring of food, etc., which at least partly compensates for the fact that his direct contributions to the rearing of the young are less.

SO-CALLED HERMAPHRODITES AMONG MAMMALS

Cases of partial hermaphroditism[2] are reported frequently in medical literature. Mammalian embryos all go through early stages in which it is difficult to be sure of the sex of the individual. That is to say, the embryos seem to have the anatomical potentiality of developing into either sex. Which way the development turns is usually determined by the sex chromosome mechanism, which normally throws the balance definitely one way or the other. Then, as differentiation proceeds, many organs develop in a way which is irreversible. Occasionally this balance is not turned so definitely, and some of the organs develop in one direction and some in the other. Almost all combinations of this can exist, but the development of some of the organs precludes the development of the organs of the other sex so that functional hermaphroditism is impossible. A few cases of birds which were functional females at one time in their lives and later became functional males have been reported. The accessory sex organs in mammals are more specialized, and the development of those organs is less reversible; therefore, functional sex-reversal seems impossible among mammals. Even in birds it is impossible for one individual to be a functional male and a functional female at the same time.

Probably the things most likely to upset the normal balance of the

[2] Colloquially called "morphadites" by many stockmen.

sex-determining mechanism are abnormal chromosome divisions or the action of definite genes on the development of the sex organs. At other times the cause of the abnormal development is a hormonal disturbance, such as exists in the case of the free-martin.

Many of the more frequent cases of supposed hermaphroditism are only incomplete embryological development of some of the sex organs. Similar embryological accidents sometimes affect parts of the body not related to sex. Such may be illustrated by the case of hare-lip or cleft palate, which occurs frequently in man. The upper lip in man develops embryologically as a center piece and two pieces from the side. These normally grow together at an early stage. The lines of their union are still more or less visible in the adult and give the human upper lip its typical slight approach to a three-lobed condition. Sometimes one or both of these unions fail to be completed. The result is an individual with an upper lip divided into two or, in extreme cases, three parts. This is known as hare-lip. Often the defective union extends back through the palate and impairs speech. In a similar way, embryological accidents in the development of the sex organs often result in superficial appearance of hermaphroditism. The condition called hypospadias, which occasionally occurs in farm animals, is a case of this. In it the individual is truly a male but bears some superficial resemblance to a female and is incapable of reproduction because the urethral groove does not complete its development to form a closed tube. Most so-called hermaphrodites among mammals are really males whose development is imperfect. The extreme cases, of course, are sterile; but some of the milder cases may be capable of reproduction.

Hermaphroditism which has its origin in definite genes or in chromosomal disturbances may be inherited to some extent. The actual evidence of this comes from goats and pigs, where hermaphroditism occurs more frequently in certain families than in others. Where the hermaphroditism has its origin in an embryological accident of some kind, it will not have any hereditary tendency unless the original cause of the embryological peculiarity was partly hereditary.

THE FREE-MARTIN

In cattle the developing fetal membranes of twins usually grow together where they come in contact. If enough of the blood vessels on the two sets of membranes grow together, the blood of the two twins is mingled and some of that from each twin actually flows in the blood vessels of the other. This does no damage when both are males or both are females; but, when one is a male and the other is a female, the hormones secreted by the male develop first and exert enough influence on the female to prevent her normal sex development. The extent to which her development is altered varies greatly, presumably depending on how early the blood vessels fuse together and on how complete the fusion is. Occasionally the membranes do not fuse at all, or least the blood vessels on them do not grow together, and the male and female twins born are both quite normal. According to data from 283 such females tested, only about 1 in 12 among heifers born twins with bulls are fertile. The bull is not affected, presumably because his sex organs start to differentiate earlier and never permit those of the female to reach the stage where they can hamper the male's development. The free-martin condition is very rare in animals other than cattle. Why the membranes of twins should so frequently fuse in cattle and so rarely in other animals is not clear.

The proportion of twin births in cattle is generally low, being something like 1 twin birth among every 200 births in "beef cattle" and 1 twin birth among every 50 to 60 in "dairy breeds." The females of the unlike-sexed twins will rarely have any other value than that of veal or beef. The occurrence of a free-martin means practically no loss to the producer of commercial beef cattle. It means at least a small loss to the producers of dairy cattle, because normally it costs them more to raise a cow than the cow will be worth for beef. It is among breeders of purebred cattle that the free-martin causes the greatest loss. When a heifer is born twin with a bull the question immediately arises as to whether it will be profitable to keep her long enough to learn whether she will be a breeder. The answer will depend in part upon how valuable she would be if she did prove fertile. If she is of an ordinary family, neither of her parents nor many of her brothers and sisters being of outstanding merit, it will probably not be worth while to raise her if that is going to cost much more than her beef value at maturity. If her value as a breeding animal would be high, then it might be wise to keep her in the hope that she will be fertile. For example, suppose it will cost about $75 to raise this heifer until she is three years old, and that her probable beef value at that time will be only $50. The loss incurred in raising each heifer which proves to be barren would therefore be $25. If she is of such valuable breeding and individuality that if fertile she would be worth $350 at the age of three years, that would be $275 more than it cost to raise her. In the long run one such heifer which did prove fertile would pay for the loss on 11 which did not. If her breeding value were still higher, it would be wise to raise her. If her probable mature breeding value would be only $125, not enough profit could be made on the occasional successful case to make up for the many where the heifer was finally proved barren. That is a rough outline of what should be considered when one is deciding whether to keep such a twin heifer to maturity. The situation may be further modified by other evidence. Many of the free-martins show evidences of abnormality, especially a distinctly enlarged clitoris or a fold of skin containing a cord along the median plane of the body just above the rear attachment of the udder, even at birth. If the individual appears to be abnormal at birth, attempting to raise it for breeding purposes is almost useless. On the other hand, if the heifer appears absolutely normal at birth, the chances of her being fertile are higher than the general average of about 1 in 12.

SCIENTIFIC ASPECTS OF HERMAPHRODITISM

To the practical breeder all hermaphrodites or partial approaches to that condition are only annoying sources of loss and are evidence of nature's blunders. To the physiologist or embryologist they are full of interest because they may throw light on the interplay of hormones and organ development and give him new information about this subject. Nature has performed for him experiments which he could not perform for himself. For example the free-martin shows him the equivalent of castration of the female embryo and partial transplantation of the male gonads at an extraordinarily early stage which he could not achieve in his laboratory. For this reason the literature of physiology and embryology has far more cases of this kind reported than corresponds to the financial importance of the subject to the practical breeder.

SUMMARY

Sexual reproduction makes it possible for each new mutation to be tried in combination with all previously existing genes. This is such an important evolutionary advantage that nearly all species have developed ways of reproducing sexually, although many of the plants and some animals as specialized as insects, retain the ability to reproduce asexually at times. Even in many of the simplest organisms, an occasional generation of conjugation or sexual reproduction may occur.

The separation of the sexes into different individuals brings advantages from a division of labor. These are small or non-existent in the case of most plants but are considerable in nearly all higher animals.

The embryos of the higher animals usually have the anatomical potentiality of developing into either sex. Normally the sex-chromosomes throw the balance definitely in one direction or the other. Many of the changes in the developing sex organs are irreversible. Sometimes the normal sex-determining balance may be upset by chromosomal abnormalities or by the effect of definite genes in such a way that development does not proceed definitely toward the one sex or the other but toward one sex in some organs and toward the other sex in others. Various grades of hermaphroditism result. In dioecious plants and in some of the lower animals this normal balance can sometimes be upset by certain environmental conditions.

Tendencies to hermaphroditism seem sometimes to be inherited. The evidence of this among the mammals comes mostly from goats and pigs.

Many of the cases of supposed hermaphroditism are merely embryological accidents, usually occurring in an individual which was originally male but develops imperfectly during its embryology.

The free-martin is an especially interesting case, sometimes regarded as hermaphroditic but really the result of partial hormonal castration during embryonic development.

REFERENCES

Allen, Edgar. 1932. Sex and internal secretions. 1,008 pp. Baltimore: William Wood & Co.

Crew, F. A. E. 1925. Animal genetics. pp. 187–253. Edinburgh: Oliver and Boyd.

Moore, Carl R. 1941. The influence of hormones on the development of the reproductive system. Jour. of Urology, 45:869–74.

Swett, W. W.; Matthews, C. A.; and Graves, R. R. 1940. Early recognition of the free-martin condition in heifers twinborn with bulls. Jour. Agr. Res. 61:587–624.

CHAPTER 36

Gestation Periods

A certain length of gestation is characteristic of each species; but, like other characteristics, the lengths of individual gestation periods vary. The actual union of the ovum and spermatozoon may be several hours after the successful service. The gestation period is measured as the number of days between the day of service and the day of birth of the young, since the exact hour of fertilization will not be known.

In most of the animal kingdom except the mammals the fertilized eggs hatch outside the body of the mother. There are exceptions to this general rule. For example, there are some snakes and some fish in which hatching takes place inside the body of the mother and others in which it takes place outside, but in either case the young hatch in a similar stage of development. Their embryology is more like that of birds than like that of mammals. Even among the mammals a few of the most primitive—for example, the duck-bill of Australia—lay eggs which hatch outside their bodies in somewhat the same way as do those of birds. In many of the primitive mammals (the "marsupials"), the eggs hatch within the body of the mother, but development proceeds only a short way before the young are born at a very immature stage. After they leave the mother's body, they are nourished and carried by the mother in a pouch. The only North American representative of these mammals is the common opossum.

In the higher mammals ("Placentalia"), the ovum not only starts its development within the body of the mother but develops a special mechanism, the placenta, by which its blood vessels come into close contact but not actual union with those of the mother. Through the placenta food and oxygen from the mother diffuse to the embryo, and the carbon dioxide and other waste products from the embryo diffuse into the blood of the mother. The placenta enables the embryo to develop far before birth, while it is still sheltered against an environment hostile in many respects. Because of the better care the mother can provide, no such enormous number of fertilized eggs need be started on the road to development in order that a few shall reach maturity safely as is the case with most lower animals, e.g., the thousands or even millions of eggs spawned per female among fishes.

All the higher mammals share the evolutionary advantages of the placental mechanism, but they differ widely in the degree of maturity which their young attain before they are born. For example, the young of rabbits are born without their hair, with their eyes closed, and are fully dependent on their mother for many days. At the other extreme, although both are rodents, the young guinea pig is born fully equipped with hair, with its eyes open, is able to eat lettuce and cabbage before it is a day old and under favorable circumstances can even survive without a foster mother if its own mother dies at its birth. Most farm animals are intermediate in this respect.

CAUSES OF VARIATION IN GESTATION LENGTH

Within the species there are breed differences which are usually slight but can be detected when large numbers of each breed are compared (see *Jour. Ani. Sci.* 2:50–52 and 4:13–14 for examples). Within the breed there are no doubt individual differences, although it is rare that one female is so different from the average of the breed and has enough gestations that one could be certain that she really differs from the breed to which she belongs. A common cause of variation is disease, such as abortion. Environmental conditions not usually classified as disease may sometimes bring about a shortening or lengthening of the gestation period. The season of the year in which pregnancy begins has a slight effect on the length of the gestation period in some cases. There is some evidence to indicate that nutritional conditions, even when not severe enough to be actually pathological, may affect the length of gestation. Within a species large litters are usually carried for a shorter time than small litters. Various workers have at times detected an influence of the sex of the offspring upon the length of time it is carried, but the evidence is contradictory. This is probably never a major cause of variation. For example, Uppenborn reports from a study of 5,600 cases that stallion foals are carried an average of $1.6 \pm .2$ days longer than mare foals—a real difference, but not a practically important one. On the same subject Mauch reports an average of $1.7 \pm .5$. The size of the offspring may influence the length of time it is carried. First gestations are often a little shorter than later ones.

"NORMAL" AND "ABNORMAL" GESTATION PERIODS

There is no entirely satisfactory test to determine whether a particular gestation is normal in length or otherwise. This problem comes up in a practical way most frequently in connection with cattle when the question is raised as to whether a given gestation terminated normally or in an abortion. Abortion may occur at all stages from the time when the embryo is so small that it escapes detection, to the stage of a gestation period practically normal in length. The "average length of gestation" is a statistical composite picture of many individual cases, no one of which may perhaps have been of exactly the average length. Some variation is normal and is just as characteristic of the species as is variation in other traits or organs.

As a practical guide to normal and abnormal gestations, Table 21 shows the standard deviation of gestation lengths for the farm animals. Approximately two-thirds of all gestation periods may be expected to differ from the average by less than the standard deviation. A useful rule is to suspect of being abnormal any gestation differing from the species average by more than the standard deviation. Where the breed or herd average is known and is based on reliable numbers, deviations should be figured from it instead of the species average, of course. Any gestation differing from the average by more than twice the standard deviation is likely to be abnormal and calls for examination as to whether there may not be some diseased condition which needs attention, or whether perhaps there may not have been some mistake in the breeding date.

The standard deviations in Table 21 were calculated from actual data which may in some cases have included errors in breeding dates or may have included as normal some gestation periods which really were terminated by abortion at such an advanced stage that the offspring was able to live. So far as they could be detected, all such cases were excluded from these data, but it is unlikely that all were detected.

TABLE 21

GESTATION PERIODS IN COMMON MAMMALS

A. Summarized from published actual counts various breeds and places:

Kind of Animal	No. of Gestations Included in Average	Average Length of Gestation	Standard Deviation of Individual Gestation Lengths
Horses	28,456	335.9	About 10 or 11 days
Ass	14	366.9	12 days
Mares bred to jacks	2,338	350	About 19 days
Cattle	27,810	282.1	About 5 days
Swine	6,535	114.3	2.2 days
Sheep	4,417	149.1	2.4 days
Goats	6,761	150.8	3.3 days
Dog	147	61.3	3.1 days
Rabbit	1,540	31.4	1.1 days
Silver fox	797	52.2	0.9 days

B. Quoted from various books:

Kind of Animal	Average Length of Gestation Periods
Farm animals:	
Jennet	About 12 months. Quite variable
Mare	About 11 months. Quite variable
Cow	280 to 285 days. 283 is most frequently stated
Ewe	147 to 150 days
Doe (goat)	149 to 154 days
Sow	112 to 114 days
Bitch	58 to 67 days, usually about 63
Cat	60 to 64 days. Some state 50 days
Laboratory animals:	
Rabbit	31 days
Chinchilla	111 days
Hamster	15 days
Guinea pig	69 days
Mouse	21 days
Rat	21 days
Other animals, about which there is less certainty:*	
Bear	6 months. Recent evidence indicates that this is too short
Beaver	4 months. Another writer states 65 days
Buffalo	10–12 months. (315 days with $\sigma = 5.5$ days in the Asiatic buffalo)
Camel	13 months
Dromedary	12 months
Elephant	20–24 months
Ferret	42 days
Fisher	352 days
Fitch	43 days
Fox	52 days
Giraffe	14 months
Lion	3½ months
Marten	267 days
Mink	51 days
Monkey	7 months
Muskrat	21 days
Nutria	140 days
Opossum	12½ days
Otter	55 days
Prjewalsky's horse	356–359 days
Puma	79 days. One writer says 15 weeks
Raccoon	65 days
Reindeer	8 months
Seal	11–12 months
Skunk	40 days. Some state 63 days
Squirrel	28 days
Tiger	22 weeks
Walrus	One year
Wolf	60 days to 63 days
Zebra	Same as the horse

* For more details see: *Breeding Data on Fur Bearing Animals*, special circular dated June, 1933, from the Department of Veterinary Science, University of Wisconsin; or: Kenneth, J. H. 1943. Gestation; a table and bibliography. Edinburgh. Oliver and Boyd.

The practical effect of including a few errors of this kind is to make the standard deviations a little larger and the means a little smaller than they would otherwise be.

PRACTICAL USES FOR KNOWING THE LENGTHS OF GESTATION PERIODS

First of all the caretaker needs to know when to prepare for the young by isolating the prospective dam and fixing things so that she can take good care of her new-born. It is not safe to rely entirely upon the breeding calendar in doing this, both because individual animals may vary distinctly from the expected date and because, after all, that animal actually may have been bred at some other date than the one recorded for it.

Another practical use for the knowledge of gestation length is in settling cases of disputed parentage where a female may have been bred to two different males at different heat periods and the young is born at a time which does not correspond exactly to either service date. Often such cases cannot be satisfactorily settled, and the only honorable thing to do is to regard the offspring as a grade or at best as a purebred not eligible to registry.

Observing gestation lengths carefully may help some in disease control by enabling the breeder to quarantine each female several days before she is due to produce her young and by calling his attention immediately to any which deviate enough from expectation that they are likely to need veterinary attention.

CHAPTER 37

Sex Ratios

The proportions of the two sexes are approximately equal in all the vertebrates. There are other animals among which sex ratios are normally very far from equality. For example, in some of the gall flies the males may be as rare as 1 or 2 per cent of the population. There are slight but consistent deviations from equality in the sex ratio even among the higher animals. A summary of the usual percentage of males among the total births for several species is shown in Table 22.

These are the sex ratios among those actually born. Sometimes these are called secondary ratios to distinguish them from the primary ratios which exist at the time of conception. The secondary sex ratio can differ from the primary one if there are differences in the prenatal mortality of the two sexes. Such differences are known to exist in several species. Apparently it is the general rule among the mammals that the prenatal and also the immediate postnatal mortality is higher among males than among females. It is usually impossible to determine the primary sex ratio directly. In practically all writings on the subject the sex ratio stated is that at birth unless otherwise specified. Some writers even distinguish a tertiary sex ratio, which would be the ratio of the sexes existing at maturity or at some other age—perhaps at weaning time for the farm animals. The tertiary sex ratio is so influenced by postnatal conditions of management that it is rarely used.

Sex ratios are usually expressed as the number of males per hundred females or as the percentage of male births among all births studied. The first method has the disadvantage that anything producing a certain effect on one sex would magnify the proportion more than if it produced the same effect on the other sex. For example, if anything occurred to destroy one-fourth of the males and the sexes had really been present in exactly equal number at the start, the sex ratio would be stated as 75 males to 100 females. On the other hand, if something had destroyed one-fourth of the females, the sex ratio would be stated as 133 males to 100 females. In the latter case the deviation would at first glance appear larger, although the amount of change is really

TABLE 22
SEX RATIOS IN SEVERAL SPECIES OF ANIMALS

Animal	Percentage of Males Among All Births	Approximate Number of Births Studied	Author, Date, or Notes
Farm animals:			
Horse	49.7	1,111,908	Düsing
"	48.9	34,497	Richter
"	49.3	11,261	Uppenborn
"	49.7	62,002	Schlechter
"	49.1	4,109	Lauprecht, 1932
"	49.9	25,560	Darwin
Mule	44.3	1,416	Craft, 1933
Cattle	50.5	3,559	Gowen and Pearl
" (dairy)	49.4	13,000	Roberts, 1930, and Roberts and Yapp, 1928
"	51.5	124,000	Johansson, 1932
"	49.9	20,579	Engeler, 1933
"	52.2	11,450	Ward, 1941
Sheep	49.5	91,640	Chapman and Lush, 1932
"	49.0	127,587	Henning, 1939
Goat (Angoras)	50.1	3,000	Lush, *et al.*, 1930
Swine	52.3	23,000	McPhee, *et al.*, 1932
"	50.6	48,000	Krallinger, 1930
"	51.1	3,639	Hetzer, *et al.*, 1933
"	50.1	5,373	Kozelhua, 1940
Dog	51.5	50,000	Whitney, 1927
"	52.8	159,304	Winzenburger, 1936
"	52.4	324,323	Druckseis, 1935
Cat	55.0	653	Jones, 1922 (embryos only)
Chicken	49.4	102,143	Jull, 1940
"	50.8	23,273	At eight weeks. Hays, 1941
Other mammals:			
Man	50.7 to 51.7	....	Various authors
"	51.4	....	At birth. Crew, 1937
"	52.4	....	Stillbirths. Crew, 1937
Rat	51.2	....	Cuénot
Mouse	50.0 to 54.1	....	....
Rabbit	51.1	....	....
Guinea pig	50.8	....	Ibsen
" "	49.4	2,014	Schott, 1930

the same. When expressed as percentages of the total births, such changes would appear of equal size, regardless of the sex in which they occurred.

CAUSES OF VARIATION IN SEX RATIOS

The deviations of sex ratios from exact equality are small, but some of them are based on too large numbers to be accidental. Hence they have aroused the interest of biologists, out of all proportion to the economic importance of such small deviations. There is an enormous literature on the subject of sex ratios and the causes of their deviations from exact equality. Crew's book[1] will serve as an introduction to that subject, but one has only to look under "sex ratios" in the indexes of *Biological Abstracts* or of the *Experiment Station Record* to note the large amount written on that subject.

The usual cause of deviations from equality where the numbers are small is chance. Among a group of 32 cows having 5 calves each, the most probable single result is that one cow will have only heifer calves and one only bull calves, five cows will each have one daughter and four sons and another five will each have four daughters and one son. The remaining 20 will each have two sons and three daughters or two daughters and three sons. It is to be expected that extreme deviations will occur just by chance. Those are sometimes impressive to persons who do not have firsthand familiarity with the wide variation which chance can produce in small samples.

The most probable causes for the slight but real differences found between sex ratios and exact equality are differential mortality of the embryos of the two sexes and differences in the motility or longevity of the two kinds of spermatozoa. The latter would lead to the initial production of more embryos of one sex, even though the two kinds of gametes were produced in equal numbers. Among the mammals mortality is a little higher among the males than among the females at all ages, but the reverse is true in birds. Sex-linked lethal genes are well known in Drosophila and lead to abnormal sex ratios. Dr. King's partial success[2] in selecting two strains of rats for a high and a low sex ratio may have been based on such lethals or on lethals which affected only one sex. In plants there is good evidence of differences in the rates at which various kinds of pollen tubes grow down through the maternal tissue to reach the ovules. It is doubtful that anything as extreme as this is important in the higher animals; yet there might be enough of it to explain the slight deviations from exact equality which actually exist.

---

[1] Crew, F. A. E., 1927, *The genetics of sexuality in animals*, Cambridge: The University Press.

[2] *Journal of Experimental Zoology*, 27:1–35.

Various investigators have found that differences in sex ratios were sometimes associated with such things as race, season of the year, year-to-year differences, excessive sexual activity, etc.; but none of these are large enough to be economically important, although they do challenge the investigator to explain them. Species crosses sometimes result in an unequal sex ratio, but such crosses are too rare to be important in practical animal breeding. The cause in these cases is the disturbed balance between two different sets of sex-determining genes. For example, Cole and Painter found 332 males but only 10 females among hybrids between the pigeon and the ring dove. Sex was not determined on 418 others which died in the first week of incubation or on 89 others which died later. Presumably high mortality among the females was the main cause of the extreme sex ratio. In pigeons the females are heterogametic.

THE POSSIBILITY OF SEX CONTROL

It seems unlikely at present that the breeder will ever be able to control the sex of the young produced. Apparently such control would require either some treatment which would destroy the fertilized eggs which are of one sex but leave unharmed those of the other sex, or it would require some sort of treatment before fertilization which would separate the two kinds of spermatozoa or would kill one kind and leave the other unharmed. The first method, even if such a highly selective treatment were discovered, would be impractical in the case of multiple births because it would merely reduce the fertility by eliminating those of the undesired sex without much if any compensating increase in those of the desired sex. The second method conceivably might be of some practical use but would require some chemical or some method of treatment so delicately balanced that it would destroy spermatozoa of one kind without harming those of the other kind or would separate one kind from the other without harming the fertilizing ability of the desired kind. The improbability of there being such a treatment, or of finding it even if it does exist, becomes evident when it is considered that in the farm animals the two kinds of spermatozoa would be exactly alike in well over 90 per cent of the material they contain.

There have been an enormous number of theories of sex control and sex determination. Drelincourt, writing in the seventeenth century, named 276 "false theories" of sex determination. It is only fair to add that his own theory was the two hundred seventy-seventh false one! Geddes and Thomson, writing in 1901, estimated that the number of published theories of this kind had doubled since Drelincourt's time. Many of these theories are so vague or mystical that it is not possible to test them experimentally. Such are the theories which invoke differences in "potency," "mental states," and the like. Many of those which have a physical basis are susceptible to experimental tests. All which have been tested so far can be explained on the chromosome theory of sex determination. Even the cases of sex reversal have only shown that certain environmental circumstances may at times be powerful enough to turn the balance in the other direction from that in which the chromosome mechanism would normally have thrown it. The same thing has been done in dioecious species of plants, such as hemp, where sex reversal can be accomplished by the proper combination of environmental circumstances.

Any theory of sex determination, no matter how absurd, will be right in about half of the cases, just as a matter of chance. The laws of chance permit small samples to deviate widely from the expected average.[3] Hence it is to be expected that even the most absurd theories will sometimes seem to fit rather well a sample consisting of a few cases. Not many years ago a man who thought he had discovered such a theory of sex determination advertised widely in livestock magazines that if breeders would pay him 50 cents per head he would tell them how to get calves of the desired sex. He guaranteed to refund the money whenever the results were not as he had predicted. Now he could be expected to be right half of the time, just as a matter of chance. If he had obtained enough business, this would have been a profitable undertaking for him, because at the most he would only have to return half the money he took in. Moreover, many probably would not ask for their money back or would lose their receipts, etc. In this particular case the man making these advertisements seemed to be sincere in his belief that he had discovered some natural law and was only trying to profit from his discovery as an inventor would from a patent. No doubt he would have been indignant if he had been accused of fraud. The critical test in such cases is to predict what the future sex will be in enough cases to give the laws of chance an opportunity to work out. If the predictions are not correct in significantly more than half the cases, there is no real evidence to support the supposed method of sex control. The subject has enough appeal to popular interest that any supposed new discovery bearing on it is almost sure to get headlines and wide publicity.[4]

[3] *Human Biology* 9:99–103.

[4] *Journal of Heredity*, 24:264–74.

CHAPTER 38

Fertility and Breeding Efficiency

The number of young produced per female per year or other unit of time is one of the most important factors in successful animal production. The cost of producing and maintaining the breeding females and breeding males must be met from the sale of their products and from the salvage value of those parents which are still alive when their breeding usefulness is ended. As long as they are kept for breeding, beef cattle and swine produce nothing for sale except their offspring. Dairy cattle, sheep, goats, horses, and poultry produce milk, wool, mohair, work, and eggs, respectively, in addition to producing their offspring. Even in these animals a considerable part of the income—in sheep, more than half of it—comes from the sale of young stock not needed for replacements. Also, the amount of milk a dairy cow produces in her lifetime depends much upon the frequency with which she freshens. Unless it has a high advertising value which the breeder is in a position to use, a phenomenally high record for one lactation may be unprofitable if that lactation is preceded by a long dry period and if the cow is not bred again until well along in her lactation. The production of wool by sheep, mohair by goats, and work by horses are almost the only returns in animal husbandry which do not depend closely upon reproductive activity.

The number of young produced could be counted at various ages for the purposes of measuring breeding efficiency. From the standpoint of profits and losses it might be most logical to count the number which reach marketing age. From the standpoint of finding the causes of high or low breeding efficiency, more information can be had by counting the number of young weaned or the number born alive among the mammals or the number of birds hatched or the number of eggs laid among poultry.

The subject of breeding efficiency receives attention in varying degrees in different branches of animal husbandry. Ranchers pay much attention to percentage calf crop, which is usually based on the number of calves branded, and to lamb crop, which is usually based on the number of lambs weaned or on the number which come back from the summer range wherever summer and winter ranges are separated. Poultrymen are much concerned with hatching percentages. Swine growers often refer to the number of pigs weaned per sow and also to the number farrowed per litter. Dairymen citing a cow's record often add that she calved again within a certain number of months or carried a calf a certain number of days during the lactation. Only a few dairymen keep definite count of their percentage calf crop, as beef raisers do, or keep track of the average length of time between calvings in their herds. Horsemen sometimes refer to their colt crop, but the number of mares on each farm is usually so small that each man thinks and speaks of his colts individually. Lambert, *et al.* report[1] 65 live colts per year per 100 mares bred. A similar figure for Austria is 52 colts.[2]

Variations in breeding efficiency are the net results of a complicated interplay of genetic and environmental circumstances. Genetic causes usually play a minor part in individual cases, being generally overshadowed in importance by circumstances of nutrition, temporary state of health of the dam, accident, and disease. Yet genetic causes often play a large part in differences between the averages of groups, such as breeds or herds.

FERTILITY

In popular usage fertility is the ability of an animal to produce large numbers of living young. The inability to produce any offspring at all is sterility. Either sex may be sterile, but stockmen usually speak of sterile females as "barren." Sterile is usually an absolute term meaning that the individual is incapable, for the time at least, of producing any young at all; but fertile is ordinarily a relative term, and high and low fertility are used to describe differences between the numbers of young per litter or differences in the frequencies of pregnancies.

In technical writings a distinction is sometimes made between fecundity, fertility, and prolificacy. In that case fecundity is the potential capacity of the female to produce functional ova, regardless of what happens to them after they are produced; fertility is the ability to produce living young or, in the case of poultry, to produce eggs which will start to develop; and prolificacy is a relative term used to express whether many or few offspring result from a given mating or from a certain individual during its lifetime. The distinction between fecundity and fertility is easily illustrated in poultry where a hen may have high fecundity but her eggs may be low in fertility or hatchability; that is, she may lay many eggs but only a few of those will start development

[1] *Proc. Amer. Soc. An. Prod.*, pp. 358–65, 1939.

[2] Züchtungskunde, 13:210.

if incubated or perhaps only a few of those which start to develop will go far enough to hatch into living chicks. Prolificacy is usually applied only to females or to groups, such as breeds, strains, or herds. These distinctions are sometimes necessary for precision in scientific writings but are unimportant in a practical way so far as concerns the mammals. Fertility is used here to mean in a comparative sense the ability of parents to produce large numbers of young.

The number of functional ova released is the first limitation on fertility, but the number actually fertilized may be much less. Failure to be fertilized may result from several circumstances. The spermatozoa may be few in numbers or low in vitality. Normally the male ejaculates millions of spermatozoa at each service; therefore, the number of these does not usually limit the number of young born. There is good evidence, however, that in occasional instances the male is so near the borderline of sterility that the number of normal spermatozoa is low enough to prevent the litter from being as large as it would otherwise be.

The spermatozoa retain their ability to fertilize the ova for only a few hours after they are released in the female genital tract. In some species the ova seem to be liberated at a certain stage in or just after the heat period. If service occurs too early, the spermatozoa are dead before the ova are liberated. If service occurs too late, the ova have passed the period when they could have been fertilized. In dairy cattle the probability of conception is highest when inseminations are made in the middle or later part of the heat period. There is some evidence[3] that the number of pigs per litter in swine may be increased by allowing two services, one early and one late in heat. Partially offsetting this is the danger of exhausting the male by so many services that the number or vitality of the spermatozoa in his future services to other sows would be diminished enough to lose more than was gained by the extra service. It is possible thus to exhaust the males, but the experimental evidence indicates that such exhaustion rarely becomes an actual fact.

Occasionally a given mating produces no results although both individuals later prove fertile in other matings. For example, a given mating is made two or three times without success; then the same female is mated to another male and conception occurs. Naturally such cases can be interpreted in several different ways. Possibly conception would have resulted if she had been mated at that time to the first male. The usual number of services per conception in herds of dairy cattle, not complicated by the presence of contagious abortion, is something like 1.8 to 2.5. With that high a percentage of unsuccessful services under what

[3] Missouri Agr. Exp. Sta., Bul. 310, p. 15.

may be regarded as average conditions, it is not surprising that there would be occasional cases where the same service is repeated three or more times without success and yet if repeated one more time would result in conception.

NUMBER OF YOUNG BORN

Many of the ova which are fertilized die before they reach the stage when they could be born normally. A few of these deaths may be the results of lethal genes for which the embryos are homozygous and which stop development at a certain stage. Others may be mere embryological accidents which prevent some vital organ in the embryo from developing as it should. A portion of those deaths may be the result of overcrowding in the uterus, or of insufficient nutrition, particularly with animals which have such large litters as swine and rabbits do. Part of the evidence for this is the fact that large litters have a higher proportion of embryos which do not complete development than small litters do. Some embryos die from the consequences of infections in the uterus.

Embryos which die before completing their normal development may gradually disintegrate and be resorbed. After they are completely resorbed, normal reproductive functioning of the female may be resumed. Sometimes, especially if bacteria are present, some putrefaction begins, and the embryo along with the residues of its membranes may be aborted. If the embryo which died is one of a large litter, many of which are still alive, as is usually the case with swine and rabbits, it will normally be expelled along with the others when they are born. If it died when it was extremely small, it may have had time to be resorbed before this takes place. Even if it is not resorbed, it is often so small that it is not noticed. If it did not die until shortly before the time of normal birth, the breeder will merely note that there is a stillborn individual in that litter. Many of the stillborn young in such animals as swine have no doubt been dead for several days before birth.

VITALITY AFTER BIRTH

The percentage of those which die between birth and marketable age varies greatly with different kinds of animals. Much of this can be controlled by sanitation and careful management. On farms where swine are fairly well managed, something like two-thirds to three-fourths of the pigs born are weaned. By far the larger proportion of those which are not weaned are dead at birth or die within the first 48 hours after birth. Reasonably exact and useful information about the average percentage of lambs, calves, and colts born which survive at least to weaning time is lacking. Ranchmen often speak of a 70 per cent calf crop, but in most regions that is more nearly a goal than an average of what is actually achieved. Western sheepmen speak of an 80 or 90 per cent lamb crop but do not so conventionally use any one figure as cattlemen do. The vitality of twin colts is especially low as compared with twins in other species. Uppenborn found that only 14 per cent of the colts born as twins lived beyond the age of one year.

As the young animal grows older its mother's influence on its fate becomes less and the effects of its own genes become relatively more important. It is common practice in some of the sow testing work in Sweden and Germany to use the weight of the litter at three or four weeks of age as a measure of the sow's productivity. By putting the weighing date as late as three or four weeks, they measure not only the sow's fertility but also her ability to care for her pigs and produce plenty of milk for them. Soon after three weeks of age the pigs begin to eat other feed, and then weights are affected more and more by their own individual abilities.

AGE AT FIRST BREEDING

Breeding efficiency can be lowered seriously by postponing the first breeding to a needlessly late age. Females bred at a very early age are apt to appear stunted, especially during the first lactation; but their size when mature is affected very little by their having been bred early. Extensive comparisons of early and delayed breeding with several classes of livestock were made at the Missouri Station many years ago and with beef cattle at the Oregon Station and with sheep at the North Dakota Station more recently. These comparisons have shown distinct advantages from breeding early with no disadvantages more serious than a more stunted appearance of the early-bred females during their first lactations. In dairy cattle, for instance, heifers bred to calve first at less than 24 months of age produce almost as much in their first five lactations as do heifers bred to calve first at more than 34 months. Moreover, the early-bred heifers finish their fifth lactations at an average age some 15 months less than that of the late-bred heifers. Thus, almost as much production is attained; and the feed and other maintenance costs for over a year are saved. Breeding could be at such an early age that difficulties at parturition would cause much trouble, but actually that rarely happens. Troubles at parturition seem about as frequent with older females as with younger ones.

INTERVALS BETWEEN PREGNANCIES

Breeding efficiency of a herd or flock as a whole may be seriously lowered by having long intervals between successive pregnancies. In some branches of animal husbandry the breeding policy in this respect is fixed primarily by the feed resources. An extreme example is the range cattleman or sheepman whose feed resources may be abundant from April to October but whose animals may have to get along on a submaintenance ration through the winter. If he cannot have his calves and lambs dropped before the middle of spring so that they can be grown largely on the abundant feed of the next few months, he may prefer to let the cows and ewes remain unbred until next year, rather than have out-of-season calves and lambs. Moreover, ewes will normally breed only at certain seasons of the year. For such men the problem of keeping breeding efficiency high is concentrated in a short season of the year and is mainly one of getting as nearly as possible 100 per cent of the cows or ewes to conceive during that time.

For dairymen and swine producers and for beef growers in the farming regions, there is also the possibility of increasing breeding efficiency by rebreeding reasonably soon after calving or farrowing. Theoretically it is possible to overdo this by breeding females so soon after the gestation period is ended that their strength and reserve vitality can be so exhausted that they will not produce strong offspring. Actually it is doubtful whether such damage is often done. Natural physiological checks and balances will guard against this danger to some extent as they do in a state of nature. The females may fail to come in heat if their reserves of health are low. Meanwhile each month that they are unnecessarily kept from reproducing adds to the overhead costs of their maintenance, which must be divided among the offspring they do produce. There is even some possibility that each additional ovulation adds to the scar tissue in the ovaries and thereby lowers fertility. Moreover, each additional month brings the animal nearer the end of its productive lifetime when one of its offspring, which might otherwise be available for sale, must be kept to replace it.

These considerations make it likely that the wisest general policy is to breed for the first time at an age early enough to favor the highest lifetime production and to rebreed at almost the earliest opportunity after each pregnancy. The data so far analyzed make it seem likely that lifetime production by dairy cows is higher with frequent breeding and many short lactations than it is with longer but fewer lactations. Even when a dairyman rebreeds promptly it is difficult to keep the average calving interval under 13 months. By raising two litters per year swine growers who are equipped to take care of fall-farrowed pigs can reduce very much the costs per pig of maintaining the breeding herd. In some of the dairy regions of northwestern Europe where there is a large surplus of skimmilk, as in Denmark, some swine producers wean their pigs at an early age and attempt to average two and one-half litters per year per sow, although few of them actually achieve that. The average number of litters produced per scored sow per year was 1.88 in the recognized swine centers of Denmark for the five years ending September 1, 1934.

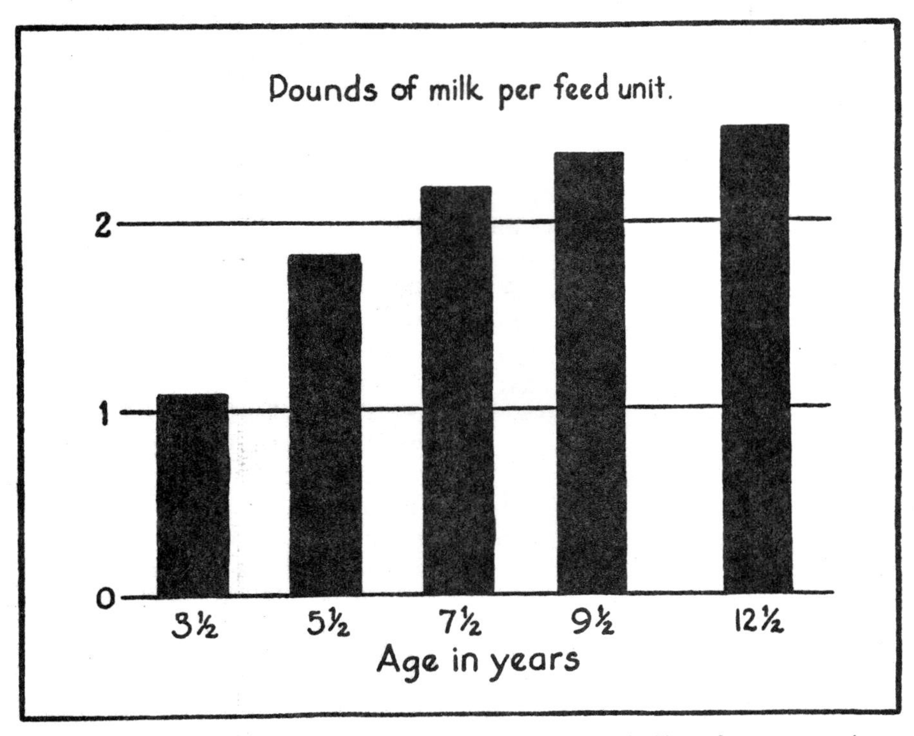

FIG. 49. "It pays to have cows which last a long time." Chart from cow-testing associations in Denmark showing how the returns in milk per unit of feed eaten by a cow during her whole lifetime increase rapidly with the increasing length of her productive life. (From *Kvaegavlen i Jylland,* 1933.)

LONGEVITY

The length of life of the parent is an important part of breeding efficiency, both from the economic standpoint and because it affects the possibilities of improving the breed. The expenses of rearing the parent until it is of breeding age are only partially met in most cases by the salvage value of the parents which are still alive to be sold when they are through producing young. Figure 49 shows a Danish analysis of cow-testing association data. The amount of milk produced per unit of feed eaten by the cow for her entire lifetime increases sharply as the cow's lifetime becomes longer. The longevity of the parents has an important effect on the possibilities of improving the breed by selection because it affects the percentage of young which must be saved merely for replacements. Several studies of dairy cattle indicate that under average conditions about 20 to 30 per cent of the cows are replaced each year. This means a productive life of four years or a little less and that something over 60 per cent of the heifer calves born must be saved for replacements. Seath reports an average of 3.78 calvings per cow's lifetime for Jerseys and Holsteins in Louisiana. Barker in Canada gives 6.1 years for the average life expectancy of dairy females and 6.0 years for registered beef cattle. The most favorable figure yet reported from studying any large body of data is that of Engeler, who found that the herdbook cows of the Brown Swiss breed averaged 4.56 calves during their entire lives, but that 69 per cent of the heifer calves born to herdbook cows were used for breeding. Range sheep in Nevada produced 57 lambs at market age per 100 ewes turned with the rams, but there was a complete turnover of the ewe flock in five years. About 53 per cent of the ewe lambs born would be needed for replacements.[4]

TWINS

Twins are a special case of fertility, particularly interesting in animals where births are normally single. In some breeds of sheep there are more twin births than single births.[5] In some other breeds less than one-fifth of the births are twins. Goats are rather similar to sheep in the frequency of twinning. Johansson summarized the records of nearly a million births in cattle and found that 1.88 per cent of all births in "dairy breeds" were twins. The corresponding figure for "beef breeds" was only 0.44 per cent. Engeler, studying 14,111 births in Brown Swiss cattle, found that 97.3 per cent were single births, 2.7 per cent were twins, .03 per cent were triplets, and there was one case of quadruplets. Sanders, studying more than a million and a quarter cattle records, found that 99.2 per cent were single births. Twins are about as rare in horses as they are in cattle. Uppenborn found that 1.5 per cent of all pregnancies among some 11,000 cases in horses were twins. Lauprecht, studying the records of about 1,500 mares, each of which had foaled several times, found that 1.5 per cent of all births were twin births. Except for the case of identical twins, which are common in man but rare in the farm animals, twins are not apt to be any more alike genetically than are full brothers and sisters born at different times. They are usually more like each other outwardly, however, because they have been subject to the same intrauterine environment before birth and often are reared under much the same environment afterward.

[4] Nevada Agr. Exp. Sta., Bul. 145.

[5] *Scientific Agriculture*, 22:11–17, 1941.

Variations in the tendency to produce twins are inherited to some extent. The result is that the general frequency of twinning is stronger in some individuals and races than in others, but things other than heredity play the major part in determining whether any particular birth shall be a twin birth.

IDENTICAL TWINS

About 50 per cent of the like-sexed twins in man are identical. These start as a single fertilized egg but very early in their embryology divide into two separate individuals. If this division occurs early enough, they may even have completely separate sets of fetal membranes. Usually they have a single chorion but separate amnions. There are cases where the division is not complete; then the result is some form of a double monster, ranging from almost complete separation (the so-called "Siamese twins") to just a faint beginning of the doubling. At large fairs there will often be some kind of a side-show exhibiting living or mounted freaks of this kind from various domestic animals.

Where the separation is complete the result is two separate individuals, of unusual genetic interest because they will be alike in all their genes. Their coefficient of relationship will be 100 per cent, although they may not be any more homozygous than the average of their breed or species. Comparisons of these twins with ordinary fraternal twins (for which the coefficient of relationship is usually 50 per cent) give us much of our information about the importance of heredity in human affairs. In most cases both kinds of twins are subjected to much the same environment, yet the resemblance between identical twins is so much greater than the resemblance between fraternal twins that there is rarely any doubt as to whether a given pair of twins is identical if more than three or four traits are considered. This remarkable similarity of identical twins has been observed from ancient times. Those rare cases where identical twins were reared apart under distinctly different environments provide interesting evidence on the importance of heredity and environment in man.

Identical twins are a subject of much popular interest but of little practical importance to the breeder. They are rare in the farm animals. Johansson estimates that 6.0 ± 1.2 per cent of all twin births in cattle are identical. Bonnier's estimate is 8.5 per cent. Although they are rare, enough have been assembled at different times and places by Kronacher in Germany, Bonnier in Sweden, and McMeekan in New Zealand to yield some interesting information on the interplay between heredity and environment. Kronacher has given detailed descriptions of many cases from cattle, including some which were double monsters, and has emphasized their usefulness as sources of information about the heritability of various characteristics. Identical twins seem to be much rarer in sheep and horses if, indeed, they occur at all. There are other animals in which they are not so rare. For instance, in the nine-banded armadillo the normal mode of reproduction is to produce four in a litter, the four litter mates being identical quadruplets.

TWINS WHICH ARE NOT FULL BROTHERS

Cases are frequently reported where two individuals are born in the same litter but are by different sires. Use is sometimes made of this in crossbreeding experiments to furnish a more adequate control for comparing the purebreds with the crossbreds. Boars of different breeds are mated to a sow which belongs to the same breed as one of the boars. Usually the resulting litter will contain some pigs by each boar. If the breeds are so chosen that the crossbreds can be distinguished from the purebreds by their color, this procedure provides nearly a perfect control of environmental conditions. Somewhat more spectacular cases are sometimes produced in animals which normally produce but one or two young at a time. Thus, Angora does have been known to produce twins, one of which was sired by an Angora buck while another was sired by a Toggenburg buck. A Guernsey nurse cow at the Iowa State College produced twins, one sired by a Hereford and the other by an Aberdeen-Angus bull.[6] Cases are on record where a mare has produced twins, one of which was sired by a horse while the other was sired by a jack. Such cases are conspicuous but illustrate no new principle. Such twins are genetically half brothers but, because of being exposed to so nearly identical environment, are apt to resemble each other more than ordinary half brothers do.

TWINS BORN AT DIFFERENT TIMES

Normally in farm animals when one pregnancy has begun, the female does not come in heat again until it is terminated. This is the general rule, but exceptions occur. When a second pregnancy begins before the preceding one is terminated, the usual consequence is that the second embryo and membranes are expelled also at the time the first pregnancy is terminated, no matter if the second embryo is still so immature that it has no chance of living outside the mother's body. Rarely it will happen that the second embryo is not expelled but continues its development normally and a second birth occurs several weeks or even months after the first one. Because such things as this are so rare and contrary to usual experience, they are apt to be reported when they do occur. Descriptions of such cases often appear in medical literature as well as in stock breeding writings.

[6] *Journal of Heredity*, 31:306.

MANAGEMENT WHICH PROMOTES BREEDING EFFICIENCY

Most individual variations in fertility within a breed are probably matters of management. The principal ways to get high fertility are the general ones of keeping the animals as free from disease as possible and in a reasonably good nutritive condition. The probability of conception and the number of young per conception can be noticeably increased in sheep and swine, and probably in cattle also, by having the females in fair condition and then distinctly improving or increasing their ration about two or three weeks before breeding begins. This is called "flushing" and is widely practiced by sheepmen and swine growers. Range cattlemen occasionally practice it where the price of a suitable concentrate, such as cottonseed meal, is not prohibitive. The percentage calf crop on the range is noticeably higher following a year when grazing was unusually good early in the breeding season. Other specific management to increase breeding efficiency includes breeding at a reasonably early age, fairly prompt rebreeding, having enough males (especially under range conditions) that they will find all females in heat and will not have their fertility impaired by overwork, giving the males extra feed during the breeding season, allowing only one or two services to each female (often practical under farm conditions but not under range conditions, of course), etc.

Artificial insemination is sometimes useful in overcoming sterility. It is sometimes used to breed a female at a distant place to some highly prized male without the expense of transporting one to the other. If artificial insemination is used extensively it permits the keeping of fewer sires and thereby effects a saving in maintenance costs. Against this must be charged the costs of the special apparatus or skilled labor required. By making it unnecessary to save so many sires, selection of the sires can be more intense. That should lead to somewhat more rapid progress although not in proportion to the number of sires which the artificial insemination permits discarding. For example, by reference to Table 12 it will be seen that if 100 per cent represents the progress resulting from selection among the males in one generation when 10 per cent of all males must be saved, then 117 per cent would represent the progress expected if only 5 per cent of the males needed to be saved, 139 per cent would represent the expected progress if only 2 per cent of the males need to be saved and 150 per cent would represent the progress to be expected if only 1 per cent needed to be saved. These figures concern only that part of the progress which is made by selecting the sires. The increasing intensity of selection among the sires would not increase the actual merit of the offspring so rapidly, however, because selection among the females contributes something to that, and this selection among females would not be altered by the use of artificial insemination. It appears, therefore, that the widespread use of artificial insemination would increase the rates of progress somewhat but not enough to change the prospects for breed improvement very radically.

GENETIC MEASURES TO PROMOTE BREEDING EFFICIENCY

So far as individual variations in fertility have a hereditary basis, they are subject to selection; and the average of the herd or flock in this respect can be changed by the same breeding methods used for changing other characteristics. That differences in fertility are often inherited is evident from a number of facts, such as the breed differences which exist in sheep and swine. Breeds of cattle also differ in the percentages of twins which they produce. Genetic literature contains many cases of inheritance of definite malformations which result in sterility. Superficially it seems a contradiction of terms to speak of the inheritance of sterility, since a sterile individual would not leave offspring. Nevertheless, sterility can be inherited just as lethal genes are; that is, carried along in heterozygous condition. Each different gene which definitely produces sterility must be of low frequency in the breed (just as lethal genes are) because of the intense natural selection against it. There might, however, be many such genes each in a different allelic series, each individually rare and yet enough of them that all together they would cause a noticeable amount of sterility in the whole breed. Data collected by Pearson on the number of children per family are interpreted by Fisher as showing that about 40 per cent of the observed variability in human family size is genetic in origin. There is conclusive evidence that part of the variation in longevity in Drosophila and in man is inherited,[7] but the direct evidence for farm animals is still fragmentary.

There is automatically some selection for high fertility, since those individuals with more offspring have more chances to get their sons or daughters saved for breeding purposes. This may be canceled by the breeder's selection for large size and excellent appearance. Individuals born in large litters may be initially handicapped by the greater competition and crowding. This is distinctly the case with twins in horses and, although less extreme, is noticeable with twins in cattle, sheep, and goats. Also, in swine the birth weights decrease and the mortality increases as litter size increases; although size of litter is by no means the major factor affecting birth weights and mortality, and its influence on weaning weight is comparatively unimportant. Moreover, the sow which has just nursed a large litter to weaning time is often thinner and

[7] Consult Pearl's *The Biology of Death*.

appears less attractive than one which has just weaned a few or none. Unless careful attention is paid to their records of production, the more productive sow is apt to be culled on account of her appearance whenever any culling is done. All this results in some tendency to select breeding animals which are below average in fertility themselves or come from parents of low fertility. In a similar way some selection for low fertility takes place in nature and may hold in equilibrium the selection which would otherwise favor the more fertile strains. Under many circumstances the most desirable condition with respect to fertility is neither the maximum nor the minimum. In the very largest litters all of the individuals are so handicapped that they cannot struggle through to maturity as successfully as those in smaller litters. On the other hand, those in the least fertile strains leave comparatively few offspring when they could have reared a larger number to maturity with almost equal success. This seems certain to cause some variation in litter size to be epistatic so far as profitability is concerned. Studies with swine have led to the general conclusion that a litter size of something like 9 to 12 pigs born alive leads to the largest net profit. If the litters are larger than this they contain too high a percentage of stillborns, and the survivors are too small at weaning time. If the litters are smaller than nine, incomplete use is made of the sow's ability to nurse and rear pigs. None of the breeds of swine in the United States seem to be in danger of exceeding this optimum as a general average for the breed, although individual sows do exceed it frequently in single litters. Litter size in swine has a rather low repeatability; the intraherd correlation between the sizes of successive litters from the same sow being of the order of one-sixth.[8] In selecting for litter size it would be unreasonable to expect actually to achieve each generation much more than about one-sixth of what is reached for in the selections, especially since the boars could express their inheritance only indirectly through the performance of their daughters.

The important practical step is to keep a reasonably complete and up-to-date record of each female's actual production and to lay heavy emphasis on that in all selections. By keeping the record in the form of lifetime production, some emphasis is automatically laid on longevity. The selection differential which can actually be made for productivity is usually small, even when the breeder thinks he is paying much attention to that. It is often an eye-opening experience to average the records on that point and see just what has been accomplished. For example, in the course of one breeding experiment at the Iowa Station (unpublished) in which much emphasis was laid on selecting for large litters, the data for 143 gilts farrowing in the first six years showed an average

[8] USDA Technical Bulletin. No. 836. 1942.

of 7.9 pigs per litter. The 92 gilts which were saved to produce at least one more litter averaged 8.4 pigs in their first litters. The selection differential actually achieved at this first culling was thus only one-half pig per litter. If the variance in litter size is about one-tenth hereditary in the simple additive manner—a reasonable assumption as far as the present evidence goes—and if selection of the boars out of prolific dams had an equal effect, this selection would increase the average litter size about one-twentieth of a pig per generation. With about 2½ years per generation, such selection would require 50 years to increase the size of the litter one pig! These computations are somewhat too pessimistic since they omit the selection which takes place after the second and later litters and the selection which takes place in saving gilts only from the larger litters in the first place. In these same data the additional selection differential achieved in culling sows after their second litters was .24 pig per litter. Selection for large litters continued all through the lives of the sows and in selecting gilts and young boars in the first place. It is within reason to expect that the improvement in litter size might be several times as rapid as indicated in the estimate above. The example illustrates two things: (1) that the selection differential actually achieved for fertility may be small even when much attention is paid to it; and (2) that such computations may be used in setting up standards of the average progress which may reasonably be expected.

SUMMARY

The number of young produced per female per year or other unit of time is important in determining profits in animal husbandry. It is also important in determining the percentage of young which must be saved to replace their parents and in thereby setting limits on the intensity of selection possible, especially among the females.

Fertility is limited in the first place by the number of ova produced. It is further limited by failure of some of the ova to become implanted or, if implanted, to develop normally enough to result in living young at birth.

Those which die between birth and weaning cause further loss in breeding efficiency. Much of this loss comes from faults of management, but some is from inherent weakness of the dam and the young. As the young grow older, their own inherent vitality and other characteristics play a larger part in their fate, and the characteristics of their dam play a correspondingly lesser part.

For mammals the first convenient stage for measuring differences in fertility is when the young are born. For economic reasons or for convenience, the number of young produced is often counted first at weaning time.

Fraternal twins or triplets are no more like each other genetically than ordinary full brothers and sisters. They are apt to be a little more like each other outwardly because of having been exposed to the same peculiarities of environment.

Identical twins, which are common in man but rare in the farm animals, are cases where a single fertilized ovum has later developed into two separate individuals. Such cases are unusually interesting to students of heredity because the two twins have identical inheritance. Their unusual likeness, as compared with that of ordinary fraternal twins, furnishes an opportunity to estimate the relative importance of heredity and environment.

Individuals born of the same mother at the same time may occasionally be half brothers, having been sired by different males.

In very rare cases a second pregnancy begins before the preceding pregnancy has terminated. Usually the younger embryo is expelled along with the older when the older one is born. In rare cases the younger embryo is retained and may be born several weeks or months later.

Longevity of the parents is important in breeding efficiency both because it spreads the cost of their rearing over a longer productive life and because it permits more intense selection of the smaller percentage of replacements needed.

Policies of management which promote breeding efficiency include such things as: keeping the animals reasonably free from disease and in good nutritive condition, flushing, breeding at a reasonably early age, fairly prompt rebreeding, not having too many females per male, giving the male extra feed during the breeding season, etc.

Genetic ways of promoting breeding efficiency consist of selecting the more fertile animals, with especial emphasis on their lifetime production of young. While it is probable that most of the ordinary individual variations in fertility are matters of management, yet some of these variations are hereditary, and some degree of success can be had by selecting for them, using the same methods as are used in breeding for other characteristics which can be manifested in one sex only. To do such selecting it is essential that up-to-date records of lifetime productivity be kept on each female.

REFERENCES

Briggs, Hilton M. 1936. Some effects of breeding ewe lambs. N. Dak. Agr. Exp. Sta., Bul. 285.

Castle, W. E. 1924. The genetics of multi-nippled sheep. Jour. of Heredity, 15:75–85.

Chapman, A. B., and Casida, L. E. 1935. Factors associated with breeding efficiency in dairy cattle. Proc. Amer. Soc. An. Prod. for 1934, pp. 57–59.

———. 1936. Length of service period in relation to productive and reproductive efficiency in dairy cows. Proc. Amer. Soc. An. Prod. for 1935, pp. 66–70.

———, and Dickerson, G. E. 1936. The relation of age at first calving to butterfat production in the first five lactations. Proc. Amer. Soc. An. Prod. for 1936.

Crew, F. A. E. 1925. Animal genetics. pp. 296–311. Edinburgh: Oliver and Boyd.

Dow, G. F. 1932. Costs and returns in producing milk, raising heifers, and keeping herd bulls in Maine. Maine Agr. Exp. Sta., Bul. 361.

Engeler, W. 1933. Die Ergebnisse statistischer Auswertung 40-jahriger Herdebuchaufzeichnungen beim schweizerischen Braunvieh. Bern: Verbandsdruckerei A. G.

Fisher, R. A. 1930. The genetical theory of natural selection. pp. 194–99. Oxford University Press.

Fleming, C. E., *et al.* 1940. Range sheep production in northeastern Nevada. Nevada Agr. Exp. Sta., Bul. 151.

Johansson, Ivar. 1931. The sex ratio and multiple births in cattle. Zeit. f. Zücht., Reihe B, 24:183–268.

———. 1933. Multiple births in sheep. Proc. Amer. Soc. An. Prod. for 1932, pp. 285–91.

Kronacher, C., and Sanders, D. 1936. Neue Ergebnisse der Zwillingsforschung beim Rind. Zeit. f. Zücht. Reihe B, 34:1–172.

Lloyd-Jones, Orren, and Hays, F. A. 1918. The influence of excessive sexual activity of male rabbits. Jour. Exp. Zool., 25:463–97.

Lovell, R., and Hill, A. Bradford. 1940. A study of the mortality rates of calves in 335 herds in England and Wales. Jour. Dairy Res., 11:225–42.

Lush, Jay L., and Lacy, M. D. 1932. The ages of breeding cattle and the possibilities of using proven sires. Iowa Agr. Exp. Sta., Bul. 290.

Lush, Robert H. 1925. Inheritance of twinning in a herd of Holstein-Friesian cattle. Jour. Heredity, 16:273–79.

Mumford, F. B. 1921. The effect on growth of breeding immature animals. Missouri Agr. Exp. Sta., Res. Bul. 45.

Newman, H. H. 1933. The effects of hereditary and environmental differences upon human personality as revealed by studies of twins. Amer. Nat., 67:193–205.

Perry, E. J. (Editor). 1945. The artificial insemination of farm animals. Rutgers University Press.

Seath, D. M., and Neasham, E. W. 1943. A study of breeding records in dairy herds. Louisiana Agr. Exp. Sta., Bul. 370.

Smith, A. D. B., and Robison, O. J. 1931. The average ages of cows and bulls in six breeds of cattle. Jour. Agr. Sci., 21:136–49.

Spielman, Arless and Jones, I. R. 1939. The reproductive efficiency of dairy cattle. Jour. Dairy Science, 22:329–34.

Uppenborn, Wilhelm. 1933. Untersuchungen über die Trächtigkeitsdauer der Stuten. Zeit. f. Zücht., Reihe B, 28:1–27.

von Patow, C. 1933. Genetische Untersuchungen am Schafen. Zeit. f. Zücht., Reihe B, 26:285–321.

Withycombe, Robert; Potter, E. L.; and Edwards, F. M. 1930. Deferred breeding of beef cows. Oregon Agr. Exp. Sta., Bul. 271.

CHAPTER 39

General Considerations

In the preceding chapters each of the methods by which the breeder can change the genetic composition of a breed has been examined to see what it will and what it will not do and hence under what circumstances it would be useful. Before passing to the individual breeder's plans for his own procedure it is appropriate to review for a moment what each of the tools available to him will do best.

Selection—i.e., differences in the number of offspring which different kinds of individuals are permitted to have—is the most effective method for changing the frequency of genes and the genetic averages of the breed for various characteristics. It is naturally the method the breeder considers first because his profit or loss will depend mostly upon the average merit of the stock which he produces for sale. So far as genes produce effects which are consistently desirable in all combinations with other genes, the changes produced by selection are permanent (unless and until equally effective counter-selection has taken place); but insofar as the effects of the genes are desirable in some combinations and undesirable in others, many of the changes which selection produces when it is first practiced are lost within a generation or two if selection ceases. Selection produces only slight changes in the homozygosity or uniformity of a population unless it continues over many generations. If there is much epistasis, selection may increase uniformity rather sharply when first practiced; but continued selection is necessary to hold that increase in uniformity. Selection does not change the uniformity of the next generation nearly as much as it changes the average. The usefulness of selection can almost be summarized by saying that it usually carries the breeder toward his goal but is almost powerless to do much toward "fixing" characteristics. In cases where there is much epistasis, the rate of progress in changing the herd or breed average slackens considerably after the first few generations.

Inbreeding is in many respects a perfect complement to selection; it can do what selection cannot do, and can do feebly or not all that which selection can do well. Inbreeding is uniquely powerful as a means of producing homozygosity and prepotency or genuine "fixity" of type. It produces uniformity within the various inbred lines but marked differences between lines. Thus, it leads to the apparent paradox of a very non-uniform breed composed of very uniform lines or families. Because of its family-forming power, inbreeding can make selection much more effective, especially in cases where there is much epistasis. Inbreeding which is directed toward keeping relationship to some admired ancestor high (i.e., linebreeding) is a combination of selection and inbreeding particularly useful for perpetuating favorable epistatic effects of certain gene combinations.

Outbreeding promotes individual merit by tending to conceal recessive genes. It remedies damage done by inbreeding and is useful for introducing desirable genes into a population which lacks them. Because it destroys family distinctness, covers recessives and scatters favorable epistatic combinations, outbreeding hinders progress in breed improvement except in those cases where a little outcrossing is necessary to introduce desired genes into a family which lacks them. Outbreeding is primarily a method for the producer of market animals.

The breeding of outwardly similar individuals together has practically no effect upon homozygosis or prepotency but does increase immediately the average resemblance between parent and offspring. That happens merely because the parents are chosen for their resemblance to each other and each offspring has a chance to inherit from both parents genes which will make it seem like them both. Mating like to like increases the proportion of extreme individuals and decreases the proportion of intermediates, thereby making the population more variable provided no accompanying selection is practiced. The effects of mating like to like are limited by the correlation between genotype and phenotype and cannot be extreme unless that correlation is high. Hence, in practice, assortive mating based on somatic resemblance and unaccompanied by selection does little but increase the variability of the population. Mating unlikes together to correct defects, where both are too extreme but in opposite directions, makes the population more uniform and keeps it nearer to an intermediate type. The mating of somatic unlikes has little effect on homozygosis. It is a very useful breeding system wherever the goal is an intermediate, particularly in a species where fertility is moderate and some of the extreme individuals must be used for parents.

PLANS FOR INDIVIDUAL BREEDERS

FOR ALL KINDS OF BREEDERS:

1. Decide what kind or type of animal and what level of production

would be ideal for the breeder's own individual circumstances and local conditions.

2. Find what living animals most nearly have the genes needed to produce that ideal animal.
   a. By judging and testing each animal.
   b. By paying some attention to the merit of recent ancestors and close collateral relatives.
   c. By studying the progeny of each animal.
3. Obtain, as far as can be done at reasonable prices, those animals which come nearest to having the ideal genes and let each have offspring in numbers proportional to the closeness with which its heredity approaches the ideal.

FOR BREEDERS OF PUREBREDS:

4. Keep the future herd closely related to the best animals of the present and of the recent past, letting the relationship to poor or ordinary animals be diluted by the natural halving effect of the processes of inheritance.
5. Outcross only when it is necessary to prevent some serious defect from being fixed on the whole flock or herd. The higher the average individual merit of the herd and the farther the breeder has gone in his linebreeding program, the milder and more tentative such outcrosses should be.

FOR BREEDERS OF MARKET ANIMALS:

4. Outbreed so far as that can be done without using animals of distinctly poorer heredity than would be available if related individuals were mated together. By using sires which are closely bred but unrelated to the females on which they are to be used, the maximum of heterosis and individual merit can be kept without losing as much in uniformity as if the sires were not closely bred themselves.

The first step in any animal breeding program is to decide what is ideal. Until a breeder knows what kind of animal he wants, he is stopped in his tracks and can neither select the best nor discard the worst. Somewhat indefinite words, such as best, worst, poorer, better, more productive, meritorious, etc., have intentionally been used in this book instead of more precise words in discussing selection and kindred problems because the purpose was to discuss ways of attaining the goals the breeder wants and not to enter into the subject of what ideal for each kind of animal would be most profitable. Each breeder needs to consider his own physical and biological resources, his own markets and his own personal inclinations to decide what characteristics his ideal animals should possess. Naturally a beginning breeder would defer somewhat to the opinions of those who have had more experience under somewhat similar conditions, but his own conditions may be different from those of the men from whom he is receiving advice. It seems likely that the matter of local adaptability will receive more attention in the future than it has in the past. Probably there will always be at least enough interchange of breeding stock to keep that from being overdone. The ideal must often be a compromise between satisfying the market and satisfying one's own local conditions most completely. Conceivably the butcher's interest in high dressing percentage and high quality of meat, if carried too far, might result in animals with vital organs too small for them to be as healthy and thrifty as the farmer wishes, while the animals which would suit the farmer best because they were healthiest, most robust, largest, and quickest growing might be too big, bony, and coarse to suit the butcher. The commercial ideal is largely dependent upon economic conditions which can change much more rapidly than the breed average can be changed. Because of this it is natural not to follow a current economic change as far as would be wise if one could be certain that the change would be permanent. Sometimes the farmer's ideal and the breeder's ideal are not quite identical. That may have a rational basis wherever the commercial ideal is an intermediate, but most of the females in the farmer herds are far to one side of that ideal. For example, in the 1930's the Danes were striving to lengthen their hogs to meet better the demands of their British market. Since most sows were too short, the breeder ideal was for even longer hogs than the bacon factories wanted. They hoped that extremely long boars would produce from the farmers' sows pigs about right in length. The breeder ideal may, however, differ only in stressing some details of breed type or in following some current fashion which has gone farther than economic conditions justify. A word of caution should be added about paying too much attention to what is said to be the customer's demand. It is difficult to be sure just what the customers do want, and many a man has gone to considerable trouble to satisfy the supposed demands of his customers only to find that an insignificant portion of them really wanted these peculiar characteristics enough to pay extra money for them.

That the ideal degree of development of a characteristic may be quite different in animals intended for different purposes is illustrated by the following quotation from a most interesting book[1] concerning the breeding of German Shepherd dogs for various kinds of service:

[1] Humphrey, Elliott, and Warner, Lucien. 1934. *Working dogs*. Baltimore: The Johns Hopkins Press. 253 pp.

"There are several characteristics which do not bear the same relationship to all forms of service. A trait may be essential to the excellent performance of certain work but nonessential or even detrimental to the proper execution of other services. The reader will find it convenient to refer to the accompanying chart:

Form of Service	Olfactory Acuity	Nose Obedience	Aggressive-ness	Distrust
Police (frontier, penitentiary, etc.)	+	+	++	−
Trailing	++	++	+	−
Liason	+	+	−	+
Blind guiding	−	++	+	−
Sanitary (Red Cross)	+	0	0	−
Herding	+	+	+	0
Companion	0	0	+	−

In this chart, ++ indicates that the trait is essential in its highest developed form; +, that it is desirable to at least a limited degree; −, that the presence of the trait is detrimental to good work; and 0, that its presence or absence is unimportant."

The second step in the breeder's plans—finding which animals most nearly have the genes he wants—has been discussed in detail in chapters 12 to 19. It is useless to institute an elaborate search for perfect animals, because in few if any cases have such animals yet been born. The breeder will always be under the necessity of compromising, getting animals which are above average as a whole but which are below average in some respects, taking care that at no time do all of his breeding animals have the same defect. Only in rare cases will the breeder know the Mendelian formula for more than a few genes in his animals. He will never see genes but can judge whether or not they are present only by the effects they produce, either in this animal itself or in some of its close relatives. In point of time, pedigree selections come first; but in most populations they are less dependable than selections based on the individual's own characteristics or on the characteristics of its progeny. Individual selection of the parents keeps the animals which would have had the worst pedigrees from being born. If individual selection has been extreme among the parents, there is only a little room left for pedigree selections among the progeny, especially among the female progeny. Individual selections are usually more accurate than selections on pedigree or selections on progeny except in the case of characteristics which can be expressed only in the other sex, but if the worst individuals have been discarded without being tried as breeding animals, the progeny test when it first becomes available brings fresh evidence from an entirely new direction and for the moment offers more possibilities for further progress than can be had by paying attention to the remaining differences between pedigrees and individualities of those which have already survived the earlier cullings on those two bases. Inbreeding seems to deserve more use than it has yet received as a means of finding which animals have the best genes. Not only does it uncover recessives more surely than any other method, but it also increases the relationship between the inbred animal and its parents and other relatives so that the animal's pedigree and the merits of the family to which it belongs become more dependable as indicators of its own genes than can be the case with animals which are not inbred. Considerable inbreeding is necessary if family selection is to be very effective. A breeder can sometimes get help in finding which of his young males have the best genes by leasing for progeny-testing in other herds those which have such good pedigrees and such good individuality that he might wish to use them himself if their progeny prove them good. There may be more of these than he can progeny-test himself, and leasing them will give him a larger number from which to choose in bringing back the ones which would improve his own herd the most. The costs of such a plan would come from some increase in the possibility of disease transmission and the possibility that the lease price might be less than could be had for these animals by outright sale. Perhaps the possibilities for business disagreements are numerous enough to be important. Leasing young sires for progeny-testing is not a new idea. Bakewell was famous for his annual ram-lettings. The plan seems to deserve wider use than it has generally received.

The third step in the practical breeder's plan is to get the animals which have the most nearly ideal heredity, so far as he can afford to buy them, and to let them reproduce at rates in proportion to how nearly ideal their heredity is. There will be some things about each animal which are not ideal, but in breeding for its good qualities one must breed for these undesired ones also. The gene is the unit of inheritance, but the animal is the smallest unit which can be chosen or rejected for breeding purposes. To breed exclusively from one or two of the best animals available would tend to fix their qualities, both good and bad, on the herd. In fact, that is the essence of what happens under extreme inbreeding. Moreover, the breeder will make at least a few mistakes in estimating which animals have the very best inheritance. Hence, in a practical program the breeder will hesitate to use too extensively even a very good sire. With as many as four or five sires in use at all times and no one of them used far more extensively than the others, the danger of fixing traits on the whole herd against selection is small. Perhaps it may be ignored for many animal generations. Where no inbreeding is practiced, this danger practically disappears because the next sire, being unrelated, will rarely have many of the same undesired genes as his predecessor.

The breeder is far from having full power to decide how many offspring each of his animals shall have. Some of the animals will die or become sterile or will be prevented in other ways from leaving as many offspring as the breeder wants. Females from which he wants a herd sire may persist in producing only daughters for several years. Consequently, some animals from which he really did not want so many offspring must leave more to make up for the offspring he does not get from the animals he prefers.

The fourth step in the plan for the breeder of purebreds is to stay with the best individuals, once he has found them. This is the essential object of linebreeding. It utilizes the laws of Mendelian inheritance to hold at a nearly constant level the probable likeness of future animals to the best proved sires and dams of the past, but to dilute the relationship of the future animals to ordinary or poor animals of the past. Such linebreeding requires planning and, if continued many generations, necessarily involves co-operation with other breeders except in those cases where the herd is large enough to maintain economically as many as two to five sires in service at all times. Without such co-operation it is likely that, sooner or later, first one undesired gene and then another will be fixed in the herd in spite of selection. It is not generally advisable to plan pedigrees very exactly for more than a generation or two in advance, because that does not give enough opportunity for selection. Instead, one can decide that he will use one of perhaps a half dozen individuals still to be born, all of which will have pedigrees which will fit into his plans reasonably well. Then there is opportunity for individual selection among that half dozen animals.

The purebred breeder whose herd is above average in merit will outcross only when that is necessary to prevent some undesirable gene from becoming entirely fixed on his herd or (which is another way of saying the same thing) to introduce into his herd some desirable genes not already there. The better his herd and the farther he has gone with his linebreeding program, the more reluctant he will be to outcross and the milder his outcrosses will be. If his herd is large enough to maintain at all times as many as five different sires, it is probable that no outcrosses will be really necessary even in a human lifetime. Perhaps that would be true with an even smaller herd. Evidence on how well selection can keep control of such mild inbreeding rates is still scanty. However, it would be a rather rare herd which already contained, at the moment the owner began his linebreeding program, absolutely all the desirable genes which exist in the breed.

The breeder of market animals will follow the policy of outbreeding continually, just as the breeder of purebreds will follow the different policy of linebreeding to his best stock. The breeder of market animals probably will find it wise to linebreed only when he can find no other stock as good as that which he already has and when the extra individual merit of his stock is enough above that of any other he might use that it will more than compensate him for the inbreeding risk involved. Probably the commercial dairyman and the man who is raising horses which are not purebreds will also follow the outbreeding policy, although that is by no means certain, especially in the case of the dairyman who naturally will be keeping most of his heifers for breeding. Present evidence indicates that a policy of using closely bred sires of good individuality, each unrelated to the females on which he is to be used, will maintain the maximum heterosis and individual merit in the breeding females and that the lack of uniformity among them and their offspring will not be extreme enough to be important, as long as each sire always comes from a closely bred strain. It will even pay to sacrifice something in individuality if that is necessary to secure a sire from a closely bred strain. This is in principle the same general idea as that underlying the "criss-crossing" of swine, which has given good results at the Minnesota Station, and is an old general idea which has been prominent in the plans of many breeders, except that many have not insisted on the sire's being closely bred. If epistatic effects are more important than is generally thought at present, there may be limits beyond which the outbreeding should not go. Practically all our breeds are so large that one can outbreed continually within the limits of those breeds, using a sire from first one and then another family. For commercial purposes wider outbreeding effects can be obtained by cross-breeding, wherever the cost of replacing the females is less than the advantages to be gained.

The ideal plan for the most rapid improvement of the breed differs from the plan for the individual breeder of purebreds chiefly in that the individual breeder dare not risk quite so much inbreeding deterioration as could be risked in every herd in the whole breed if the object of all breeders were to improve the breed with little regard to their own immediate financial benefit. The inbreeding deterioration which would be produced under the ideal breeding system for rapid improvement of the breed would be different from group to group and would disappear in most of the resulting outcrosses. However, for the individual breeder who is managing his plan by himself, such a defect when it began to appear might interfere too much with his sales. It might force him to outcross before the full benefits of his linebreeding had occurred.

IDEAL BREEDING SYSTEM FOR RAPID IMPROVEMENT OF THE WHOLE BREED

An ideal breeding system for the most rapid improvement of the breed as a whole would be about as follows: Each breed would be divided into many small groups, each such group rarely introducing any breeding animals from other groups and then only with caution. Each group would be large enough for the use of two or three breeding males at all times and, of course, would include a much larger number of females. The smaller the group the higher will be the rate of fixation of genes on account of the inevitable inbreeding, and the more frequently will there be need to outcross the better groups with each other. If the groups were much larger than this, progress toward uniformity within each group and toward the distinctness from group to group which is necessary for effective intergroup selection would be needlessly slow. Such a system is pictured diagrammatically in Figure 50, where the large area represents a whole breed and each small area within it means a partially isolated subgroup of the breed into which individuals from other subgroups are rarely introduced. Naturally the few introductions which are made would usually be from the neighboring subgroups and only rarely from a distant subgroup. Groups of subgroups or major geographical subdivisions might thus tend to form within the breed.

The consequence of such a separation into groups, each breeding very largely within itself, would be that each such group would quickly become more uniform than herds are today and that each group would become different from other groups. Selection between the groups would then be effective to an extent impossible today and unattainable in the selection of individual animals for moderately or slightly heritable characteristics, no matter how much the animal is studied nor how skilled is the man who does the selection.

Many of these subgroups would begin to show undesired traits varying in severity. Side by side with these, they would show other highly desired traits more uniformly than present herds do. Groups showing many desired and few undesired traits would be outcrossed mildly to neighboring groups which were strong where they themselves were weak. Then by renewed linebreeding with rigid selection for the traits they wished to introduce by the outcross, the breeders would attempt to fix the introduced desired traits without losing the desired traits they already had. Groups showing few desired and many undesired traits would either be discarded altogether or would be graded up by the continued use of sires from the most successful groups until their individual merit was restored or even exceeded that of the most successful group. Then the breeders would start breeding within this group to find and fix some one of the almost infinite number of desirable new combinations of genes which would be possible. The general rule would be that the more successful each subgroup was, the less readily would any outcrossing be done and the milder such outcrossing would be.

The general picture thus presented is an alternation of mild linebreeding with tentative outcrosses, both accompanied at all times by intense selection. In this way more emphasis is placed on family formation and inter-family selection. The most of the linebreeding would be done in the best of the herds with extreme outcrossing being confined to the poorer herds.

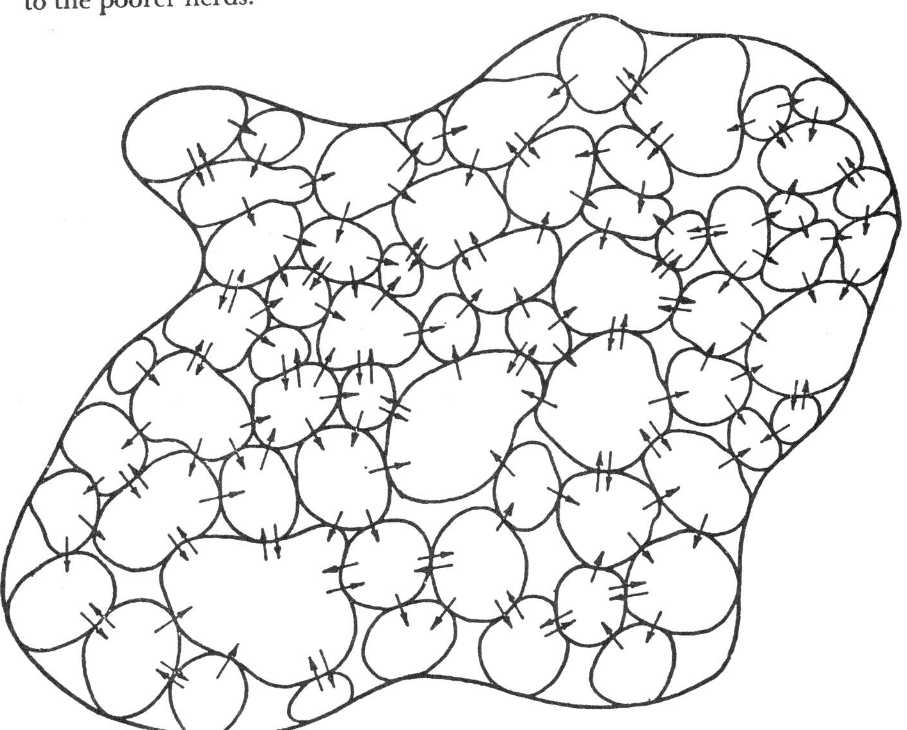

FIG. 50. Subdivision of a breed into small local groups which exchange breeding stock only at rare intervals and then only with neighboring local groups.

IMPORTANT PROBLEMS STILL UNSOLVED

How generally important is epistasis? Do most gene substitutions tend to produce the same effect when made in all kinds of individuals, almost regardless of the other genes which are present? If so, inbreeding systems are not so necessary for progress as has been implied here. But the reverse may be true. It may be that nearly all genes, except lethals and others which produce distinct breakdowns in the functioning of vital organs, are epistatic in their effects. If so, inbreeding is even more necessary for producing permanent changes than has been implied here, and most of the progress which appears to be made by intensifying selection is temporary and will disappear whenever the selection is relaxed. There are only scraps of actual evidence about this in animals. The results of Sprague and Tatum with corn indicate that most of the hereditary differences in an unselected population are additive, but that the differences which still remain between the survivors in an intensely selected population are mostly epistatic. (Jour. Amer. Soc. Agron. 34:923–32.)

Does each gene generally have many different effects, some good and some bad, or do most genes merely affect one organ or one characteristic of an animal? If the former is true, each gene will have some desirable and undesirable effects and it may be impossible ever to establish strains absolutely homozygous for all the desired traits. In such an event, breeding plans must be pointed toward the ultimate condition of closely bred but not individually meritorious purebreds to be continually outbred in the most extreme fashion for the production of market stock. If most genes have but one effect, we may reasonably hope to build purebreds which will not only be far more homozygous than market stock but which will also be fully as meritorious individually as their crossbreds could be.

Standards of the progress which might reasonably be expected under various conditions and with various kinds of selection would be useful in a number of ways. So far as concerns the purely additive portion of the variance, the information needed is fairly simple, consisting of the size of the standard deviation in the population being considered, the percentage of the variance which is additively genetic, and the selection differential which can be attained by the plan of selection being considered. But for the epistatic portion of the variance it does not appear that there is any way to predict what possibilities of desirable combinations exist. Wright has given[2] a generalized description of the way in which epistatic variance contributes to the correlation between relatives and of the general consequences of selection where epistatic interactions are involved; but it does not appear that this provides any way to predict when an epistatic interaction of considerable importance may result from bringing together genes which individually give no hint of the effects they will have when combined. Probably that must remain a matter of trial and error.

Simple and reasonably complete objective ways of measuring the practical merit of each animal would be very useful. For dairy cattle and poultry there is an approach to that in weighing and testing the milk and in counting and weighing the eggs, but other things also need

[2] *Journal of Genetics,* 30:243–66. Also *Proc. Sixth International Cong. Genetics,* 1:356–66.

to be taken into account for these animals. For the other classes of animals the standards for measuring practical merit are not even this well developed. Considerably more needs to be done in developing practical selection indexes which will pay attention to each practically important characteristic without the risk of overemphasizing it.

Machinery for co-operating in animal breeding can doubtless be made more efficient and useful. The dairy herd improvement associations are an example of what can be done in this direction; but, if there is to be any large increase in linebreeding in small herds or in community breeding, closer co-operative organization aimed directly at that will be essential. Bull circles may foreshadow the pattern which such efforts will take.

OPPORTUNITIES IN ANIMAL BREEDING

The number of combinations of existing genes which have never yet been brought together is practically infinite. In every breed there are enough unfixed genes available to make possible the production of animals more extreme than have ever yet existed in almost any direction that the breeder might desire. All our breeds are still exceedingly plastic, and the breeder's opportunities to mold them to his own desire are so great that there is no occasion to mourn his inability to produce new mutations at will and probably no important reason to regret that the established breeding systems make it impossible for a breeder to use blood from outside the breed unless he wishes to form a new breed of his own. There is reason to think that a new breed could be formed from crossing two or more of the existing breeds, but the plans and specifications for doing that will probably require that the herd be at least large enough to keep in service at all times three to five sires and a much larger number of females. Otherwise, the inbreeding consequent on the small number of animals which one man could manage would be certain sooner or later to fix on the whole herd some undesired combinations of parental traits. Also, in combining some of the extreme characteristics of one breed with other extreme characteristics of another breed it will be necessary to allow for several generations to permit the desirable new combinations of genes to come together. How many generations would be required to make a breed at least as uniform as the breeds from which it was derived will, of course, depend upon how many genes are involved in the differences between the breeds, how many animals there are from which to select, how accurate the selections are, how much linebreeding is done to those which appear to come closest to the new combination desired and upon how homozygous the parental breeds were. For example, if eight pairs of equally important genes differ in the two parental breeds, the average breeding value of the second crossbred generation (the F_2 generation) will differ from the desired true-breeding combinations by four standard deviations. It would be exceedingly unlikely that one could reach the goal in as few as four more generations. If several different characteristics, each dependent on several genes, are to be combined, it may well require eight to ten generations of breeding after the original cross to come reasonably close to the goal. This is not at all to say that making such a breed is impossible but merely to call attention to the time which will probably be required and to the need of budgeting that in the plans. Partly offsetting this is the fact that the breeds which are to be used in the cross are far from homozygous and that, by making more use of inbreeding than is usually done, one might within two or three generations from the first cross surpass the degree of homozygosity which characterizes the parents.

Where the new ideal is an intermediate between the two breeds in nearly all characteristics, rather than a mosaic of some characteristics from one and other characteristics from the other, the first cross or second cross generation may already average near the desired ideal. In that case the generations required for selection to move the average to the desired point would be unnecessary, and it would only remain to linebreed intensively enough to increase homozygosity and uniformity to the point that would warrant calling the new group a breed. These considerations probably explain why most deliberate attempts to found a breed have failed. The founders have not had large enough numbers to carry on their own breeding plans without dangerously high inbreeding, or else they have not had enough time to reach their goal before they died. Not often were their heirs interested in continuing these plans. Theoretically there seems to be no bar to forming a new breed in this way if one has animals enough and time enough, but it is a rather impressive fact that practically no breeds were formed thus. Extensive mixing of races was involved in the founding of many breeds, but that seems to have been undertaken for other reasons and was profitable as it went along. Only incidentally and after some time had elapsed was it observed that somewhere out of the welter of crossbreeding there emerged a group which seemed to have merit enough that their owners recognized them as a breed and sought to perpetuate them as such.

Currently the most active interest in forming new breeds is in tropical and subtropical regions for which none of the established breeds seem well suited. The Santa Gertrudis cattle in Texas; dairy cattle suitable to Brazil, to the West Indies, and to India; and beef cattle for the more tropical parts of South Africa, the Gold Coast and Kenya, are among the more striking examples. Generally the settlers in temperate regions were able to find in Europe improved breeds which were already fairly well suited to their needs. However, Corriedale sheep in New Zealand, Columbia sheep in the United States, lard breeds of hogs in the United States, Morgan, Standardbred, and American Saddle Horses in the United States, all illustrate that even the temperate regions have sometimes found it profitable to make their own breeds. It is not likely that all of this which should be done has already been accomplished, but it is clear that the difficulties in the way of forming a new breed of farm animals are greater than in the way of forming a new variety of plants. Recently formed breeds, such as the Hereford hog and Palomino and Quarter horses, indicate that the process of breed formation is not entirely ended but several breeds have been launched and become popular and then have disappeared. Examples in the United States are the Sapphire hog of about 1914 and the Mulefoot hog of a decade earlier. Other examples of crossing which were aimed from the very beginning at forming a new breed but did not succeed commercially are the Bowlker herd of crosses between Guernseys and Holstein-Friesians in the United States and the Tranekjaer herd of crosses between Jerseys and Red Danish cattle in Denmark.

Index